T0269546

CAMBRIDGE LIBRARY COLLECTION

Books of enduring scholarly value

Darwin, Evolution and Genetics

More than 150 years after the publication of On the Origin of Species, Darwin's 'dangerous idea' continues to spark impassioned scientific, philosophical and theological debates. This series includes key texts by precursors of Darwin, his supporters and detractors, and the generations that followed him. They reveal how scholars and philosophers approached the evidence in the fossil record and the zoological and botanical data provided by scientific expeditions to distant lands, and how these intellectuals grappled with topics such as the origins of life, the mechanisms that produce variation among life forms, and heredity, as well as the enormous implications of evolutionary theory for the understanding of human identity.

Scientific Papers of Asa Gray

Born in the state of New York, Asa Gray (1810–88) abandoned a medical career to pursue his true interest in botany. He sought the mentorship of the influential American botanist John Torrey and their collaborative efforts in classifying North American flora according to biological similarities paved the way for Gray's professorship at Harvard University after years of research. Gray was also one of the few scientists to whom Charles Darwin revealed his early ideas of evolutionary theory. After Gray's death, his fellow botanist Charles Sprague Sargent (1841–1927) compiled the lesser-known writings of a prolific author whose user-friendly *Manual of the Botany of the Northern United States* and other works inspired generations of botany enthusiasts. The two-volume collection appeared in 1889. Covering the period from 1841 to 1886, Volume 2 contains essays on botanical topics and biographical sketches of influential naturalists.

Cambridge University Press has long been a pioneer in the reissuing of out-of-print titles from its own backlist, producing digital reprints of books that are still sought after by scholars and students but could not be reprinted economically using traditional technology. The Cambridge Library Collection extends this activity to a wider range of books which are still of importance to researchers and professionals, either for the source material they contain, or as landmarks in the history of their academic discipline.

Drawing from the world-renowned collections in the Cambridge University Library and other partner libraries, and guided by the advice of experts in each subject area, Cambridge University Press is using state-of-the-art scanning machines in its own Printing House to capture the content of each book selected for inclusion. The files are processed to give a consistently clear, crisp image, and the books finished to the high quality standard for which the Press is recognised around the world. The latest print-on-demand technology ensures that the books will remain available indefinitely, and that orders for single or multiple copies can quickly be supplied.

The Cambridge Library Collection brings back to life books of enduring scholarly value (including out-of-copyright works originally issued by other publishers) across a wide range of disciplines in the humanities and social sciences and in science and technology.

Scientific Papers of Asa Gray

VOLUME 2

EDITED BY
CHARLES SPRAGUE SARGENT

CAMBRIDGE
UNIVERSITY PRESS

CAMBRIDGE
UNIVERSITY PRESS

University Printing House, Cambridge, CB2 8BS, United Kingdom

Cambridge University Press is part of the University of Cambridge.

It furthers the University's mission by disseminating knowledge in the pursuit of education, learning and research at the highest international levels of excellence.

www.cambridge.org
Information on this title: www.cambridge.org/9781108083676

This edition first published 1889
This digitally printed version 2015

ISBN 978-1-108-08367-6 Paperback

SCIENTIFIC PAPERS

OF

ASA GRAY

SELECTED BY

CHARLES SPRAGUE SARGENT

VOL. II.

ESSAYS; BIOGRAPHICAL SKETCHES

1841–1886

London

MACMILLAN AND CO.

1889

The Riverside Press, Cambridge, Massachusetts, U. S. A.
Printed by H. O. Houghton and Company.

CONTENTS.

ESSAYS.

ESSAYS.

EUROPEAN HERBARIA.[1]

THE vegetable productions of North America, in common with those of most other parts of the world, have generally been first described by European botanists, either from the collections of travelers, or from specimens communicated by residents of the country, who, induced by an enlightened curiosity, the love of flowers, or in some instances by no inconsiderable scientific acquirements, have thus sought to contribute, according to their opportunities, to the promotion of botanical knowledge. From the increase in the number of known plants, it very frequently happens that the brief descriptions, and even the figures, of older authors are found quite insufficient for the satisfactory determination of the particular species they had in view ; and hence it becomes necessary to refer to the herbaria where the original specimens were preserved. In this respect, the collections of the early authors possess an importance far exceeding their intrinsic value, since they are seldom large, and the specimens often imperfect.

With the introduction of the Linnæan nomenclature, a rule absolutely essential to the perpetuation of its advantages was also established, namely, that the name under which a genus or species is first published shall be retained, except in certain cases of obvious and paramount necessity. An accurate determination of the Linnæan species is therefore of the first importance ; and this, in numerous instances, is only attained with certainty by the inspection of the herbaria of Linnæus and those authors upon whose descriptive phrases or figures

[1] American Journal of Science and Arts, xl. 1. (1841.)

he established many of his species. Our brief notices will therefore naturally commence with the herbarium of the immortal Linnæus, the father of that system of nomenclature to which botany, no less than natural history in general, is so greatly indebted.

This collection, it is well known, after the death of the younger Linnæus, found its way to England, from whence it is not probable that it will ever be removed. The late Sir James Edward Smith, then a young medical student, and a botanist of much promise, was one morning informed by Sir Joseph Banks that the heirs of the younger Linnæus had just offered him the herbarium with the other collections and the library of the father, for the sum of 1000 guineas. Sir Joseph Banks not being disposed to make the purchase, recommended it to Mr. Smith; the latter, it appears, immediately decided to risk the expectation of a moderate independence, and to secure, if possible, these treasures for himself and his country; and before the day closed had actually written to Upsal, desiring a full catalogue of the collection, and offering to become the purchaser at the price fixed, in case it should answer his expectations.[1]

[1] The next day Mr. Smith wrote as follows to his father, informing him of the step he had taken and entreating his assistance : —

"HONORED SIR: You may have heard that the young Linnæus is lately dead : his father's collections and library, and his own, are now to be sold ; and the whole consists of an immense hortus siccus, with duplicates, insects, shells, corals, materia medica, fossils, a very fine library, all the unpublished manuscripts, in short, everything they were possessed of relating to natural history and physic ; the whole has just been offered to Sir Joseph Banks for 1000 guineas, and he has declined buying it. The offer was made to him by my friend Dr. Engelhart, at the desire of a Dr. Acrel of Upsal, who has charge of the collection. Now, I am so ambitious as to wish to possess this treasure, with a view to settle as a physician in London, and read lectures on natural history. Sir Joseph Banks and all my friends to whom I have entrusted my intention, approve of it highly. I have written to Dr. Acrel, to whom Dr. Engelhart has recommended me, for particulars and the refusal, telling him if it was what I expected, I would give him a very good price for it. I hope, my dear sir, you and my good mother will look on this scheme in as favorable a light as my friends here do. There is no time to be lost, for the affair is now talked of in all companies, and a number of people wish to be purchasers. The Empress

His success, as soon appeared, was entirely owing to his promptitude, for other and very pressing applications were almost immediately made for the collection, but the upright Dr. Acrel, having given Mr. Smith the refusal, declined to entertain any other proposals while this negotiation was pending. The purchase was finally made for 900 guineas, excluding the separate herbarium of the younger Linnæus, collected before his father's death, and said to contain nothing that did not exist in the original herbarium; this was assigned to Baron Alstrœmer, in satisfaction of a small debt. The ship which conveyed these treasures to London had scarcely sailed, when the king of Sweden, who had been absent in France, returned home and dispatched, it is said, an armed vessel in pursuit. This story, though mentioned in the "Memoir and Correspondence of Sir J. E. Smith," and generally received, has, we believe, been recently controverted. However, the king and the men of science in Sweden were greatly offended, as indeed they had reason to be, at the conduct of the executors, in allowing these collections to leave the country; but the disgrace should perhaps fall more justly upon the Swedish government itself and the University of Upsal, which derived its reputation almost entirely from the name of Linnæus. It was, however, fortunate for science that they were transferred from such a remote situation to the commercial metropolis of the world, where they are certainly more generally accessible. The late Professor Schultes, in a very amusing journal of a botanical visit to England in the year 1824, laments indeed

of Russia is said to have thoughts of it. The manuscripts, letters, etc., must be invaluable, and there is, no doubt, a complete collection of all the inaugural dissertations which have been published at Upsal, a small part of which has been published under the title of "Amænitates Academicæ," a very celebrated and scarce work. All these dissertations were written by Linnæus, and must be of prodigious value. In short, the more I think of this affair the more sanguine I am, and earnestly hope for your concurrence. I wish I could have one half hour's conversation with you; but that is impossible." (Correspondence of Sir James Edward Smith, edited by Lady Smith, vol. i. p. 93.)

The appeal to his father was not in vain; and, did our limits allow, we should be glad to copy, from the work cited above, the entire correspondence upon this subject.

that they have fallen to the lot of the " toto disjunctos orbe Britannos " ; yet a journey even from Landshut to London may perhaps be more readily performed than to Upsal.

After the death of Sir James Edward Smith the herbarium and the other collections, and library of Linnæus, as well as his own, were purchased by the Linnæan Society. The herbarium still occupies the cases which contained it at Upsal, and is scrupulously preserved in its original state, except that, for more effectual protection from the black penetrating dust of London, it is divided into parcels of convenient size, which are closely wrapped in covers of strong paper lined with muslin. The genera and covers are numbered to correspond with a complete manuscript catalogue, and the collection, which is by no means large in comparison with modern herbaria, may be consulted with great facility.

In the negotiation with Smith, Dr. Acrel stated the number of species as 8000, which probably is not too low an estimate. The specimens, which are mostly small, but in excellent preservation, are attached to half-sheets of very ordinary paper, of the foolscap size [1] (which is now considered too small), and those of each genus covered by a double sheet, in the ordinary manner. The names are usually written upon the sheet itself, with a mark or an abbreviation to indicate the source from which the specimen was derived. Thus those from the Upsal garden are marked *H. U.*, those given by Kalm, *K.*, those received from Gronovius, *Gron.*, etc. The labels are all in the handwriting of Linnæus himself, except a few later ones by the son, and occasional notes by Smith, which are readily distinguished, and indeed are usually designated by his initials. By far the greater part of the North American plants which are found in the Linnæan herbarium were received from Kalm, or raised from seeds collected by

[1] Upon this subject Dr. Acrel, giving an account of the Linnæan collections, thus writes to Smith: " Ut vero vir illustrissimus, dum vixit, nihil ad ostentationem habuit, omnia vero sua in usum accommodata ; ita etiam in hoc herbario, quod per XL. annos sedulo collegit, frustra quæsiveris papyri insignia ornamenta, margines inauratas, et cet. quæ ostentationis gratia in omnibus fere herbariis nunc vulgaria sunt."

him. Under the patronage of the Swedish government, this enterprising pupil of Linnæus remained three years in this country, traveling throughout New York, New Jersey, Pennsylvania, and Lower Canada; hence his plants are almost exclusively those of the northern States.[1]

Governor Colden, to whom Kalm brought letters of introduction from Linnæus, was then well known as a botanist by his correspondence with Peter Collinson and Gronovius, and also by his account of the plants growing around Coldenham, New York, which was sent to the latter, who transmitted it to Linnæus for publication in the "Acta Upsalensia." At an early period he attempted a direct correspondence with Linnæus, but the ship by which his specimens and notes were sent was plundered by pirates;[2] and in a letter sent by Kalm, on the return of the latter to Sweden, he informs Linnæus that this traveler had been such an industrious collector as to leave him little hopes of being himself farther useful. It is not probable therefore that Linnæus received any plants from Colden, nor does his herbarium afford any such indication.[3]

[1] "Ex his Kalmium, naturæ eximium scrutatorem, itinere suo per Pennsylvaniam, Novum Eboracum, et Canadum, regiones Americæ ad septentrionem vergentes, trium annorum decursu dextre confecto, in patriam inde nuper reducem læti recipimus: ingentem enim ab istis terras reportavit thesaurum non conchyliorum solum, insectorum, et amphibiorum sed herbarum etiam diversi generis ac usus, quas, tam siccas quam vivas, allatis etiam seminibus eorum recentibus et incorruptis, adduxit." (Linn. Amæn. Acad. vol. iii. p. 4.)

[2] See Letter of Linnæus to Haller, September 24, 1746.

[3] The *Holosteum succulentum* of Linnæus (*Alsine foliis ellipticis carnosis* of Colden) is, however, marked in Linnæus's own copy of the "Species Plantarum" with the sign employed to designate the species he at that time possessed; but no corresponding specimen is to be found in his herbarium. This plant has long been a puzzle to American botanists; but it is clear from Colden's description that Dr. Torrey has correctly referred it, in his "Flora of the Northern and Middle States" (1824); to *Stellaria media*, the common Chickweed. Governor Colden's daughter seems fully to have deserved the praise which Collinson, Ellis, and others have bestowed upon her. The latter, in a letter to Linnæus (April, 1758), says : "Mr. Colden of New York has sent Dr. Fothergill a new plant, described by his daughter. It is called Fibraurea, gold-thread. It is a small creep-

From Gronovius, Linnæus had received a very small number of Clayton's plants, previous to the publication of the "Species Plantarum"; but most of the species of the "Flora Virginica" were adopted or referred to other plants on the authority of the descriptions alone.

Linnæus had another American correspondent in Dr. John Mitchell,[1] who lived several years in Virginia, where he collected extensively; but the ship in which he returned to England having been taken by pirates, his own collections, as well as those of Governor Colden, were mostly destroyed. Linnæus, however, had previously received a few specimens, as, for instance, those on which Proserpinaca, Polypremum, Galax, and some other genera were founded.

There were two other American botanists of this period, from whom Linnæus derived either directly or indirectly much information respecting the plants of this country, namely, John Bartram and Dr. Alexander Garden of Charleston, South Carolina.

ing plant, growing on bogs; the roots are used in a decoction by the country people for sore mouths and sore throats. The root and leaves are very bitter, etc. I shall send you the characters as near as I can translate them." Then follows Miss Colden's detailed generic character, prepared in a manner which would not be discreditable to a botanist of the present day. It is a pity that Linnæus did not adopt the genus, with Miss Colden's name, which is better than Salisbury's Coptis. "This young lady merits your esteem and does honor to your method. She has drawn and described four hundred plants in your system: she uses only English terms. Her father has a plant called after him Coldenia; suppose you should call this (alluding to a new genus of which he added the characters) Coldenella, or any other name which might distinguish her among your genera." (Ellis, Letter to Linnæus.)

[1] To him the pretty *Mitchella repens* was dedicated. Dr. Mitchell had sent to Collinson, perhaps as early as the year 1740, a paper in which thirty new genera of Virginian plants were proposed. This Collinson sent to Trew at Nuremberg, who published it in the "Ephemerides Acad. Naturae Curiosorum" for 1748; but in the mean time most of the genera had been already published, with other names, by Linnæus or Gronovius. Among Mitchell's new genera was one which he called Chamædaphne: this Linnæus referred to Lonicera, but the elder (Bernard) Jussieu, in a letter dated February 19, 1751, having shown him that it was very distinct both from Lonicera and Linnæa, and in fact belonged to a different natural order, he afterwards named it Mitchella.

The former collected seeds and living plants for Peter Collinson during more than twenty years, and even at that early day extended his laborious researches from the frontiers of Canada to southern Florida and the Mississippi. All his collections were sent to his patron Collinson,[1] until the death

[1] Mr. Collinson kept up a correspondence with all the lovers of plants in this country, among whom were Governor Colden, Bartram, Mitchell, Clayton, and Dr. Garden, by whose means he procured the introduction of great numbers of North American plants into the English gardens. "Your system," he writes Linnæus, "I can tell you, obtains much in America. Mr. Clayton and Dr. Colden at Albany, on Hudson's River, in New York, are complete professors, as is Dr. Mitchell at Urbana, on Rappahannock River in Virginia. It is he that has made many and great discoveries in the vegetable world." "I am glad you have the correspondence of Dr. Colden and Mr. Bartram. They are both very indefatigable, ingenious men. Your system is much admired in North America." Again, "I have but lately heard from Mr. Colden. He is well, but what is marvellous, his daughter is perhaps the first lady that has so perfectly studied your system. She deserves to be celebrated." "In the second volume of 'Edinburg Essays' is published a Latin botanic dissertation by Miss Colden ; perhaps the only lady that makes a profession of the Linnæan system, of which you may be proud." From all this, botany appears to have flourished in the North American colonies. But Dr. Garden about this time writes thus to his friend Ellis : "Ever since I have been in Carolina, I have never been able to set my eye upon one who had barely a regard for botany. Indeed, I have often wondered how there should be one place abounding with so many marks of the divine wisdom and power, and not one rational eye to contemplate them ; or that there should be a country abounding with almost every sort of plant, and almost every species of the animal kind, and yet that it should not have pleased God to raise up one botanist. Strange, indeed, that the creature should be so rare ! " But to return to Collinson, the most amusing portion of whose correspondence consists of his letters to Linnæus shortly after the publication of the "Species Plantarum," in which (with all kindness and sincerity) he reproves the great Swedish naturalist for his innovations, employing the same arguments which a strenuous Linnæan might be supposed to advance against a botanist of these latter days. "I have had the pleasure," Collinson writes, "of reading your 'Species Plantarum,' a very useful and laborious work. But, my dear friend, we that admire you are much concerned that you should perplex the delightful science of botany with changing names that have been well received, and adding new names quite unknown to us. Thus, botany, which was a pleasant study and attainable by most men, is now become, by alterations and new names, the study of a man's life, and

of that amiable and simple-hearted man, in 1768; and by
him many seeds, living plants, and interesting observations
were communicated to Linnæus, but few, if any, dried specimens. Dr. Garden, who was a native of Scotland, resided in
Charleston, South Carolina, from about 1745 to the commencement of the American Revolution, devoting all the time
he could redeem from an extensive medical practice to the
zealous pursuit of botany and zoölogy. His chief correspondent was Ellis at London, but through Ellis he commenced a
correspondence with Linnæus, and to both he sent manuscript
descriptions of new plants and animals with many excellent
critical observations. None of his specimens addressed to the
latter reached their destination, the ships by which they were
sent having been intercepted by French cruisers; and Linnæus
complained that he was often unable to make out many of Dr.
Garden's genera for want of the plants themselves. Ellis was
sometimes more fortunate, but as he seems usually to have contented himself with the transmission of the descriptions alone,
we find no authentic specimens from Garden in the Linnæan
herbarium.

We have now probably mentioned all the North American
correspondents of Linnæus; for Dr. Kuhn, who appears only
to have brought him living specimens of the plant which bears
his name, and Catesby, who shortly before his death sent a
few living plants which his friend Lawson had collected in
Carolina, can scarcely be reckoned among the number.[1]

none now but real professors can pretend to attain it. As I love you I
tell you our sentiments." (Letter of April 20, 1754.) "You have begun
by your 'Species Plantarum'; but if you will be forever making new
names, and altering good and old ones, for such hard names that convey
no idea of the plant, it will be impossible to attain to a perfect knowledge
in the science of botany." (Letter of April 10, 1755: from Smith's Selection of the Correspondence of Linnæus, etc.)

[1] In a letter to Haller, dated Leyden, January 23, 1738, Linnæus
writes: "You would scarcely believe how many of the vegetable productions of Virginia are the same as our European ones. There are Alps in
the country of New York, for the snow remains all summer long on the
mountains there. I am now giving instructions to a medical student
here, who is a native of that country, and will return thither in the course
of a year, that he may visit those mountains, and let me know whether

The Linnæan Society also possesses the proper herbarium of its founder and first president, Sir James E. Smith, which is a beautiful collection and in perfect preservation. The specimens are attached to fine and strong paper, after the method now common in England. In North American botany, the chief contributors are Menzies, for the plants of California and the Northwest coast; and Muhlenberg, Bigelow, Torrey, and Boott, for those of the United States. Here also we find the Cryptogamic collections of Acharius, containing the authentic specimens described in his works on the Lichens, and the magnificent East Indian herbarium of Wallich, presented some years since by the East India Company.

The collections preserved in the British Museum are scarcely inferior in importance to the Linnæan herbarium itself, in aiding the determination of the species of Linnæus and other early authors. Here we meet with the authentic herbarium of the " Hortus Cliffortianus," one of the earliest works of Linnæus, which comprises some plants which are not to be found in his own proper herbarium. Here also is the herbarium of Plunkenet, which consists of a great number of small specimens crowded, without apparent order, upon the pages of a dozen large folio volumes. With due attention, the originals of many figures in the " Almagestum " and " Amaltheum Botanicum," etc., may be recognized, and many Linnæan species thereby authenticated. The herbarium of Sloane, also, is not without interest to the North American botanist, since many plants described in the " Voyage to Jamaica," etc., and the " Catalogue of the Plants of Jamaica," were united by Linnæus, in almost every instance incorrectly, with species peculiar to the United States and Canada. But still more important is the herbarium of Clayton, from whose notes and specimens Gronovius edited the " Flora Virginica." [1] Many Linnæan species are founded on the plants

the same alpine plants are found there as in Europe." Who can this American student have been? Kuhn did not visit Linnæus until more than fifteen years after the date of this letter.

[1] " Flora Virginica, exhibens plantas quas J. Clayton in Virgini col-

here described for which this herbarium is alone authentic;
for Linnæus, as we have already remarked, possessed very
few of Clayton's plants. The collection is nearly complete,
but the specimens were not well prepared, and are not therefore always in perfect preservation. A collection of Catesby's plants exists also in the British Museum, but probably
the larger portion remains at Oxford. There is besides,
among the separate collections, a small but very interesting
parcel selected by the elder Bartram, from his collection
made in Georgia and Florida almost a century ago, and presented to Queen Charlotte, with a letter of touching simplicity. At the time this fasciculus was prepared, nearly all the
plants it comprised were undescribed, and many were of entirely new genera; several, indeed, have only been published
very recently, and a few are not yet recorded as natives of
North America. Among the latter we may mention *Petive-
ria alliacea* and *Ximinea Americana*, which last has again
recently been collected in the same region. This small parcel
contains the Elliottia, Muhl., Polypteris, Nutt., Baldwinia,
Nutt., Macranthera, Torr., Glottidium, Mayaca, Chaptalia,
Befaria, *Eriogonum tomentosum*, *Polygonum polygamum*,
Vent., *Gardoquia Hookeri*, Benth., *Satureia (Pycnothymus)
rigida*, Cliftonia, *Hypericum aureum*, *Galactia Elliottii*,
Krameria lanceolata, Torr., *Waldsteinia (Comaropsis) lobata*, Torr. & Gr., the *Dolichos? multiflorus*, Torr. & Gr.,
the Chapmannia, Torr. & Gr., *Psoralea Lupinellus*, and others
of almost equal interest or rarity, which it is much to be regretted were not long ago made known from Bartram's discoveries.

The herbarium of Sir Joseph Banks, now in the British
Museum, is probably the oldest one prepared in the manner
commonly adopted in England, of which, therefore, it may
serve as a specimen. The plants are glued fast to half-sheets
of very thick and firm white paper of excellent quality (similar to that employed for merchants' ledgers, etc.), all care-

legit." Ludg. Bat. 8vo, 1743. Ed. 2, 4to, 1762. The first edition is cited
in the "Species Plantarum" of Linnæus; the second, again, quotes the
specific phrases of Linnæus.

fully cut to the same size, which is usually 16½ inches by 10¾, and the name of the species is written on the lower right-hand corner. All the species of a genus if they are few in number, or any convenient subdivision of a larger genus, are enclosed in a whole sheet of the same quality, and labelled at the lower left-hand corner. These parcels, properly arranged, are preserved in cases or closets, with folding-doors made to shut as closely as possible, being laid horizontally into compartments just wide enough to receive them, and of any convenient depth. In the Banksian herbarium, the shelves are also made to draw out like a case of drawers. This method is unrivalled for elegance, and the facility with which the specimens may be found and inspected, which to a working botanist with a large collection is a matter of the greatest consequence. The only objection is the expense, which becomes very considerable when paper worth at least ten dollars per ream is employed for the purpose, which is the case with the principal herbaria in England; but a cheaper paper, if it be only sufficiently thick and firm, will answer nearly as well. The Banksian herbarium contains authentic specimens of nearly all the plants of Aiton's "Hortus Kewensis," in which many North American species were early established. It is hardly proper indeed, that either the elder or younger Aiton should be quoted for these species, since the first edition was prepared by Solander, and the second revised by Dryander, as to vol. 1 and 2, and the remainder by Mr. Brown. Many American plants from the Physic garden at Chelsea, named by Miller, are here preserved, as also from the gardens of Collinson, Dr. Fothergill (who was Bartram's correspondent after Collinson's death), Dr. Pitcairne, etc. There are likewise many contributions of indigenous plants of the United States, from Bartram, Dr. Mitchell, Dr. Garden, Fraser, Marshall, and other early cultivators of botany in this country. The herbarium also comprises many plants from Labrador and Newfoundland, a portion of which were collected by Sir Joseph Banks himself; and in the plants of the northern and arctic regions it is enriched by the collections of Barry, Ross, and Dr. Richardson. Two sets of plants collected by the venerable Menzies in Vancouver's voyage, are preserved at the British Museum, the one incorporated with the Banksian herbarium, the other forming a separate collection. Those of this country are of the Northwest Coast, the mouth of the Oregon River, and from California. Many of Pursh's species were described from specimens preserved in this herbarium, especially the Oregon plants of Menzies and those of Bartram and others from the more southern United States, which Pursh had never visited, although he often adds the mark *v. v.* (*vidi vivam*) to species which are only to be met with south of Virginia.

The herbarium of Walter still remains in the possession of the Fraser family, and in the same condition as when consulted by Pursh. It is a small collection, occupying a single large volume. The specimens, which are commonly mere fragments, often serve to identify the species of the "Flora Caroliniana," although they are not always labelled in accordance with that work.

The collections of Pursh, which served as the basis of his "Flora Americæ Septentrionalis," are in the possession of Mr. Lambert, and form a part of his immense herbarium. These, with a few specimens brought by Lewis and Clark from Oregon and the Rocky Mountains, a set of Nuttall's collections on the Missouri, and also of Bradbury's so far as they are extant, with a small number from Fraser, Lyon, etc., compose the most important portion of this herbarium, so far as North American botany is concerned. There is also a small Canadian collection made by Pursh subsequently to the publication of his Flora, a considerable number of Menzies's plants, and other minor contributions. To the general botanist, probably the fine herbarium of Pallas, and the splendid collection of Ruiz and Pavon (both acquired by Mr. Lambert at a great expense) are of the highest interest; and they are by no means unimportant in their relations to North American botany, since the former comprises several species from the Northwest Coast and numerous allied Siberian forms, while our California plants require in some instances to be compared with the Chilian and Peruvian plants of the latter.

Besides the herbaria already mentioned, there are two others in London of more recent formation, which possess the highest interest as well to the general as to the American botanist, namely that of Professor Lindley, and of Mr. Bentham. Both comprise very complete sets of the plants collected by Douglas in Oregon, California, and the Rocky Mountains, as well as those raised from seeds or bulbs, which he transmitted to England, of which a large portion have from time to time been published by these authors. Mr. Bentham's herbarium is probably the richest and most authentic collection in the world for *Labiatæ*, and is perhaps nearly unrivalled for *Leguminosæ*, *Scrophularineæ*, and the other tribes to which he has devoted especial attention; it is also particularly full and authentic in European plants. Professor Lindley's herbarium, which is very complete in every department, is wholly unrivalled in Orchidaceous plants. The genus-covers are made of strong and smooth hardware paper, the names being written on a slip of white paper pasted on the lower corner. This is an excellent plan, as covers of white paper, in the herbarium of an active botanist, are apt to be soiled by frequent use. The paper employed by Dr. Lindley is $8\frac{1}{2}$ inches in length and $11\frac{1}{2}$ inches wide, which, as he has himself remarked, is rather larger than is necessary, and much too expensive for general use.

The herbarium of Sir William J. Hooker, at Glasgow, is not only the largest and most valuable collection in the world, in the possession of a private individual, but it also comprises the richest collection of North American plants in Europe. Here we find nearly complete sets of the plants collected in the arctic voyages of discovery, the overland journeys of Franklin to the polar sea, the collections of Drummond and Douglas in the Rocky Mountains, Oregon, and California, as well as those of Professor Scouler, Mr. Tolmie, Dr. Gairdner, and numerous officers of the Hudson's Bay Company, from almost every part of the vast territory embraced in their operations from one side of the continent to the other. By an active and prolonged correspondence with nearly all the botanists and lovers of plants in the United States and Canada, as well as by the collections of travelers, this herbarium is rendered unusually rich in the botany of this country; while Drummond's Texan collections, and many contributions from Mr. Nuttall and others, very fully represent the flora of our southern and western confines. That these valuable materials have not been buried, or suffered to accumulate to no purpose or advantage to science, the pages of the "Flora Boreali-Americana," the "Botanical Magazine," the "Botanical Miscellany," the "Journal of Botany," the "Icones Plantarum," and other works of this industrious botanist, abundantly testify; and no single herbarium will afford the student of North American botany such extensive aid as that of Sir William Hooker.

The herbarium of Dr. Arnott of Arlary, although more especially rich and authentic in East Indian plants, is also interesting to the North American botanist, as well for the plants of the "Botany of Captain Beechy's Voyage," etc., published by Hooker and himself, as the collections of Drummond and others, all of which have been carefully studied by this sagacious botanist.

The most important botanical collection in Paris, and, indeed, perhaps the largest in the world, is that of the Royal Museum at the Jardin des Plantes or Jardin du Roi. We cannot now devote even a passing notice to the garden and magnificent new conservatories of this noble institution, much less to the menagerie and celebrated museum of zoölogy and anatomy, of the cabinet of mineralogy, geology, and fossil remains, which, newly arranged in a building recently erected for its reception, has just been thrown open to the public. The botanical collections occupy a portion of this new building. A large room on the first floor, handsomely fitted up with glass cases, contains the cabinet of fruits, seeds, sections of stems, and curious examples of vegetable structure from every part of the known world. Among them we find an interesting suite of specimens of the wood, and another comprising the fruits, or nuts, of nearly all the trees of this country; both collected and prepared by the younger Michaux. The herbaria now occupy a large room or hall, immediately over the former, perhaps eighty feet long and thirty wide above the galleries, and very conveniently lighted from the roof. Beneath the galleries are four or five small rooms on each side, lighted from the exterior, used as cabinets for study and for separate herbaria, and above them the same number of smaller rooms or closets, occupied by duplicate or unarranged collections. The cases which contain the herbaria occupy the walls of the large hall and of the side rooms. Their plan may serve as a specimen of that generally adopted in France. The shelves are divided into compartments in the usual manner : but instead of doors the cabinet is closed by a curtain of thick and coarse brown linen, kept extended by a heavy bar attached to the bottom, which is counterpoised by concealed weights, and the curtain is raised or dropped by a pulley. Paper of very ordinary quality is generally used, and the specimens are attached, either to half sheets or to double sheets, by slips of gummed paper, or by pins, or sometimes the specimen itself is glued to the paper. Genera or other divisions are separated by interposed sheets, having the name written on a projecting slip.

According to the excellent plan adopted in the arrangement of these collections, which is due to Desfontaines, three kinds of herbaria have been instituted, namely : 1. The general herbarium. 2. The herbaria of particular works or celebrated authors, which are kept distinct, the duplicates alone being distributed in the general collection. 3. Separate herbaria of different countries, which are composed of the duplicates taken from the general herbarium. To these, new accessions from different countries are added, which from time to time are assorted and examined, and those required for the general herbarium are removed to that collection. The ancient herbarium of Vaillant forms the basis of the general collection ; the specimens, which are all labelled by his own hand, are in excellent preservation, and among them plants derived from Cornuti or Dr. Surrasin may occasionally be met with. This collection, augmented to many times its original extent, by the plants of Commerson, Dombey, Poiteau, Leschenault, etc., and by the duplicates from the special herbaria, probably contains at this

time thirty or forty thousand species. Of the separate herbaria, the most interesting to us is that made in this country
by the elder Michaux, from whose specimens and notes the
learned Richard prepared the " Flora Boreali-Americana."

Michaux himself, although an excellent and industrious
collector and observer, was by no means qualified for authorship; and it is to L. C. Richard that the sagacious observations, and the elegant, terse, and highly characteristic specific
phrases of this work are entirely due. There is also the very
complete Newfoundland collection of La Pylaie, comprising
about 300 species, and a set of Berlandier's Texan and Mexican
plants, as well as numerous herbaria less directly connected
with North American botany, which we have not room to
enumerate. Here, however, we do not find the herbaria of
several authors which we should have expected. That of
Lamarck, for instance, is in the possession of Professor Rœper
at Rostock, on the shores of the Baltic; that of Poiret belongs
to Moquin-Tandon of Toulouse; that of Bosc, to Professor
Moretti of Pavia; and the proper herbarium of the late Desfontaines, which however still remains at Paris, now forms a
part of the very large and valuable collection of Mr. Webb.
The herbarium of Mr. Webb, although of recent establishment,
is only second to that of Baron Delessert; the two being far
the largest private collections in France, and comprising not
only many older herbaria, but also, as far as possible, full sets
of the plants of recent collectors. The former contains many
of Michaux's plants (derived from the herbarium of Desfontaines), a North American collection, sent by Nuttall to the
late Mr. Mercier of Geneva, a full set of Drummond's collections in the United States and Texas, etc. The latter also
comprises many plants of Michaux, derived from Ventenat's
herbarium, complete sets of Drummond's collections, etc.
But a more important, because original and perhaps complete,
set of the plants of Michaux is found in the herbarium of the
late Richard, now in possession of his son, Professor Achille
Richard, which even contains a few species which do not exist in the herbarium at the Royal Museum. The herbarium
of the celebrated Jussieu, a fine collection, which is scrupu-

lously preserved in its original state, by his worthy son and successor, Professor Adrien Jussieu, comprises many North American plants of the older collectors, of which several are authentic for species of Lamarck, Poiret, Cassini, etc. The herbarium of De Candolle at Geneva, accumulated throughout the long and active career of this justly celebrated botanist, and enriched by a great number of correspondents, is surpassed by few others in size, and by none in importance. In order that it may remain as authentic as possible for his published works, especially the "Prodromus," no subsequent accessions to families already published are admitted into the general herbarium, but these are arranged in a separate collection. The proper herbarium, therefore, accurately exhibits the materials employed in the preparation of the "Prodromus," at least so far as these were in Professor De Candolle's own possession. As almost twenty years have elapsed since the commencement of this herculean undertaking, the authentic herbarium is of course much less rich in the earlier than in the later orders. The *Compositæ*, to which seven years of unremitted labor have been devoted, form themselves an herbarium of no inconsiderable size. It is unnecessary to enumerate the contributors to this collection (which indeed would form an extended list), since the author, at least in later volumes of the "Prodromus," carefully indicates, as fully as the work permits, the sources whence his materials have been derived. The paper employed is of an ordinary kind, somewhat smaller than the English size, perhaps fifteen inches by ten ; and the specimens are attached to half sheets by loops or slips of paper fastened by pins so that they may readily be detached, if necessary, for particular examination. Several specimens from different sources or localities, or exhibiting the different varieties of a species, are retained when practicable ; and each species has a separate cover, with a label affixed to the corner, containing the name and a reference to the volume and page of the "Prodromus" where it is described. The limits of genera, sections, tribes, etc., are marked by interposed sheets, with the name written on projecting slips. The parcels which occupy each compartment of the well-filled shelves are protected by pieces of binder's board, and secured by a cord, which is the more necessary as the doors are not closed by doors or curtains.

The royal Bavarian herbarium at Munich is chiefly valuable for its Brazilian plants, with which it has been enriched by the laborious and learned Martius. The North American botanist will, however, be interested in the herbarium of Schreber, which is here preserved and comprises the authentic specimens described or figured in his work on the grasses, the American species mostly communicated by Muhlenberg. The *Gramineæ* of this and the general herbarium have been revised by Nees von Esenbeck, and still later by Trinius. It was here that the latter, who for many years had devoted himself to the exclusive study of this tribe of plants, and had nearly finished the examination of the chief herbaria of the continent, preparatory to the publication of a new Agrostographia, was suddenly struck with a paralysis, which has probably brought his scientific labors to a close.

The imperial herbarium at Vienna, under the superintendence of the accomplished Endlicher, assisted by Dr. Fenzl, is rapidly becoming one of the most valuable and extensive collections in Europe. The various herbaria of which it is composed have recently been incorporated into one, which is prepared nearly after the English method. It, however, possesses few North American plants, except a collection made by Enslin (a collector sent to this country by Prince Lichtenstein, from whom Pursh obtained many specimens from the southern States), and some recent contributions by Hooker, etc. There is also an imperfect set of plants collected by Hænke (a portion of which are from Oregon and California), so far as they are yet published in the " Reliquæ Hænkeanæ " of Presl, in whose custody, as curator of the Bohemian museum at Prague, the original collection remains.

The herbarium of the late Professor Sprengle still remains in the possession of his son, Dr. Anthony Sprengle, at Halle, but is offered for sale. It comprises many North American plants communicated by Muhlenberg and Torrey. The herbarium of Schkuhr was bequeathed to the university of Wittenberg, and at the union of this university with that of Halle was transferred to the latter, where it remains under the care of Professor von Schlechtendal. It contains a large portion of the Carices described and figured in Schkuhr's work, and is therefore interesting to the lovers of that large and difficult genus. The American specimens were mostly derived from Willdenow, who obtained the greater portion from Muhlenberg.

The royal Prussian herbarium is deposited at Schöneberg (a little village in the environs of Berlin), opposite the royal botanic garden, and in the garden of the Horticultural Society. It occupies a very convenient building erected for its reception, and is under the superintendence of Dr. Klotzsch, a very zealous and promising botanist. It comprises three separate herbaria, namely, the general herbarium, the herbarium of Willdenow, and the Brazilian herbarium of Sello. The principal contributions of the plants of this country to the general herbarium, garden specimens excepted, consist of the collections of the late Mr. Beyrich, who died in western Arkansas, while accompanying Colonel Dodge's dragoon expedition, and a collection of the plants of Missouri and Arkansas by Dr. Engelmann, now of St. Louis ; to which a fine selection of North American plants, recently presented by Sir William Hooker, has been added. The botanical collections made by Chamisso, who accompanied Romanzoff in his voyage round the world, also enrich this herbarium; many are from the coast of Russian America and from California ; and they have mostly been published conjointly by the late Von Chamisso and Professor Schlechtendal in the "Linnæa" edited by the latter.

The late Professor Willdenow enjoyed for many years the correspondence of Muhlenberg, from whom he received the greater part of his North American specimens, a considerable portion of which are authentic for the North American plants of his edition of the "Species Plantarum." In addition to these we find in his herbarium many of Michaux's plants communicated by Desfontaines, several from the German collector Kinn, and perhaps all the American species described by

Willdenow from the Berlin garden. It also comprises a portion of the herbarium of Pallas, the Siberian plants of Stephen, and a tolerable set of Humboldt's plants. This herbarium is in good preservation, and is kept in perfect order and extreme neatness. As left by Willdenow, the specimens were loose in the covers, into which additional specimens had sometimes been thrown and the labels often mixed, so that much caution is requisite to ascertain which are really authentic for the Willdenovian species. To prevent farther sources of error, and to secure the collection from injury, it was carefully revised by Professor Schlechtendal while under his management, and the specimens attached by slips of paper to single sheets, and all those that Willdenow had left under one cover, as the same species, are enclosed in a double sheet of neat blue paper. These covers are numbered continuously throughout the herbarium, and the individual sheets or specimens in each are also numbered, so that any plant may be referred to by quoting the number of the cover and that of the sheet to which it is attached. The arrangement of the herbarium is unchanged, and it precisely accords with this author's edition of the " Species Plantarum." Like the general herbarium, it is kept in neat portfolios, the back of which consists of three pieces of broad tape, which, passing through slits near each edge of the covers, are tied in front: by this arrangement their thickness may be varied at pleasure, which, though of no consequence in a stationary herbarium, is a great convenience in a growing collection. The portfolios are placed vertically on shelves protected by glass doors, and the contents of each are marked on a slip of paper fastened to the back. The herbaria occupy a suite of small rooms distinct from the working rooms, which are kept perfectly free from dust.

Another important herbarium at Berlin is that of Professor Kunth, which is scarcely inferior in extent to the royal collection at Schöneberg, but it is not rich or authentic in the plants of this country. It comprises the most extensive and authentic set of Humboldt's plants, and a considerable number of Michaux's which were received from the younger Richard. As the new " Enumeratio Plantarum " of this industrious botanist proceeds, this herbarium will become still more important.

For a detailed account of the Russian botanical collections and collectors, we may refer to a historical sketch of the progress of botany in Russia, etc., by Mr. Bongard, the superintendent of the Imperial Academy's herbarium at St. Petersburg, published in the "Recueil des Actes" of this institution for 1834. An English translation of this memoir is published in the first volume of Hooker's "Companion to the Botanical Magazine."

NOTES OF A BOTANICAL EXCURSION TO THE MOUNTAINS OF NORTH CAROLINA.[1]

THE peculiar interest you[2] have long taken in North American botany, and your most important labors in its elucidation, indicate the propriety of addressing to yourself the following remarks, relating, for the most part, to the hasty collections made by Mr. John Carey, Mr. James Constable, and myself, in a recent excursion to the higher mountains of North Carolina. Before entering upon our own itinerary, it may be well to notice very briefly the travels of those who have preceded us in these comparatively unfrequented regions. The history of the botany of the Alleghany Mountains would be at once interesting and on many accounts useful to the cultivators of our science in this country; but with my present inadequate means, I can only offer a slight contribution towards that object.

So far as I can ascertain, the younger (William) Bartram was the first botanist who visited the southern portion of the Alleghany Mountains. Under the auspices of Dr. Fothergill, to whom his collections were principally sent, and with whom his then surviving father had previously corresponded, Mr. Bartram left Philadelphia in 1773, and after traveling in Florida and the lower part of Georgia for three years, he made a transient visit to the Cherokee country, in the spring of 1776. In this journey he ascended the Seneca or Keowee River, one of the principal sources of the Savannah, and crossing the mountains which divide its waters from those of the Tennessee, continued his travels along the course of the latter to the borders of the present State of Tennessee. Finding that his researches could not safely be extended in

---

[1] American Journal of Science and Arts, xlii. 1. (1842.)
[2] Sir W. J. Hooker.

this direction, after exploring some of the higher mountains in the neighborhood, he retraced his steps to the Savannah River, proceeding thence through Georgia and Alabama to Mobile. His well-known and very interesting volume of travels [1] contains numerous observations upon the botany of these regions, with occasional popular descriptions, and in a few cases Latin characters of some remarkable plants; as, for example, the *Rhododendron punctatum* (which he calls *R. ferrugineum*), *Stuartia pentagyna* (under the name of *S. montana*), *Azalea calendulacea* (which he terms *A. flammea*), Trautvetteria, which he took for a new species of Hydrastris, *Magnolia auriculata*, etc. He also notices the remarkable intermixture of the vegetation of the north and south, which occurs in this portion of the mountains where Halesia, Styrax, Stuartia, and Gelseminum (although the latter " is killed by a very slight frost in the open air in Pennsylvania ") are seen flourishing by the side of Birches, Maples, and Firs of Canada.

I should next mention the name of André Michaux, who at an early period, amid difficulties and privations of which few can now form an adequate conception, explored our country from Hudson's Bay to Florida, and westward to the Mississippi, more extensively than any subsequent botanist. A few of his plants have not yet been rediscovered, and a considerable number remain among the rarest and least known species of the United States; it may therefore be useful to give a particular account of his peregrinations, especially through the mountain region which he so diligently explored, and in which he made such important discoveries. For this purpose I am fortunately supplied with sufficient materials, having had the opportunity of consulting the original journals of Michaux, presented by his son to the American Philosophical Society. I am indebted for this privilege to the kindness of John Vaughn, Esq., the secretary of the society, who directed my attention to these manuscripts, and permitted me

---

[1] " Travels through North and South Carolina, Georgia, East and West Florida, the Cherokee Country," etc. By William Bartram. Philadelphia, 1791.

to extract freely whatever I deemed useful and interesting. The first fasciculus of the diary is wanting; but we learn from a chance record, as well as from published sources,[1] that he embarked at L'Orient on the 29th of September, 1785, and arrived in New York on the 13th of November. The private journal from which the following information is derived commences in April, 1787; prior to which date he had established two gardens, or nurseries, to receive his collections of living plants until they could be conveniently transported to France: one in New Jersey, near the city of New York; the other about ten miles from Charleston, South Carolina. Into the latter it appears he introduced some exotic trees, which he thought suitable to the climate; and the younger Michaux, who visited this garden several years afterwards, mentions two Ginkgos (*Salisburia adiantifolia*), which in seven years had attained an elevation of thirty feet; also some fine specimens of *Sterculia platanifolia*, and a large number of young plants of *Mimosa Julibrissin*, propagated from a tree which his father had brought from Europe. From this stock, probably, the latter has been disseminated throughout the southern States, and is beginning to be naturalized in many places.

I have no means of ascertaining what portions of the country Michaux had visited previously to April, 1787, when he set out from Charleston on his first journey to the Alleghany Mountains, by way of Savannah, ascending the river of that name to its sources in the Cherokee country, and following very nearly the route taken by Bartram eleven years before.[2] He reached the sources of the Keowee River on the

---

[1] See Michaux, "Flora Boreali-Americana"; Introduction. See also "A Sketch of the Progress of Botany in Western America," by Dr. Short, in the "Transylvania Journal of Medicine," No. 35; and in Hooker's "Journal of Botany" for November, 1840. I am informed that an interesting notice of Michaux is contained in the 8th volume of the "Dictionnaire Encyclopedique de Botanique" (under the head of *Voyageurs*); a work which unfortunately I am not able at this moment to consult.

[2] In this journey he was accompanied by his son, who shortly afterwards returned to Europe. Before they reached Augusta, their horses

14th of June, and was conducted by the Indians across the mountains to the head of the Tugaloo (the other principal branch of the Savannah), and thence to the waters of the Tennessee. After suffering much inconvenience from 'unfavorable weather and the want of food, he returned to the Indian village of Seneca by way of Cane Creek, descended along the Savannah to Augusta, and arrived at Charleston on the 1st of July. His notes in this as well as in subsequent journeys to the mountains often contain remarks upon the more interesting plants he discovered; and in some cases their localities are so carefully specified that they might still be sought with confidence. On the 16th of July he embarked for Philadelphia, which he reached on the 27th; and, after visiting Mr. Bartram, traveled to New York, arriving at the garden he had established in New Jersey about the 1st of August. Returning by water to Charleston the same month, he remained in that vicinity until February, 1788, when he embarked for St. Augustine, and was busily occupied, during this spring, in exploring east Florida. His journal mentions several sub-tropical plants, now well known to be indigenous to Florida, but which are not noticed in his Flora: such as the Mangrove *Guilandina Bonduc*, *Sophora occidentalis*, two or three Ferns, and especially the Orange.[1] Leaving Florida at the beginning of June, he returned by land to Savannah and Charleston, where he was confined by sickness the remainder of the summer. Late in the autumn, however, he made a second excursion to the sources of the Savannah, chiefly to obtain the roots and seeds of the remarkable plants he had previously discovered. He pursued the same route as before, except that he ascended the Tugaloo, instead of the

were stolen, a misfortune which, it appears from Michaux's remarks, was of no uncommon occurrence in those days ; and they were obliged to pursue their journey to that place on foot. On the way he discovered "a shrubby Rumex," which he terms *Lapathum occidentale ;* doubtless the *Polygonella parvifolia* of his Flora, and also the *Polygonum polygamum* of Ventenat.

[1] "Les bois etoient remplis d'oranges aigres," etc. Michaux, MSS. See also Bartram's "Travels"; and Torrey & Gray, "Flora of North America," i. p. 222.

Seneca or Keowee River, crossing over to the latter; and, climbing the higher mountains about its sources in the inclement month of December, when they were mostly covered with snow, he at length found some trees of *Magnolia cordata*, to obtain which was the principal object of this arduous journey. Retracing his steps, he reached Charleston at the end of December, with a large collection of living trees, roots, and seeds. The remainder of the winter Michaux passed in the Bahama Islands, returning to Charleston in the month of May. Early in June he set out upon a journey to a different portion of the mountains of North Carolina, by way of Camden, Charlotte (the county seat of Mecklenburg), and Morganton, reaching the higher mountains at "Turkey Cove, thirty miles from Burke Court House" (probably the head of Turkey Creek, a tributary of the Catawba), on the 15th of June. From this place he made an excursion to the Black Mountain, in what is now Yancey County, and afterwards to the Yellow Mountain, which Michaux at that time considered to be the highest mountain in the United States. If the Roan be included in the latter appellation, as I believe it often has been, this opinion is not far from the truth; since the Black Mountain alone exceeds it, according to Professor Mitchell's recent measurements. Descending this elevated range on the Tennessee side, and traveling for the most part through an unbroken wilderness, near the end of June he reached the Block House on the Holston, famous in the annals of border warfare. Several persons had been killed by the Indians during the preceding week, and general alarm prevailing, Michaux abandoned his intention of penetrating into Kentucky, and resolved to botanize for a time in the mountains of Virginia. He accordingly entered that State, and arrived on the 1st of July at "Washington Court House, première ville dans la Virginie que l'on trouve sur la cote occidentale des montagnes, en sortant de la Carolinie Septentrionale." To this he adds the following note: "Première ville, si l'on peut nommer ville une Bourgade composée de douze maisons (log-houses). Dans cette ville on ne mange que de pain de Mays. Il n'y a viande fraiche, ni cidre, mais seulement du mauvais Rum." Abingdon, the county seat of Washington County, is now a flourishing town; but Michaux's remarks are still applicable to more than one *première ville* in this region. From this place he continued his course along the valley of Virginia throughout its whole extent, crossing New River, the Roanoke, and passing by Natural Bridge, Lexington, Staunton, and Winchester; thence by way of Frederick in Maryland, and Lancaster, Pennsylvania, he arrived at Philadelphia on the 21st of July, and at New York on the 30th. In August and September he returned to Charleston by way of Baltimore, Alexandria, Richmond, and Wilmington, North Carolina. In November he revisited the mountains explored early in the preceding summer, passing through Charlotte, Lincolnton, and Morganton, to his former headquarters at Turkey Cove; from whence he visited the north branch of Catawba (North Cove, between Linville Mountain and the Blue Ridge?), the Black Mountain, Toe River, etc.; and returned to Charleston in December, with two thousand five hundred young trees, shrubs, and other plants. From January until April, 1791, this indefatigable botanist remained in the vicinity of Charleston; but his memoranda for the remainder of that year are unfortunately wanting. The earliest succeeding date I have been able to find is March 22, 1792, when he sold the "Jardin du Roi" at Charleston, and going shortly afterwards by water to Philadelphia, he botanized in New Jersey and around New York until the close of May. In the beginning of June he visited Milford, Connecticut, to procure information from a Mr. Peter Pound, who had traveled far in the northwest; and at New Haven took passage in a sloop for Albany, where he arrived on the 14th of June (having botanized on the way at West Point, Poughkeepsie, etc.); on the 18th he was at Saratoga; on the 20th he embarked at Skenesborough (Whitehall), botanized more or less on the shores of Lake Champlain, reaching Montreal on the 30th of June, and Quebec on the 16th of July.[1] The remainder of this season was

---

[1] Among the plants collected in this journey, he particularly mentions having found *Aconitum uncinatum* near Quebec; but in the Flora no

devoted to an examination of the region between Quebec and
Hudson's Bay, the botany of which, as is well known, he was
the first to investigate. His journal comprises a full and
very interesting account of the physical geography and vege-
tation of that inclement district. Leaving Quebec in October, and returning by the same
route, we find our persevering traveler at Philadelphia early
in December. It appears that he now meditated a most formidable journey, and made the following proposition to the
American Philosophical Society : " Proposé à plusieurs mem-
bres de la Société Philosophique les avantages pour les Etats-
Unis d'avoir des informations geographiques des pays a l'ouest
de Mississippi, et demandé qu'ils aient à endosser mes traites
pour la somme de £3600, si je suis disposé a voyager aux
sources du Missouri, et même rechercher les rivières qui
coulent vers l'ocean Pacifique. Ma proposition ayant été
accepté, j'ai donné a Mr. Jefferson, Secretaire d'Etat, les
conditions auxquels je suis disposé à entreprendre ce voyage.
. . . J'offre de communiquer toutes les connoisances et informations geographiques à la Société Philosophique ; et je reserve à mon profit toutes les connoisances en histoire naturelle
que j'acquirerai dans ce voyage." Remaining in Philadelphia
and its vicinity until the following summer, he set out for
Kentucky in July, 1793, with the object of exploring the
western States (which no botanist had yet visited), and also
of conferring with General Clarke (at Mr. Jefferson's request) on the subject of his contemplated journey to the
Rocky Mountains. He crossed the Alleghanies in Pennsylvania, descended the Ohio to Louisville, Kentucky, traversed
that State and western Virginia to Abingdon, and again traveled through the valley of Virginia to Winchester, Harper's
Ferry, etc., arriving at Philadelphia on the 12th of December
of the same year. Conferences respecting his projected expedition were now renewed, in which Mr. Genet, the envoy from
the French republic, took prominent part ; but here the mat-

other locality is given than the high mountains of North Carolina. Major
Le Conte found it several years ago in the southwestern part of New
York, and Mr. Lapham has recently detected it in Wisconsin.

ter seems to have dropped, since no further reference is made to the subject in the journal; and Michaux left Philadelphia in February, 1794, on another tour to the southern States. In July of that year he again visited the mountains of North Carolina, traveling from Charleston to Turkey Cove by his old route. On this occasion he ascended the Linville Mountain, and the other mountains in the neighborhood; but having "differé a cause du manque des provisions," he left his old quarters (at Ainsworth's), crossed the Blue Ridge, and established himself at Crab Orchard on Doe River. From this place he revisited the Black Mountain, and, accompanied by his new guide, Davenport, explored the Yellow Mountain, the Roan, and finally the Grandfather, the summit of which he attained on the 30th of August.[1] Returning to the house of his guide, he visited Table Mountain on the 5th of September, and proceeded (by way of Morganton, Lincolnton, Salisbury, and Fayetteville, North Carolina) to Charleston, where he passed the winter.

On the nineteenth day of April, 1795, our indefatigable traveler again set out, reached the Santee River at Nelson's Ferry, ascended the Wateree, or Catawba, to Flat Rock Creek, visited Flat Rock,[2] crossed Hanging-Rock Creek, and ascended

---

[1] His earlier journals are full of expressions of loyalty to the king under whose patronage his travels were undertaken; but now transformed into a republican : "Monté au sommet de la plus haute montagne de toute l'Amerique Septentrionale, chanté avec mon compagnon-guide l'hymne de Marseillois, et crié, ' Vive la Liberté et la Republique Française.' " If this enthusiasm were called forth by mere elevation, he should have chanted his pæans on the Black Mountain and the Roan, both of which are higher than the Grandfather.

[2] I believe this is the only instance in which the name of Flat Rock occurs in Michaux's journal ; it is in South Carolina, not far from Camden. Here, without doubt, he discovered *Sedum pusillum* (Diamorpha, Nutt.), the habitat of which is said to be "in Carolina Septentrionali, loco dicto Flat Rock." Mr. Nuttall, who subsequently collected the plant at the same locality, inadvertently continued this mistake, by assigning the habitat, "Flat Rock near Camden, *North* Carolina," as well in his "Genera of North American Plants," as in a letter to Dr. Short on this subject. (Vide Short on Western Botany, in the "Transylvania Journal of Medicine," and in Hooker's "Journal of Botany " for November, 1840, p. 103.)

the Little Catawba to Lincolnton. In the early part of May he revisited Linville Mountain, the Yellow Mountain, the Roan, and some others, and then descended Doe River and the Holston to Knoxville, Tennessee. Thence, crossing the Cumberland Mountains, and a wilderness one hundred and twenty miles in extent, he arrived at Nashville on the 16th of June, at Danville, Kentucky, on the 27th, and at Louisville on the 20th of July. In August he ascended the Wabash to Vincennes, crossed the country to the Illinois River, and devoted the months of September, October, and November to diligent herborizations along the course of that river, the Mississippi, the lower part of the Ohio, and throughout the country included by these rivers. In December he descended the Mississippi in a small boat to the mouth of the Ohio, and ascended the latter and the Cumberland to Clarksville, which he reached on the 10th of January, 1796, after a perilous voyage in the most inclement weather. Leaving that place on the 16th, he arrived at Nashville on the 19th of January; and after making a journey to Louisville and back again, he started for Carolina at the close of February, crossed the Cumberland Mountains early in March, reached Knoxville on the 8th, Greenville on the 18th, Jonesborough on the 19th, and on the 22d crossed the Iron Mountains into North Carolina, descended Cane Creek (which rises in the Roan), and spent several days in exploring the mountains in the vicinity, with his former guide, Davenport. In April he returned to Charleston by his usual route; and on the 13th of August embarked for Amsterdam in the ship Ophir. This vessel was wrecked on the coast of Holland, on the 10th of October, and Michaux lost a part of the collections he had with him; on the 23d of December, 1796, he arrived at Paris

Hence some confusion has arisen respecting the locality of this interesting plant, since there is both a Flat Rock and a village named Camden in North Carolina, although the two are widely separated. After all, Pursh's habitat, "on flat rocks in North Carolina, and elsewhere," proves sufficiently correct, since Mr. Nuttall himself, and also Mr. Curtis and others, have subsequently obtained it in such situations near Salisbury in that State, and Dr. Leavenworth found it abundantly throughout the upper district of Georgia.

with the portion he had saved. This notice of the travels of Michaux on this continent will suffice to show with what untiring zeal and assiduity his laborious researches were prosecuted ; it should, however, be remarked, that greater facilities were afforded him, in some important respects, than any subsequent botanist has enjoyed ; the expenses of his journey having been entirely defrayed by the French government, under whose auspices and direction they were undertaken.

The name of Fraser, so familiar in the annals of North American botany, ought, perhaps, to have preceded that of Michaux in our brief sketch ; since the elder Mr. Fraser, who had visited Newfoundland previous to the year 1784, commenced his researches in the southern States as early as 1785 ; and Michaux, on his first expedition to the mountains in 1787, speaks as having traveled in his company for several days. We believe, however, that he did not explore the Alleghany Mountains until 1789. Under the patronage of the Russian government, he returned to this country in 1799, accompanied by his eldest son, and revisited the mountains, ascending the beautiful Roan, where, " on a spot which commands a view of five States, namely, Kentucky, Virginia, Tennessee, North Carolina, and South Carolina, the eye ranging to a distance of seventy or eighty miles when the air is clear, it was Mr. Fraser's good fortune to discover and collect living specimens of the new and splendid *Rhododendron Catawbiense*, from which so many beautiful hybrid varieties have since been obtained by skilful cultivators." [1] The father and son revisited the southern States in 1807 ; and the latter, after the decease of the father in 1811, returned to this country, and continued his indefatigable researches until 1817.

Many of the rarest plants of these mountains were made

[1] "Biographical Sketch of John Fraser, the Botanical Collector," in Hooker's "Companion to the Botanical Magazine," vol. ii. p. 300 : an article from which I have derived nearly all the information I possess respecting the researches of the Frasers in this country, and to which the reader is referred for more particular information. A full list of the North American plants introduced into England by the father and son, is appended to that account.

known especially to English gardens and collections, by Mr.
John Lyon, whose indefatigable researches are highly spoken
of by Pursh, Nuttall, and Elliott. It is very probable that he
had visited the mountains previous to his assuming the charge
of Mr. Hamilton's collections near Philadelphia, which he
resigned to Pursh in 1802. At a later period, however, he
assiduously explored this region, from Georgia as far north at
least as the Grandfather Mountain; and died at Asheville in
Buncombe County, North Carolina, some time between 1814
and 1818. I am informed by my friend, the Rev. Mr. Curtis,
that his journals and a portion of his herbarium were preserved
at Asheville for many years, and that it is probable that they
may yet be found.

Michaux the younger, author of the "Sylva Americana,"
who accompanied his father in some of his earlier journeys,
returned to this country in 1801, and crossed the Alleghany
Mountains twice; first in Pennsylvania on his way to the
western States, and the next year in North Carolina on his
return to the seaboard. In crossing from Jonesboro', Tennessee, to Morganton, by way of Toe River (not Doe River as is
stated in his Travels), he accidentally stopped at the house of
Davenport, his father's guide in these mountains. The observations of the younger Michaux on this part of the Alleghany
Mountains, in a chapter of his Travels devoted to that subject,
are mainly accurate.

"In the beginning of 1805," Pursh, as he states in the
preface to his Flora, " set out for the mountains and western
territories of the southern States, beginning at Maryland and
extending to the Carolinas (in which tract the interesting
high mountains of Virginia and Carolina took my particular
attention), and returning late in the autumn through the
lower countries along the seacoast to Philadelphia." This
plan, however, was not fully carried out, since he does not
appear to have crossed the Alleghanies into the great western
valley, nor to have botanized along these mountains farther
south than where the New River crosses the valley of Virginia.
At any rate, it is certain that the original tickets of his specimens in the herbarium of the late Professor Barton, under

whose patronage he traveled, as well as those in Mr. Lambert's herbarium, furnish no evidence that he extended his researches into the mountainous portion of North Carolina; but it appears probable (from some labels marked Halifax or Mecklenburg, Virginia) that he followed the course of the Roanoke into the former State. His most interesting collections were made at Harper's Ferry, Natural Bridge, the Peaked Mountains (which separate the two principal branches of the Shenandoah), the Peaks of Otter, in the Blue Ridge; also, Cove Mountain, Salt Pond Mountain, and Parnell's Knob (with the situation of which I am unacquainted), the region around the Warm Sulphur Springs, Capon Springs, the Sweet Springs, and the mountains of Monroe and Greenbrier counties.

Early in the present century, Mr. Kin, a German nurseryman and collector, resident at Philadelphia, traveled somewhat extensively among the Alleghany Mountains, chiefly for the purpose of obtaining living plants and seeds. He also collected many interesting specimens, which may be found in the herbaria of Muhlenberg and Willdenow, where his tickets may be recognized by the orthography, and the amusing mixture of bad English and German (with occasionally some very singular Latin) in which his observations are written.

In the winter of 1816, Mr. Nuttall crossed the mountains of North Carolina from the west, ascending the French Broad River (along the banks of which he obtained his *Philadelphus hirsutus*, etc.) to Asheville, passing the Blue Ridge, and exploring the Table Mountain, where he discovered *Hudsonia montana*, etc., and collected many other rare and interesting plants.[1]

As early as 1817, the mountains at the sources of the Saluda River were visited by the late Dr. MacBride, the friend

---

[1] The spur of the Blue Ridge from which the picturesque Table Mountain rises like a tower, is called by Mr. Nuttall the Catawba Ridge. I am informed, however, by my friend Mr. Curtis, who is intimately acquainted with this interesting region, that it is not known by that name, but is called the Table Mountain Ridge. Its base is not washed by the Catawba River, but by its tributary the Linville.

and correspondent of Elliott; who, in the preface to the second volume of his "Sketch," renders an affecting and deserved tribute to his memory, and acknowledges the important services which he had rendered to that work during its progress.

The name of Rafinesque should also be mentioned in this connection; since that botanist crossed the Alleghanies four or five times between 1818 and 1833 (in Pennsylvania, Maryland, and the north of Virginia), and also explored the Cumberland Mountains.

A few years since, the Peaks of Otter, in Virginia, were visited by Mr. S. B. Buckley; and still more recently the same botanist has explored the mountains in the upper part of Alabama and Georgia, and the adjacent borders of North Carolina. Among the interesting contributions which the authors of the "Flora of North America" have received from this source, I may here mention the *Coreopsis latifolia* of Michaux, which had not been found by any subsequent botanist until it was observed by Mr. Buckley in the autumn of 1840.

No living botanist, however, is so well acquainted with the vegetation of the southern Alleghany Mountains, or has explored those of North Carolina so extensively, as the Rev. Mr. M. A. Curtis; who, when resident for a short time in their vicinity, visited, as opportunity occurred, the Table Mountain, Grandfather, the Yellow Mountain, the Roan, the Black Mountain, etc., and subsequently (although prevented by infirm health from making large collections) extended his researches through the counties of Haywood, Macon, and Cherokee, which form the narrow southwestern extremity of North Carolina. To him we are indebted for local information which greatly facilitated our recent journey, and, indeed, for a complete itinerarium of the region south of Ashe County. But, as the latter county had not been visited by Mr. Curtis, nor so far as we are aware by any other botanist, and being from its situation the most accessible to the traveler from the north, we determined to devote to its examination the principal part of the time allotted to our own excursion.

Intending to reach this remote region by the way of the Valley of Virginia, we left New York on the evening of the 22d of June, and traveling by railroad, reached Winchester, a distance of three hundred miles, before sunset of the following day. At Harper's Ferry, where the Potomac, joined by the Shenandoah, forces its way through the Blue Ridge, in the midst of some of the most picturesque scenery in the United States, we merely stopped to dine, and were therefore disappointed in our hope of collecting *Sedum telephoides, S. pulchellum, Paronychia dichotoma,* and *Draba ramosissima,* all of which grow here upon the rocks. We observed the first in passing, but it was not yet in flower. On the rocky banks of the Potomac below Harper's Ferry, we saw for the first time the common Locust-tree (*Robinia Pseudacacia*) decidedly indigenous. It probably extends to the southern confines of Pennsylvania; and from this point south it is everywhere abundant, but we did not meet with it east of the Blue Ridge. From Winchester, the shiretown of Frederick County, we proceeded by stage-coach directly up the Valley of Virginia, as that portion of the State is called which lies between the unbroken Blue Ridge and the most easterly ranges of the Alleghanies. From the Potomac to the sources of the Shenandoah it is, strictly speaking, a valley, from twenty to thirty miles in width, with a strong, chiefly limestone soil of great fertility. It is scarcely interrupted, indeed, up to where the Roanoke rises; but a branch of the Alleghanies intervenes between the latter and New River, as the upper part of the Great Kenawha is termed, from which point it loses its character in some degree, and is exclusively traversed by the western waters. The same valley extends to the north and east through Maryland and Pennsylvania, and even into the State of New York, preserving throughout the same geological character and fertile soil. Our first day's ride was to Harrisonburg, in Rockingham County, a distance of sixty-nine miles from Winchester. From the moment we entered the valley, we observed such immense quantities of *Echium vulgare,* that we were no longer surprised at the doubt expressed by Pursh whether it was really an introduced plant. This " vile foreign weed," as Dr. Darlington, agriculturally speaking, terms this showy plant, is occasionally seen along the roadside in the northern States; but here, for the distance of more than a hundred miles, it has taken complete possession even of many cultivated fields, especially where the limestone approaches the surface, presenting a broad expanse of brilliant blue. It is surprising that the farmers should allow a biennial like this so completely to overrun the land. Another plant much more extensively introduced here than in the north (where it scarcely deserves the name of a naturalized species) is *Bupleurum rotundifolium*, which in the course of the day we met with abundantly. The *Marubium vulgare* is everywhere naturalized; and *Euphorbia Lathyris* must also be added to the list of naturalized plants. The little *Verbena angustifolia* is also a common weed. We collected but a single indigenous plant of any interest, and one which we by no means expected to find, namely, *Carex stenolepsis* of Torrey,[1] which here, as in western States, to which we supposed it confined, takes the place of the northern *C. retrorsa*. We searched for its constant companion, *C. Shortii*, and the next day we found the two growing together. During the day's ride we observed that the bearded wheat was almost exclusively cultivated, and were informed that it had been found less subject to the ravages of the " Fly " than the ordinary varieties; which may be owing to recent introduction of the seed of the bearded variety from districts unmolested by this insect.

The following day we traveled only sixteen miles on our route, but from Mount Sidney made an interesting excursion on foot to Weyer's Cave, one of the largest, and certainly the most remarkable grotto in the United States. It has been so often described as to render any account on our part super-

---

[1] It is the *C. Frankii* of Kunth (1837), and of Kunze's Supplement to Schkuhr's "Caricography," where it is well figured. It was also distributed among Dr. Frank's plants under the name of *C. atherodes*, and with the locality of Baltimore in Pennsylvania! I had always supposed it to be derived from some part of the western States; but since it abounds in the Valley of Virginia, it may have been collected near Baltimore, Maryland.

fluous. Near the cave we saw some trees of *Tillia hetero-phylla*, Vent. (*T. alba*, Michx. f.), and collected a few specimens with unopened flower-buds. It appears to be the most abundant species along the mountains.

Our ride next day offered nothing of interest. Near Staunton, we saw some patches of *Delphinium consolida*, where it was pretty thoroughly naturalized in the time of Pursh. We did not observe *Spiræa lobata*, which Michaux first met with in this vicinity, and which Pursh, as well as later botanists, found in various parts of the valley. Passing the town of Lexington in the evening, we arrived at the Natural Bridge towards morning, where we remained until Monday, and had an opportunity of botanizing for a short time before we left. On the rocks we found plenty of *Asplenium Ruta-muraria, Sedun ternatum*, and *Draba ramosissima* with ripe fruit. In the bottom of the ravine, directly under the stupendous natural arch (the point which affords the most impressive view of this vast chasm), we collected specimens of *Heuchera villosa*, Michx., in fine flower on the 28th of June; although, in the higher mountains of North Carolina, where it also abounds, the flowers did not appear until near the end of July. This species is excellently described by Michaux, to whose account it is only necessary to add that the petals are very narrow, appearing like sterile filaments. Although a smaller plant than *H. Americana*, the leaves are larger, and vary considerably in the depths of the lobes. It is both the *H. villosa* and *H. caulescens* of Pursh, who probably derived the latter name from the strong elongated rhizoma, often projecting and appearing like a suffrutescent stem, by which the plant is attached to the rocks; since he does not describe the scape as leafy, nor is this at all the case in the original specimens. The *H. caulescens* of Torrey and Gray's Flora with the synonym, must also be united with *H. villosa*, which in that work is chiefly described from specimens collected by Dr. Short in Kentucky, where everything seems to grow with extraordinary luxuriance. With these, the plant we collected entirely accords except that the leaves are mostly smaller, and more deeply lobed;

but this character is not constant.[1] Soon after leaving Natural Bridge, we observed indigenous trees of the Honey Locust (*Gleditschia triacanthos*), also *Æsculus Pavia?* and, in crossing the valley of the James River, we noticed the Papaw (*Uvaria triloba*) and Negundo. The roadside was almost everywhere occupied with *Verbesina Siegesbeckia* not yet in flower; and in many places with *Melissa* (*Calamintha*) *Nepeta*, which Mr. Bentham has not noticed as an American plant, although Pursh has it as a native of the country. It was, however, doubtless introduced from Europe, but is completely naturalized in the valley of Virginia, in Tennessee, and in North Carolina east of the Blue Ridge.

On Tuesday, the 29th of June, we crossed the New River, arrived at Wytheville, or Wythe Court House, towards evening; and at Marion, or Smythe Court House, on the Middle Fork of the Holston, early the next morning. The vegetation of this elevated region is almost entirely similar to that of the northern States. The only herbaceous plants we noticed, as we passed rapidly along, which we had not seen growing before, were *Galax aphylla*, and *Silene Virginica:* the showy, deep-red flowers of the latter, no less than the different habitus, caused us to wonder how it could ever have been confounded with the northern *S. Pennsylvanica.* The only forest tree with which we were not previously familiar was the large Buckeye, *Æsculus flava*, which abounds in this region, and attains the height of sixty to ninety feet, and the diameter of two or three feet or more at the base.

At Marion we determined to leave the valley road, and to cross the mountains into Ashe County, North Carolina; the morning was occupied in seeking a conveyance for this pur-

---

[1] Much to our disappointment we did not meet with *Heuchera hispida*, although I have since learned from an inspection of Barton's herbarium, that we passed within a moderate distance from the place where Pursh discovered it. The habitat given on the original ticket, " High Mountains between Fincastle and the Sweet Springs, and some other similar places," we here cite, with the hope that it may guide some botanist to its rediscovery. The habitat in Pursh's Flora, " High Mountains of Virginia and Carolina," is probably a mere guess, so far as relates to the latter State.

pose. With considerable difficulty we at length procured a carry-all (a light, covered wagon with springs, drawn by a single horse), capable of conveying our luggage and a single person besides the driver, a simple shoemaker who had never before undertaken so formidable a journey, and who accordingly proved entirely wanting in the skill and tact necessary for conducting so frail a vehicle over such difficult mountain tracks, for roads they can scarcely be called. We had first to ascend the steep ridge interposed between the middle and south Forks of the Holston, called Brushy Mountain, during the ascent of which we commenced botanizing in earnest. The first interesting plant we met with was *Saxifraga erosa* of Pursh, but only with ripe fruit, and even with the seeds for the most part fallen from the capsules. The same locality also furnished us with specimens of the pretty *Thalictrum filipes*, Torr. & Gray (to which the name of *T. clavatum*, DC. must be restored), a plant which abounds along all the cold and clear brooks throughout the mountains of North Carolina; where it could not well have escaped the notice of Michaux, in whose herbarium De Candolle found the specimen (with no indication of its habitat) on which his *T. clavatum* was established. The authors of the " Flora of North America," having only an imperfect fruiting specimen of their *T. filipes*, and not sufficiently remarking the discrepancies between the *T. clavatum*, Hook. " Fl. Bor.-Am." and the figure and description of De Candolle's plant, in regard to the length of the styles, assumed the former to be the true *T. clavatum*, and described their own plant as a new species. But our specimens accord so perfectly with the figure of Delessert (except in the greater but variable length of the stipes to the fruit, and in the veining of the carpels, which, doubtless by an oversight of the artist, is omitted in the figure) as to leave no doubt of their identity. The subarctic plant may be appropriately called *T. Richardsonii*, in honor of its discoverer. The flowers of this species are uniformly perfect, as indeed they are figured by Delessert, although De Candolle has otherwise described them. It is a slender, delicate plant, from eight to twelve, or rarely exceeding eighteen inches in height, with pure white flowers. During this ascent we collected *Galium latifolium*, Michx., just coming into flower; and we subsequently found this species so widely diffused throughout the mountains of North Carolina, that we were much surprised at its remaining so little known since the time of Michaux. On a moist, rocky bank by the road-side, we gathered some specimens of a Scutellaria, which did not again occur to us. It proves to be a species mentioned by Mr. Bentham under *S. serrata*, and subsequently described by Dr. Riddell with the name of *S. saxatilis*, which apparently is not of uncommon occurrence westward of the Alleghany Mountains. It is a slender plant, from six to twenty inches high; and the stems often produce slender sub-terranean runners from their base. We here also collected *Asarum Virginicum*, Linn. in similar situations. In the higher mountains the northern *A. Canadense* takes the place of the former species, while *A. arifolium*, Michx. seems to be confined to the lower country. The banks of the shady and cool rivulets, which we crossed every few minutes during our ascent, were in many places covered by the prostrate or creeping *Hedyotis serpyllifolia*, Torr. & Gray (*Houstonia serpyllifolia*, Michx.), which continues to flower sparingly throughout the summer. This pretty plant has quite the habit of *Arenaria Balearica;* and the root is certainly per-ennial. We found it very abundant in similar situations throughout this mountain region. Towards the summit of this ridge we first met with the *Magnolia Fraseri* (*M. auriculata*, Bart.), which resembles the Umbrella-tree (*Magnolia Umbrella*) in the disposition of its leaves at the extremity of the branches. This, as well as *M. acuminata* (the only other species of Magnolia that we observed), is occasionally termed Cucumber-tree; but the people of the country almost uniformly called the former Wahoo, a name which in the lower part of the southern States is applied to *Ulmus alata*, or often to all the Elms indifferently. The bitter and somewhat aromatic infusion of the green cones of both these Magnolias in whiskey or apple-brandy is very extensively employed as a preventive against intermittent fevers; a use which, as the younger Michaux remarks, would doubtless be much less frequent, if, with the same medical properties, the aqueous infusion were substituted.

Nearly at the top of this mountain we overtook our awkward driver, awaiting our arrival in perfect helplessness, having contrived to break his carriage upon a heap of stones, and to overthrow his horse into the boughs of a prostrate tree. So much time was occupied in extricating the poor animal and in temporary repairs to the wagon, that we had barely time to descend the mountain on the opposite side, and to seek lodgings for the night in the secluded valley of the South Fork of the Holston. In moist, shady places along the descent of this mountain, and in similar situations throughout the mountains of North Carolina, we found plenty of the northern *Listera convallarioides,* in fine state, entirely similar to the plant from Vermont, Canada, Newfoundland, and the Northwest Coast, and agreeing completely with the figure of Swartz (in Weber & Mohr, " Beiträge zur Naturkunde," I., 1805, p. 2, t. I.), and the recent one of Hooker's " Flora Boreali Americana." It is difficult to conceive why Willdenow should cite the *Ophrys cordata* of Michaux under the *Epipactis convallarioides* of Swartz, while there is so little accordance in their characters; but this has not prevented Pursh from combining the specific phrase of the two authors into one, while he assigns a locality for the plant (New Jersey), where the *Listera convallarioides* certainly does not grow. The Rev. Mr. Curtis, I believe, first detected the plant in these mountains.

The next day (July 1) we crossed the Iron Mountains (the great chain which divides the States of North Carolina and Tennessee, and which here forms the northwestern boundary of Grayson County, Virginia) by Fox-Creek Gap, and traversing the numerous tributaries of the North Fork of the New River, which abundantly water this sequestered region, we slept a few miles beyond the boundary of North Carolina, after a journey of nearly thirty miles. It must not be imagined that we found hotels or taverns for our accommodation ; as, except at Ashe Court House, we saw no house of public entertainment from the time we left the valley of Virginia until we finally crossed the Blue Ridge and quitted the mountain region. Yet we suffered little inconvenience on this account, as we were cordially received at the farm-houses along the road, and entertained according to the means and ability of the owners ; who seldom hesitated either to make a moderate charge, or to accept a proper compensation for their hospitality, which we therefore did not hesitate to solicit from time to time. On the Iron Mountains we met with nearly all the species we had collected during the previous day, and with a single additional plant of much interest, namely, the *Boykinia aconitifolia*, Nutt. We found it in the greatest abundance and luxuriance on the southern side of the mountain, near the summit, along the rocky margins of a small brook, which for a short distance were completely covered with the plant. It here attains the height of two feet or more ; the stems, rising from a thick rhizoma (and clothed below, as well as the petioles, with deciduous rusty hairs), are terminated by a panicle of small cymes, which at first are crowded, but at length are loose, with the flowers mostly unilateral. The rather large, pure white petals are deciduous after flowering, not marcescent as in Saxifraga and Heuchera. We did not again meet with this plant ; but Mr. Curtis collected it several years ago near the head of Linville River, and Mr. Buckley obtained it in the mountains of Alabama. It also extends further north than our own locality ; for, although not described in his Flora, Pursh collected it on the Salt-Pond Mountain in Virginia.[1] I have little doubt that the *Saxifraga Richardsonii* would be more correctly transferred to Boykinia, as well as the *S. ranunculifolia ;* and, since the *S. elata* of Nuttall, in Torrey and Gray's

---

[1] The specimen in Professor Barton's herbarium (in fruit) is ticketed by Pursh : " *Heuchera villosa*, Michx. ? Salt-Pond Mountain, under the naked knob, near a spring. This spring is the highest I have seen." I know not the exact situation of this mountain from which Pursh obtained many interesting plants. The *Boykinia aconitifolia*, I may remark, would be a very desirable plant in cultivation, and might be expected to endure the winter of New York and Philadelphia ; it would certainly flourish in England.

"Flora," is referred to *Boykinia occidentalis*, in the supplement to that work, no pentandrous Saxifrage remains, except the ambiguous *S. Sullivantii*, Torr. & Gr. But the authors of the Flora having received fruiting specimens of this interesting plant, do not hesitate to remove it from the genus to which it was provisionally appended, and to dedicate it to their esteemed correspondent, the promising botanist who discovered it.[1]

While descending the mountain on the opposite side, we met with *Clethra acuminata*, a very distinct and almost arborescent species, which is well characterized by Michaux. The flowers were not yet expanded; but towards the end of July we obtained from other localities specimens in full flower, while the racemes and capsules of the preceding year were still persistent. The conspicuous bracts, it may be remarked, are as caducous in the wild as they are said to be in the cultivated plant; usually falling before the flower-buds have attained their full size. We also saw *Campanula divaricata*, Michx., not yet in flower; and obtained fruiting specimens of the *Convallaria umbellulata*, Michx. (Clintonia, Raf., not of Dougl.). While the character in Michaux is drawn from

[1] *Sullivantia*. Torrey & Gray, "Fl. N. Amer." suppl. ined. — Calyx inferne imo ovario adnatus, limbo quinquefido. Petala 5 spathulata, unguiculata, integra, summo calycis tubo inserta, marcescentia. Stamina 5, laciniis calycinis breviora : antheræ biloculares. Styli, 2, breves ; stigmatibus simplicibus. Capsula calyce inclusa, bilocularis, birostris, polysperma, rostris intus longitudinaliter dehiscentibus. Semina adscendentia, scobiformia ; testa membracea, relaxata, utrinque ultra nucleum ovalem alatim producta. Embryo cylindricus albumine vix brevior. — Herba humilis, in rupibus calcareis Ohionis vigens ; radice fibrosa perenni ; foliis plerisque radicalibus, rotundato-reniformibus, inciso-dentatis sublobatisve, longe petiolatis ; scapo gracili, decumbente ; floribus parvis (corolla conspicua, alba), cymuloso-paniculatis, post anthesin in apicem pedicellorum arcte deflexis. *S. Ohionis.* — *Saxifraga?* *Sullivantii*, Torrey & Gray, "Fl. N. Amer." i. 575. — Genus a Saxifraga præcipue diversum staminibus petalis isomeris, et seminibus scobiformibus : a Boykinia calyce fere libero, atque seminibus ; ab Heuchera capsula biloculari, etc. ; a Leptarrhena staminibus 5, antheris bilocularibus, et seminibus alato-marginatis, nec utrinque subulatis.

this species, the "planta Canadensis" there mentioned is the nearly allied *Dracæna borealis* of the "Hortus Kewensis." The two species are mixed in Michaux's herbarium; and, although the latter is almost exclusively a northern plant, we found the two species growing together on the Grandfather, Roan, and other high mountains of North Carolina. Towards the base of the mountain we saw for the first time the Pyrularia of Michaux (Oil-nut, Buffalo-tree, etc.; *Hamiltonia oleifera*, Muhl.) : a low shrub which is not of unfrequent occurrence in rich, shady soil. Its geographical range extends from the Cherokee country on the confines of Georgia (where the elder Michaux discovered it on his earliest visit to the mountains, and where Mr. Curtis has recently observed it), to the western ranges of the Alleghanies of Pennsylvania in lat. 40°, where it was found by the younger Michaux.[1] It flowers early in the season, and the oleaginous fruit in the specimens we collected had attained the size of a musket-ball.

In wet places, on the very borders of North Carolina, but still within Virginia, we first met with *Trautvetteria palmata* and *Diphylleia cymosa*; the former in full flower, the latter in fruit. Trautvetteria, which I doubt not is more nearly allied to Thalictrum than to Cimicifuga or Actæa, was collected by Pursh in Virginia, both on the Salt-Pond Mountain and on the Peaks of Otter. The Diphylleia is confined to springy places, and the margins of the shaded mountain brooks, and the rich and deep alluvial soil which is so general throughout these mountains, never occurring perhaps at a lower elevation than three thousand feet above the level of the sea. It is a more striking plant than we had supposed; the cauline leaves (generally two, but sometimes three in number) being often two feet in diameter, and the radicle, which is obicular and centrally peltate as in Podophyllum, frequently still larger; so that it is not easy (at this season) to obtain manageable specimens. The branches of the cymes are usually reddish or purple, and the gibbous deep blue and glaucous berries are almost dry when ripe. The latter often

---

[1] "Travels to the Westward of the Alleghany Mountains," etc. English edition, p. 57, etc.

contain as many as four perfect seeds; and it is proper to
remark that the embryo is not "very minute," as described
in the "Flora of North America"; but, in the ripe seeds
recently examined, is one third the length of the albumen,
as stated by Decaisne, or even longer. The cotyledons are
elliptical, flattish, and nearly the length of the thick, slightly
club-shaped radicle. The whole embryo is also somewhat
flattened; so that when the seed is longitudinally divided in
one direction, the embryo, examined in place, appears to be
very slender, and to agree with De Candolle's description.
The albumen is horny when dry, and has a bitter taste.
Along the roadside we shortly afterwards collected the equivocal *Vaccinium erythrocarpum* of Michaux, or *Oxycoccus erectus* of Pursh; a low, erect, dichotomously branched shrub,
with the habit, foliage, and fruit of Vaccinium, but the flowers
of Oxycoccus. It here occurred at a lower elevation than
usual, scarcely more than three thousand feet above the level
of the sea, and in a dwarfish state (about a foot high): subsequently we only met with it on the summit of the Grandfather and other mountains which exceed the altitude of five
thousand feet, where it is commonly three or four feet high.
We were too early for the fruit, a small, red or purplish
berry, which does not ripen until August or September. It
has an exquisite flavor, according to Pursh, who found the
plant on the mountains of Virginia; but our friend Mr.
Curtis informs us that it is rather insipid, and entirely destitute of the fine acidity of the cranberry.

On the 2d of July we continued our journey (eleven miles)
to Jefferson or Ashe Court House, a hamlet of twenty or
thirty houses, and the only village in the county. Intending
to make this place our headquarters while we remained in the
region, we had the good fortune to find excellent accommodations at the house of Colonel Bower, who evinced every disposition to further our inquiries, and afforded us very important assistance. We may remark indeed, that during our
residence amongst the mountains we were uniformly received
with courtesy by the inhabitants; who for the most part
lacked the general intelligence of our obliging host at Jeffer-

son, and could scarcely be made to comprehend the object of
our visit, or why we should come from a distance of seven
hundred miles to toil over the mountains in quest of their
common and disregarded herbs. Curiosities as we were to
these good folks, their endless queries had no air of imper-
tinence, and they entertained us to the best of their ability,
never attempting to make unreasonable charges. A very fastidious palate might occasionally be at a loss; but good corn-
bread and milk are everywhere abundant; the latter being
used from preference quite sour, or even curdled. Sweet
milk appears to be very generally disliked, being thought less
wholesome, and more likely to produce the "milk sickness,"
which is prevalent in some very circumscribed districts; so
that our dislike of sour and fondness for sweet milk was regarded by this simple people as one of our very many oddi-
ties. Nearly every farmer has a small dairy-house built over
a cold brook or spring, by which the milk and butter are kept
cool and sweet in the warmest weather.

We botanized for several days upon the mountains in the
immediate neighborhood of Jefferson, especially the Negro
Mountain, which rises abruptly on one side of the village, the
Phœnix Mountain, a sharp ridge on the other side, and the
Bluff, a few miles distant in a westerly direction. The altitude of the former is probably between four and five thousand feet above the sea; the latter is apparently somewhat
higher. They are all composed of Mica-slate; and we should
remark that we entered upon a primitive region immediately
upon leaving the Valley of Virginia. The mountain sides,
though steep or precipitous, are covered with a rich and deep
vegetable mould, and are heavily timbered, chiefly with
Chestnut, White Oak, the Tulip-tree, the Cucumber-tree, and
sometimes the Sugar Maple. Their vegetation presents so
little diversity, that it is for the most part unnecessary to distinguish particular localities. Besides many of the plants
already mentioned, and a very considerable number of northern species which we have not room to enumerate, we collected
or observed on the mountain sides, *Clematis Viorna* in great
abundance; *Tradescantia Virginica; Iris cristata* in fruit;

Hedyotis (*Amphiotis*) *purpurea*, which scarcely deserves the name, since the flowers are commonly almost white ; *Phlox paniculata ?* *Aristolochia Sipho*, without flowers or fruit ; *Ribes Cynosbati*, *rotundifolium*, Michx. (*R. triflorum*, Willd.) and *prostratum*, L'Her. ; *Allium cernuum* and *tricoccum ;* [1] *Galax aphylla ; Ligusticum actæifolium*, the strong-scented roots of which are eagerly sought and eaten by boys and hogs ; [2] The Ginseng, here called " sang " (the roots of which are largely collected, and sold to the country merchants when fresh for about twelve cents per pound, or when dry for triple that price) ; *Menziesia globularis*, mostly in fruit ; and the showy *Azalea calendulacea*, which was also out of flower, except in deep shade. [3] In the latter situations

---

[1] The latter is known throughout this region by the name of " Ramps " ; doubtless a corruption of " Ramsons," the popular appellation of *A. ursinum* in England.

[2] It is here termed " Angelico " ; while in Virginia it is called " Nondo." Bartram (Travels, p. 45, and p. 367), who found it in Georgia, notices it under the name of *Angelica lucida*, or " White-root " of the Creek and Cherokee traders. " Its aromatic carminative root is in taste much like that of ginseng, though more of the taste and scent of aniseseed : it is in high estimation with the Indians as well as white inhabitants, and sells at a great price to the southern Indians of Florida, who dwell near the sea-coast, where this never grows spontaneously." (Bartram, *l. c.*)

[3] Bartram well describes this species, under the name of *Azalea flammea*, or fiery Azalea. " The epithet fiery I annex to this most celebrated species of Azalea, as being expressive of the appearance of its flowers ; which are in general of the color of the finest red-lead, orange and bright gold, as well as yellow and cream-color. These various splendid colors are not only in separate plants, but frequently all the varieties and shades are seen in separate branches on the same plant ; and the clusters of the blossoms cover the shrubs in such incredible profusion on the hillsides, that suddenly opening to view from dark shades, we are alarmed with apprehension of the woods being set on fire. This is certainly the most gay and brilliant flowering shrub yet known ; they grow in little copses or clumps, in open forests as well as dark groves, with other shrubs, and about the bases of hills, especially where brooks and rivulets wind about them ; the bushes seldom rise above six or seven feet in height, and generally but three, four, or five, but branch and spread their tops greatly ; the young leaves are but very small whilst the shrubs are in bloom, from which circumstance the plant exhibits a greater show of splendor." (Bartram's Travels, p. 323.)

we found an arborescent tetramerous species of Prinos (in
fruit only), with large and membranaceous ovate leaves.
The same species has been collected on the Pokono Mountains in Pennsylvania by Mr. Wolle, and on the Catskills by
Mr. S. T. Carey. We should deem it the *P. lœvigatus* of
Pursh (not of Torrey, Fl. Northern States), on account of
the solitary and subsessile fertile flowers, as well as the
habitat, were not the flowers of that species said to be hex-
amerous.

In damp, very shady places high up the Negro Mountain
we saw an Aconitum not yet in flower; and on moist rocks
near the summit, obtained a few fruiting specimens of a Saxifraga which was entirely new to us. In a single very secluded
spot on the north side of the mountain, near the summit, the
rocks were covered with a beautiful small Fern, which proves
to be the *Asplenium Adiantum-nigrum* of Michaux, the *A.
montanum*, Willd., an extremely rare plant. It is certainly
distinct from the *A. Adiantum-nigrum;* being not only a
much smaller and more delicate species (two to four inches
high), but the fronds are narrower, the pinnæ ovate and
much shorter, 3–5 parted, with the pinnulæ toothed or in-
cised at the apex.

The *Veratrum parviflorum*, Michx., is of frequent occurrence throughout this region, but was not yet fully in flower,
so that our specimens were not collected until near the end of
July. The plant is excellently described in the Flora of
Michaux, where it is, probably with justice, referred to Veratrum rather than to Melanthium; since the divisions of the
perianth (yellowish-green from the first) are wholly destitute
of glands, and only differ from Veratrum in being stellate,
and tapering at the base. I may here remark that the name
Melanthium must undoubtedly be retained for *M. Virginicum* and *M. hybridum.* Some years since, in rearranging the
North American species of this family, I followed Rœmer
and Schultes in adopting the genus Leimanthium of Willdenow, without considering that Melanthium was established by
Clayton and Gronovius on *M. Virginicum*, and thus taken
up by Linnæus, with the addition of a Siberian plant, which

belongs to Zigadenus.[1] The *Melanthium Capense* (Andro-
cymbium, Willd.) was added some time afterwards.

The rocky summits of the mountains afforded us *Sedum telephoides ; Heuchera villosa ; Paronychia argyrocoma,* which
forms dense silvery tufts on the highest and most exposed
peaks ; *Veronica officinalis, serpyllifolia,* and *agrestis* (all
certainly native) ; *Lycopodium rupestre,* in a very beautiful
state, and on the Phœnix Mountain we found a solitary specimen of *L. Selago ; Arabis lyrata,* with perfectly accumbent
cotyledons ; *Potentilla tridentata,* which we only saw on the
Bluff Mountain; *Woodsia Ilvensis ; Saxifraga leucanthemifolia,* which not unfrequently attains the height of two feet,
with a large and slender effuse panicle; *Diervilla trifida,*
entirely resembling the northern plant ; *Pyrus melanocarpa ;
Sorbus Americana, β. microcarpa ; Rhododendron Catawbiense,* just out of flower, while *R. maximum,* extremely
abundant along the streams and mountain sides, was only beginning to expand its blossoms.[2] In such situations also we
found a marked dwarfish variety of *Hedyotis purpurea,* growing somewhat in tufts, and scarcely exceeding four or five
inches in height. The flowers, which are deep pink, while in
the ordinary form in this region they are nearly white, present the dimorphism which obtains in several sections of the
genus ; the stamens in some specimens being inserted in the
throat of the corolla and exsert, while in others they are inserted near the base of the tube and included ; in the former
the style is uniformly short and included, and in the latter
long and somewhat exserted. These two forms were often
seen growing side by side, and appeared to be equally fertile. The *Amianthium muscœtoxicum,* which is common in
the low country of the southern States, we here found only

[1] The *Helonias glaberrima,* " Botanical Magazine," t. 1680, on which *Zigadenus commutatus* of Schultes is founded, is *Z. glaucus ;* the specimens
came from Fraser's nursery, but doubtless were not derived from the
southern States. *Helonias bracteata,* " Botanical Magazine," t. 1703, is
Z. glaberrimus, Michx., not fully developed.

[2] These shrubs bear the name of " Laurel " ; while the *Kalmia latifolia* is universally called " Ivy," or " Ivy-bush."

in the rich open woods of the Bluff Mountain, and in similar places further south. The flowers are pure white or cream-color, in a dense and very showy raceme, at length changing to green. The cattle which roam in the woods for a great part of the year are sometimes poisoned by feeding, as is supposed, on the foliage of this plant during the autumn : hence its name of " Fall-poison." The wild Pea-vine, which is so highly prized as an autumnal feed for cattle, is the Amphi-carpæa.¹ The Lily of the Valley (*Convallaria majalis*), which we occasionally met with in fruit, appears to be identical with the European plant. It extends from the mountains of Virginia to Georgia, where it was long ago noticed by the younger Bartram. We also collected a handsome Phlox, of frequent occurrence in rich woods, which differs from *P. Carolina* (with which it has perhaps been confounded) in its perfectly smooth stem, and broader, less pointed calyx-teeth. The leaves are sometimes an inch in width, and four or five in length; the uppermost often ovate-lanceolate, and more or less cordate at the base.

A species of Carex, nearly allied to *C. gracillima*, occurs in the greatest abundance on all the higher mountains of North Carolina, forming tufts on the earth or on rocks, and flowering throughout the summer. On this account it is called *C. æstivalis* by Mr. Curtis, who discovered it several years since, and pointed out its characters.² We also met

¹ In the large woods the surface of the soil is covered with a species of wild peas, which rise three feet above the earth, and of which the cattle are very greedy. They prefer this pasture to every other, and when removed from it they fall away, or make their escape to return to it." (Michaux, F. A., Travels, p. 316.)

² *C. æstivalis* (M. A. Curtis, ined.) : spicis 3–5 gracilibus laxifloris suberectis, infirma pedunculata, cæteris subsessilibus, suprema androgyna inferne mascula, bracteis inferioribus foliaceis vix vaginantibus superioribus setaceis, perigyniis ovoideis trigonis basi apiceque acutiusculis obsolete nervosis glabris ore subintegro squamam ovatam obtusam (nunc mucronatam) duplo superantibus, stigmatibus tribus, vaginis foliorum inferiorum pubescentibus.

Hab. in montibus altioribus Carolinæ Septentrionalis ubique. Julio-Augusto floret. — *C. gracillimæ* nimis affinis; at diversa, culmis foliisque gracilioribus, vaginis infirmis pubescentibus ; bracteis vix vaginantibus ;

with *C. canescens*, Linn. ex Boott (*C. Buxbaumii*, Wahl.) and *C. conoidea*, Schk., on the moist grassy brow of a precipice of the Bluff ; and towards the base of the Negro Mountain we observed *C. virescens* and *C. digitalis*, Willd. In a cool, sequestered brook, we found the true *Cardamine rotundifolia*, Michx., growing like the Water-cress (for which it might be substituted, as its leaves have exactly the same taste), but producing numerous stolons two or three or

spicis angustioribus et laxifloris erectis, superioribus brevissime pedunculatis ; acheniis oblongo-ovoideis magis stipitatis.

The figure of *C. gracillima*, in Professor Kunze's Supplement to Schkuhr's Carices, is excellent, except that the immature perigynia are represented with more distinct beaks than I have ever seen. To this genus, already perhaps the most extensive in the vegetable kingdom, after Senecio, Mr. Sullivant has recently added another species, an account of which may be appended to this note. As Dr. Boott had already dedicated it to the zealous discoverer, without being aware that he had distributed it under another name, I trust I may be allowed to publish the notes of this sedulous caricographer unchanged :

" *C. Sullivantii* (Boott) : spica mascula solitaria cylindrica, fœmineis 3–5 cylindricis erectis gracilibus pedunculatis laxifloris, superioribus contiguis, infirma remota longe pedunculata basi attenuata, stigmatibus tribus, perigyniis ellipticis brevi-rostratis emarginatis pellucido-punctatis apice marginibusque piloso-hispidis squamam ovatam ciliatam hispido-mucronatam subæquantibus.

"Culmus bipedalis, gracilis, triqueter, pilis albis sparsis longis scabriusculus, pars spicas gerens 2–9-uncialis. Folia 2 lin. lata, culmo breviora, marginibus nervisque scabris. Bractea infirma vaginans, foliacea, culmum adæquans, reliquæ sensim breviores, superiores evaginatæ demum setaceæ. Spica mascula uncialis, vix lineam lata, sessilis vel brevi-pedunculata : squamæ, muticæ, obtusæ, apice ciliobatæ, nemo scabro, pallidæ castaneæ spicæ fœmieæ 3–5, laxifloræ, 1–1½ uncias longæ, 1–1½ lineas latæ ; superiores contiguæ ; infirma remota (uno exemplo basi composita) ; squamæ pellucidæ ciliolatæ, nervo viridi scabro, hispido-mucronatæ. Pedunculi scabri, superiores sensim breviores. Perigynium (vix maturum) 1⅔ lin. longum, ⅞ lin. latum, viride, enervium ? apice hispidulum, ciliatum, brevistipitatum, squamam subæquans vel eo paululum longius. Achenium immaturum." (Boott in litt.)

Hab. in sylvaticis prope Columbum, Ohionis, ubi detexit W. S. Sullivant, cum C. pubescente, C. gracillima, etc. vigens. Affinis *C. arctatæ* (C. sylvaticæ, auct. Amer.) ex cl. Boott. — In exemplis nuperrime receptis, perigynia satis matura sunt ovato-elliptica, lata, compresso-plana, enervia (marginibus exceptis), apice vix rostrata.

more feet in length. These runners arise not only from the
base of the stem, but from the axils of the upper leaves, and
very frequently from the apex of the weak ascending raceme
itself, which is thus prolonged into a leafy stolon, hanging
down into the water or mud, where it takes root. Its habit
and appearance are so unlike even the summer state of our
northern *C. rhomboidea*, that we could not hesitate to consider it a distinct species. The subjoined diagnostic character
will doubtless suffice for its discrimination.[1]

On the 7th of July we started for the high mountains
farther south, having hired a cumbrous and unsightly, but
convenient tilted wagon, with a pair of horses and a driver
(who rode one of the horses, according to the usual custom of
this region), for the conveyance of our luggage, and which
afforded us, at intervals, the luxury of reposing on straw at
the bottom, while we were dragged along at the rate of two or
three miles an hour.

Our first day's journey of about twenty-four miles was
somewhat tedious, as we found no new plants of any interest.
We saw, however, a variety of *Lonicera parviflora?* with
larger leaves and flowers than ordinary, the latter dull purplish; probably the *Caprifolium bracteosum*, var. *floribus
violaceo-purpureis* of Michaux. The following morning we
reached the Watauga River (a tributary of the Holston);
and leaving our driver to follow up the banks of the stream
to the termination of the road at the foot of the Grandfather,
we ascended an adjacent mountain, called Hanging-rock, and
reached our quarters for the night by a different route. The
fine and close view of the rugged Grandfather amply rewarded

---

[1] *Cardamine rotundifolia* (Michx.) : glaberrimer decumbens, stolonibus
repentibus, radice fibrosa, foliis omnibus conformibus (radicalibus sæpe
trisectis, segmentis lateralibus parvis), petiolatis rotundatis plerumque
subcordatis integriusculis vel repando sinuatis, siliquis parvis stylo subulatis, stigmate minuto, seminibus ovalibus. *C. rotundifolia*, Michx., Fl. 2,
p. 30 ; Hook., Bot. Misc., 3, t. 109, (statu vernali : in exemplis Caroli-
nianis folia caulinia magis petiolata;) Darlingt., Fl. Cest., ed. 2, p. 384.
C. rotundifolia, Torr. & Gray, Fl. N. Amer., i. p. 88.

Hab. in rivulis fontibusque opaculis montium Carolinæ, Virginiæ,
Kentucky, et in Pennsylvania.

the toil of ascending this beetling cliff, where we also obtained the *Geum* (*Sieversia*) *radiatum*, probably the most showy species of the genus. The brilliant golden flowers have a disposition to double, even in the wild state, in which we often found as many as eight or nine petals. This tendency would doubtless be fully developed by cultivation. Around the base of these mountains we saw *Blephilia nepetoides*, and another Labiate plant not yet in flower, which we took for *Pycnanthemum montanum*, Michx.

The next day (July 9th) we ascended the Grandfather, the highest as well as the most rugged and savage mountain we had yet attempted; although by no means the most elevated in North Carolina, as has generally been supposed.[1] It is a sharp and craggy ridge, lying within Ashe and Burke counties, very near the northeast corner of Yancey, and cutting across the chain to which it belongs (the Blue Ridge) nearly at right angles. It is entirely covered with trees except where the rocks are absolutely perpendicular; and towards the summit, the Balsam Fir of these mountains, *Abies balsamifera*, partly, of Michaux's Flora (but not of the younger Michaux's Sylva), the *A. Fraseri*, Pursh, prevails, accompanied by the *Abies nigra* or Black Spruce. The earth, rocks, and prostrate decaying trunks, in the shade of these trees, are carpeted with mosses and lichens; and the whole present the most perfect resemblance to the dark and sombre forests of the northern parts of New York and Vermont, except that the trees are here much smaller. The resemblance extends to the whole vegetation; and a list of the shrubs and herbaceous plants of this mountain would be found to include a large portion of the common plants of the extreme northern States and Canada.[2] Indeed the vegetation is essentially Canadian,

---

[1] According to Professor Mitchel's barometrical measurements, the Grandfather attains the altitude of five thousand five hundred and fifty-six feet above the level of the sea; the Roan, six thousand and thirty-eight feet; and the highest peak of the Black Mountain, six thousand four hundred and seventy-six feet, which exceeds Mount Washington in New Hampshire (hitherto accounted the highest mountain in the United States) by more than two hundred feet.

[2] Among the northern species which we had not previously observed

with a considerable number of peculiar species intermixed. Under the guidance of Mr. Levi Moody we followed the Watauga, here a mere creek, for four or five miles along the base of the Grandfather, until we reached a ridge which promised a comparatively easy ascent. In the rich soil of this ridge, at an elevation of about four hundred feet above the Watauga, we found one of the plants which of all others we were desirous of obtaining, namely, *Carex Fraseriana.* Mr. Curtis had made diligent but ineffectual search for this most singular and rarest of Carices, "along the Catawba near Morganton," and "near Table Mountain," where Fraser is said to have discovered it; and we believe that no subsequent botanist has ever met with it, except Mr. Kin, whose specimen in Muhlenberg's herbarium is merely ticketed " Deigher walli in der Wilternus." Muhlenberg assigns the habitat, " Tiger Valley, Pennsylvania; " but Kin probably obtained his plant in Tygart's Valley, Virginia, a secluded vale among the western ranges of the Alleghanies (in Randolph County), not far from Greenbrier Mountains, and other localities visited by this collector, as his tickets prove. Kin cultivated the plant for some time at Philadelphia, where it was seen by several botanists, and among them by Pursh, who took it for the *Mapinia sylvatica* of Aublet; a mistake which he did not discover whilst writing his Flora in Europe, although he had the cultivated *Carex Fraseriana* before him. We were too late for good specimens, but succeeded in obtaining a considerable number with the fruit still adherent. The plant grows in tufts, after the manner of *C. plantaginea;* the evergreen leaves are a foot or more in length, and often an inch and a half in width, with singularly undulate margins; the slender scapes are naked except towards the root, where they are sheathed by the convulate bases of the leaves. To the description of the spike, fruit, etc., we have nothing of any consequence to add.

in this region, we may mention *Carex flexuosa, C. plantaginea, C. scabrata, C. intumescens, Oxalis Acetosella, Streptopus roseus, Viburnum lantanoides,* and *Platanthera orbiculata* in the finest condition, and in greater profusion than we ever before met with this, the most striking of North American *Orchidaceæ.*

Long before we reached the summit we again met with the new Saxifraga,[1] which we had previously gathered on the mountains near Jefferson; but we now found it in great abundance, both in flower and with mature fruit. It grew in the greatest profusion on the dripping face of a rocky precipice near our encampment for the night, on the northwestern side of the mountain, five or six hundred feet beneath the highest summit. The vegetation here is so backward that the *Saxifraga leucanthemifolia* growing on the brow of this precipice was not yet in blossom, and the *Saxifraga erosa*, Pursh, in the wet soil at its base was scarcely out of flower, while at the foot of the mountain it had long since shed its seeds. We were therefore enabled to satisfy ourselves that *S. erosa* belongs to the section Hydatica, and that the *S. Wolleana*, Torr. & Gray, from a mountain near Bethlehem in Pennsylvania, is only a variety of this species. Pursh gathered his plant in Virginia, "out of a run near the road from the Sweet Springs to the Union Springs, five miles from the former." But if this species be the *Robertsonia micranthifolia* of Haworth's Succulent Plants, as is most probable, and consequently the *Aulaxis micranthifolia* of this author's

---

[1] *Saxifraga Careyana* (spec. nov.): foliis radicalibus longe petiolatis glabris (tenuibus) ovato-rotundis grosse crenato-dentatis basi truncatis vel subcordatis, scapo gracili nudo apice paniculato-cymoso, floribus effusis, pedicillis filiformibus, petalis lanceolato-oblongis sessilibus sepala recurva plus duplo superantibus, carpellis discretis turgidis demum divaricatis calyce liberis. Variat 1, scapo petiolisque glabriusculis : 2, scapo pedicillis nec non pagina foliorum pilis viscosis pubescentibus : 3, scapo foliis aut bracteis foliaceis 1–2 instructo : 4, foliis ovalibus oblongisve, nunc argute dentatis, in petiolum plus minus attenuatis.

Crescit in rupibus humidis opacis altissimorum montium comitatus Ashe, præsertim ad montem Grandfather dictum, alt. 3500–5000 pedes: Junio floret. Herba spithamæa, rarius pedalis. Flores parvi. Petala consimilia, sessilia, subtriplinervia, alba, immaculata. Filamenta subulato-filiformis. Carpella ovoidea, stylis brevibus apiculata, (stigmatibus subincrassatis,) basi vix aut ne vix coalita ad maturitatem per totam suturam ventralem dehiscentia, ut in pleris Saxifragis plus minus apocarpeis. Semina ovalia, striis elevatis denticulatis (per lentem augentem) longitudinaliter notata. Species distinctissima, habitu ad sect. Hydaticam, sed characteribus ad Micrantheum accedens.

subsequent " Enumeration of Saxifragaceous Plants," it must have been introduced into the English gardens by Fraser as early as 1810.[1] We know not how such a common plant could have escaped the notice of Michaux. Under the name of Lettuce the leaves are eaten by the inhabitants as a salad. At this same place we also saw an Umbelliferous plant not yet in flower, which we believe to be *Conioselinum Canadense,* Torr. & Gr. (*Selinum Canadense,* Michx.), a very rare plant in the extreme northern States and Canada, to which we had supposed it exclusively confined. We found plenty of *Cimicifuga Americana,* Michx., but were obliged to content ourselves with specimens not yet in flower, and with vestiges of the last year's fruit. It should be collected in September.

We were also too early in the season for *Chelone Lyoni,* Pursh, which we found in abundance between the precipice mentioned above and the summit of the mountain, with the flower-buds just beginning to appear. Mr. Curtis remarks that Mr. Nuttall could not have met with this exclusively mountain flower near Wilmington; and also, that the *C. Lyoni* of Pursh and the *C. latifolia* of Muhlenberg and Elliott, are doubtless founded upon one and the same species. Both, indeed, are said to have been collected by Lyon, and the leaves vary from ovate-lanceolate or oval with an acute base, to ovate with a rounded but scarcely cordate base. Pursh's character is drawn from a cultivated specimen. Here we again met with the *Aconitum* previously observed in similar situations on the Negro Mountain, and which, being then only in bud, we took for the *A. uncinatum,* a species collected in this region by Michaux, and recently by Mr. Curtis and other botanists. We were greatly surprised, therefore, to find that

---

[1] The only important discrepancy respects Haworth's character, " Corolla irregularis, petalis 2 inferioribus elongatis divaricantibus gracilioribus," and " Flores albi, rubro minute punctati " ; while the petals in our plant are very nearly equal and similar, and pure white, except the yellow spot at the base. *Aulaxis nuda,* Haworth, *l. c.* (of unknown origin), appears to be the more ordinary and nearly glabrous form of this species. Mr. Don's description of *S. erosa,* probably drawn from the cultivated plant, also differs from our plant in several minor points.

our plant, here just coming into blossom, had cream-colored flowers, very different from those of *A. uncinatum*, and more nearly resembling those of *A. Lycoctonum*.[1] On our return to Jefferson, we obtained good specimens at our original locality, where it is very abundant. The weak stems at first ascending, become prostrate when the plant is in flower, and frequently attain the length of seven or eight feet. As the stem does not climb, and its flowers are so different from those of *A. uncinatum*, it can hardly be the plant mentioned by Pursh under that species, which he saw at the foot of the Peaks of Otter and about the Sweet Springs in Virginia. It may be remarked that the ovaries of *A. uncinatum* are often nearly glabrous, and the claws of the petals entirely so. The seeds are strongly plicate-rugose, with a wing-like margin on one side.

Near the summit of the mountain we saw immense quantities of a low but very large-leaved Solidago, not yet in flower, which I take to be the *S. glomerata* of Michaux, who could not have failed to observe such a conspicuous and abundant plant, especially as it must have been in full blossom at the

---

[1] *Aconitum reclinatum* (spec. nov. § Lycoctonum) : caule elongato decumbente foliisque palmatifidis glabris, lobis divaricatis cuneatis apicem versus incisis, racemis paniculisve divergentibus laxifloris (floribus albidis), bracteolis minimis, galea horizontali conico-cylindracea ore obliquo, labio cucullorum obcordato ab ungue distante, calcare adunco, filamentis edentulis, carpellis glabris 2–4–spermis, seminibus (immaturis) squamoso-rugosis.

Hab. in opacissimis sylvis ad montes Negro Mountain et Grandfather dictos, alt. 4000–5000 pedes. Julio-Augusto floret. Caulis flaccidus, adscendens vel declinatus, denique procumbens, 3–8 pedalis, ramis gracilibus, seu paniculis laxifloris, divaricatis. Folia flaccida ; inferiora longe petiolata (circumscriptione suborbiculari), profunde 5-7 fida ; segmentis interdum 2-3 lobatis apice incisodentatis dentibus mucronatis; summa subsessilia, 3–5 partita ; venis et pagina quandoque superiori tenuissime pubescentibus. Pedicelli sparsi (pedunculique puberuli) flore longiores, bracteolis 2–3 minimis stipati. Flores minores quam in *A. Lycoctono*, albi vix flavidis tincti (in siccis leviter purpurascentes); sepalis intus pilis aureis barbatis. Galea primum adscendens, mox horizontalis, rostello brevi ractiusculo. Unguis petalorum medium cuculli adfixus ; saccus angustus, ore valde obliquo in labium obcordatum expanso. Ovaria tria, 4–6–ovulata.

time he ascended this mountain. It does not, however, altogether accord with Michaux's description, nor does that author notice the size of the heads, which in our plant are among the largest of the genus. Specimens in flower were procured by Mr. Curtis, who visited this mountain at a more favorable season. With the latter we found a Geum, which Mr. Curtis had formerly observed on the Roan Mountain (where we afterwards met with it in great abundance), and referred, I think correctly, to *G. geniculatum*, Michx., although that species is said to have been collected in Canada. The lower portion of the style is less hairy in our specimens than in Michaux's plant, a difference which, if constant, is not perhaps of specific importance. In the subjoined character I have supplied an inadvertent omission in the " Flora of North America," where the sessile head of carpels, which so readily distinguishes this species from *G. rivale*, is not mentioned.[1] Here we again found *Vaccinium erythrocarpum*, as already mentioned ; and obtained beautiful flowering specimens of *Menziesia globularis*, a straggling shrub which in this place attains the height of five or six feet.

[1] *Geum geniculatum* (Michx.): capitulo carpellorum sessili, articulo styli superiore plumoso inferiorem pubescentem excedente, achenio hirsuto petalis cuneato-obovatis (nunc emarginatis aut leviter obcordatis) exunguiculatis calycem æquantibus; floribus mox erectis.

β *Macreanum*, articulo styli inferiore sursum glabrescente. *G. Macreanum*, M. A. Curtis, in litt.

Crescit in Canada ex Michaux : an recte ? Var. β in umbrosis ad montes Grandfather et Roan, Carolinæ Septentrionalis, alt. 5500-6000 pedes, ubi imprimis detexit cl. Curtis. Julio floret. Caulis 2-3-pedalis, gracilis, foliosus, inferne pilis rigidiusculis retrorsis, superne pilis mollibus patentibus crebrioribus villosus. Folia membranacea ; radicalia nunc palmatim 3-secta, nunc interrupte pinnatisecta, haud rariusque indivisa vel sublobata in eodem stirpe ; caulinia trisecta trilobatave, lobis acutis ; superiora sessilia. Flores minores et numerosiores quam in *G. rivali ;* petala albida, venis purpurascentibus. Styli pars inferior portione plumosa primum multo, postremum modice brevior, in exemplo Michx. manifeste, at juxta apicem parce piloso-pubescens ; in var. β superne glabrata.

Should the Carolina plant hereafter prove to be a distinct species, it will of course retain the name proposed by Mr. Curtis, in honor of his friend and former associate in botanical labors, Dr. James F. McRee of Wilmington, North Carolina.

The only unwooded portion of the ridge which we ascended, an exposed rock a few yards in extent, presents a truly Alpine aspect, being clothed with lichens and mosses, and with a dense mat of the mountain Leiophyllum, a stunted and much branched shrub (five to ten inches high), with small coriaceous leaves greatly resembling *Azalea procumbens*.[1] The much denser growth, and the broader, more petiolate, and perhaps uniformly opposite leaves, as well as the very different habitat, would seem to distinguish the mountain plant from the *L. buxifolium* of the Pine Barrens of New Jersey, etc.; but although I think the learned De Candolle has correctly separated the former, under the name of *L. serpyllifolium* (*Ledum serpyllifolium*, L'Her. ined.), it is not easy to find sufficient and entirely constant distinctive characters; since the sparse scabrous puburluence of the capsule may also be observed upon the ovary of the low-country plant, in which the leaves are likewise not unfrequently opposite; and no reliance can be placed on the length of the pedicels. The synonymy requires some correction; the *Ledum buxifolium* of Michaux ("in summis montibus excelsis Carolinæ"), and of Nuttall, so far as respects the plant which "is extremely abundant on the highest summits of the Catawba Ridge," (that is, on Table Mountain,) as well as the *Leiophyllum buxifolium* of Elliott (from the mountains of Greenville district, South Carolina), must be referred to *L. serpyllifolium*, DC. We were too late to obtain the plant in blossom, excepting one or two straggling specimens ; but we were so fortunate as to obtain a few flowering specimens of *Rhododendron Catawbiense*.

I should have remarked, that so much time was occupied in the ascent of this mountain as nearly to prevent us from herborizing around the summit for that day; since we had to descend some distance to the nearest spring of water, and prepare our encampment for the night. The branches of the

[1] We are confident that the latter does not grow on the Grandfather Mountain, as is stated by Pursh, on the authority of a specimen collected by Lyon ; and have little doubt that he mistook for it this species of Leiophyllum. Vide Pursh, "Flora Amer. Sept." i. pp. 154, 301.

Balsam afforded excellent materials for the construction of our lodge; the smaller twigs with large mats of moss stripped from the rocks furnished our bed, and the dead trees supplied us with fuel for cooking our supper and for the large fire we were obliged to keep up during the night. We re-ascended the summit the next morning, and devoted several hours to its examination; but the threatening state of the weather prevented us from visiting the adjacent ridges, or the southern and eastern faces of the mountain, and we were constrained to descend towards evening to the humble dwelling of our guide, which we reached before the impending storm commenced.

Our next excursion was to the Roan Mountain, a portion of the elevated range which forms the boundary between North Carolina and Tennessee, distant nearly thirty miles southwest from our quarters at the foot of Grandfather by the most direct path, but at least sixty by the nearest carriage road. We traveled for the most part on foot, loading the horses with our portfolios, paper, and some necessary luggage, crossed the Hanging-rock Mountain to Elk Creek, and thence over a steep ridge to Cranberry Forge, on the sources of Doe River, where we passed the night. On our way we cut down a Service-tree (as the *Amelanchier Canadensis* is here called), and feasted upon the ripe fruit, which throughout this region is highly, and indeed justly prized, being sweet with a very agreeable flavor; while in the northern States, so far as our experience goes, this fruit, even if it may be said to be edible, is not worth eating. As "Sarvices" are here greedily sought after, and are generally procured by cutting down the trees, the latter are becoming scarce in the vicinity of the "plantations," as the mountain settlements are universally called. Along the streams we met with the mountain species of Andromeda (Leucothoë), doubtless Pursh's *A. axillaris*; but whether the original *A. axillaris* of the "Hortus Kewensis" pertains to this or to the species of the low country, I cannot at this moment ascertain. A portion of Pursh's character seems also to belong to the low country rather than to the mountain species, and the two are by no means clearly distinguished in subsequent works. The leaves in our specimens are oblong-lanceolate, finely acuminate, the margins closely beset throughout with spinulose - setaceous teeth; and the rather loose spicate racemes (the corolla having fallen) are nearly half the length of the leaves.

Hitherto we had searched in vain for the *Astilbe decandra;* but we first met with this very interesting plant in the rich and moist mountain woods between Elk Creek and Cranberry Forge, and subsequently in similar situations, particularly along the steep banks of streams, quite to the base of the Roan. Mr. Curtis found it abundantly near the sources of the Linville River, and at the North Cove, where it could not have escaped the notice of Michaux ; and it is doubtless the *Spiræa Aruncus* var. *hermaphrodita* of that author. It indeed greatly resembles *Spiræa Aruncus*, and at a distance of a few yards is not easily distinguished from that plant, but on a closer approach the resemblance is much less striking. Michaux appears to have been the original discoverer of this plant, and from him the specimens cultivated in the Malmaison Garden, and described by Ventenat under the name of *Tiarella biternata*, were probably derived. It was afterwards collected by Lyon,[1] and described by Pursh from a specimen cultivated in Mr Lambert's garden at Boynton. We noticed a peculiarity in this plant, which explains the discrepancy between Ventenat and Pursh (the former having figured it with linear-spatulate petals, while the latter found it apetalous), and perhaps throws some additional light upon the genus. The flowers are diœcio-polygamous, the two forms differing from each other in aspect much as the staminate and pistillate plants of *Spiræa Aruncus*. In one form, the filaments are exserted to twice or thrice the length of the calyx, and the spatulate-linear petals, inconspicuous only on account of their narrowness, are nearly as long as the stamens ; the ovaries

---

[1] Muhlenberg's specimen was also received from Lyon. The only habitat cited in this author's catalogue is Tennessee, and we ourselves collected it within the limits, as well as on the borders of that State. The late Dr. Mcbride found it in South Carolina, near the sources of the Saluda.

are well formed and filled with ovules, which, however, so far as I have observed, are never fertilized ; and the stigmas are smaller than in the fertile plant, and not papillose. In the other or fertile form, both the stamens and the petals are in an abortive or rudimentary state, and being shorter than the sepals, and concealed by them in dried specimens, are readily overlooked ; the stigmas are large, truncate, and papillose, and a portion of the ovules become fertile. The Japanese species (*Hoteia Japonica*, Morr. & Decaisne, the *Spiræa Aruncus* of Thunberg) appears to have uniform and perfect flowers ;[1] but the species from Nepal (*Astilbe rivularis*, Don, the *Spiræa barbata* of Wallich, but not of Lindley) is probably polygamo-diœcious, like our own species ; at least, the flowers are apetalous in a fragment given me by Professor Royle, and the stamens mostly equal in number to the sepals. I have no doubt that these three species belong to a single and very natural genus, for which the name of Astilbe must be retained ; for I see neither justice nor reason in superseding the prior name, as suggested by Endlicher,[2] on account of the incompleteness of the character, which correctly describes one state, at least, of the plant intended, by the subsequent Hoteia, the character of which is equally incomplete, when applied to the whole genus. The number of genera which are either divided between North America, Japan, and the mountain region of central Asia, or have nearly allied species in these countries or in the two former, is very considerable : in other cases a North American genus is replaced by a nearly allied one in Japan, etc., as Decumaria by Schizophragma, Schizandra by Sphærostemma, Hamamelis by Corylopsis, etc. I have elsewhere alluded to this subject and shall probably consider it more particularly on some future occasion.

Our next day's journey was from Cranberry Forge to Crab

---

[1] " Flores in meo Japonico specimine omnes inveni hermaphroditos, nec ullos polygamos." (Thunberg, " Flora Japonica," p. 212, sub *Spiræa Arunco.*)

[2] " Si, quod nunc asserunt auctores, Hoteia et Astilbe, Don, revera plantæ congeneres, posterius incomplete ad auctore suo descriptum supprimendum, et prius egregie stabilitum servandum erit." (Endlicher, " Genera," Suppl. p. 1416.)

Orchard on Doe River, in Tennessee, and up Little Doe River to Squire Hampton's, where we took a guide and ascended the Roan. While ascending the Little Doe River, about three miles from the junction with the large stream of that name, at one of the numerous places where the road crosses this rivulet, we again met with *Carex Fraseriana.* The plant did not appear to be so abundant in this Tennessee locality as at the Grandfather, but it is doubtless plentiful on the mountain side just above. We ascended the north side of the Roan, through the heavy timbered woods and rank herbage with which it is covered ; but found nothing new to us excepting *Streptopus lanuginosus*, in fruit, and among the grove of *Rhododendron maximum* towards the summit, we also collected *Diphyscium foliosum*, a moss which we had not before seen in a living state. In more open moist places near the summit, we found the *Hedyotis (Houstonia) serpyllifolia*, still beautifully in flower, and the *Geum geniculatum*, which we have already noticed. It was just sunset when we reached the bald and grassy summit of this noble mountain, and after enjoying for a moment the magnificent view it affords, had barely time to prepare our encampment between two dense clumps of *Rhododendron Catawbiense*, to collect fuel, and make ready our supper. The night was so fine that our slight shelter of Balsam boughs proved amply sufficient; the thermometer, at this elevation of about six thousand feet above the level of the sea, being 64° Fahr. at midnight, and 60° at sunrise. The temperature of a spring just under the brow of the mountain below our encampment we found to be 47° Fahr. The Roan is well characterized by Professor Mitchell as the easiest of access and the most beautiful of all the high mountains of that region. " With the exception of a body of (granitic) rocks, looking like the ruins of an old castle, near its southwestern extremity, the top of the Roan may be described as a vast meadow (about nine miles in length, with some interruptions, and with a maximum elevation of six thousand and thirty-eight feet), without a tree to obstruct the prospect ; where a person may gallop his horse for a mile or two, with Carolina at his feet on one side, and Tennessee on the other, with a great ocean of mountain raised into tremendous billows immediately about him. It is the pasture ground for the young horses of the whole country about it during the summer. We found the strawberry here in the greatest abundance and of the finest quality, in regard to both size and flavor, on the 30th of July." [1]

At sunrise we had fine weather and a most extensive view of the surrounding country. In one direction we could count from eight to twelve successive ranges of mountains, and nearly all the higher peaks of this whole region were distinctly visible. Soon, however, we were enveloped in a dense fog which continued for several hours, during which time we traversed the southwestern summit and made a list of the plants we saw. The herbaceous plants of this bald and rounded summit are chiefly *Aira flexuosa, Juncus tenuis, Carex intumescens, festucacea, æstivalis* of Mr. Curtis, and a narrow-leaved variety of *C. Pennsylvanica,* the latter constituting the greater part of the grassy herbage, *Luzula campestris, Lilium Philadelphicum* and *Canadense,* which here only attain the height of from four to eight inches, *Sisyrinchium anceps, Smilacina bifolia, Habenaria (Platanthera) peramœna, Veratrum viride, Helonias (Chamœlirium) dioica, Osmunda Claytoniana,* Linn. (*O. interrupta,* Michx.), *Athyrium asplenioides, Pedicularis Canadensis,* mostly with purplish-brown flowers, now just in blossom, *Trautvetteria palmata, Ranunculus repens, Thalictrum dioicum* just in flower, *Geum radiatum* in the greatest profusion (it was here that Michaux obtained this species), *Potentilla tridentata* and *Canadensis, Fragaria Virginiana,* the fruit just ripe and of the finest flavor, *Rubus villosus* now in flower, *Castilleja coccinea, Geranium maculatum, Clematis Viorna* about eight inches high, *Sanicula Marilandica, Zizia aurea, Heracleum lanatum, Hypericum corymbosum,* with larger flowers than usual, a more upright and larger-leaved variety of *Hedyotis serpyllifolia, Œnorthera glauca β, Senecio*

---

[1] Professor Mitchell of the Chapel Hill University, in the " Raleigh Register " of November 3, 1835, and in the " American Journal of Science and Arts " for January, 1839.

Balsamitæ, Rudbeckia triloba, and a dwarf variety of *R. laciniata, Liatris spicata, Cacalia atriplicifolia, Cynthia Virginica, Aster acuminatus, Solidago bicolor, S. spithamea*, (Curtis in Torr. & Gray, Fl. ined.,) a very distinct dwarf species, *S. Curtisii*, Torr & Gr. *l. c.*, not yet in flower, and *S. glomerata* in the same state as at Grandfather Mountain ; also *Saxifraga leucanthemifolia, Sedum telephioides, Heuchera villosa, Polypodium vulgare*, the dwarf variety of *Hedyotis purpurea* previously noticed, *Scirpus cæspitosus*, and *Agrostis rupestris!* which are confined to the rocky precipice already mentioned. The only tree is *Abies Fraseri*, a few dwarf specimens of which extend into the open ground of the summit; and the following are all the shrubs which we observed, namely, *Diervilla trifida, Menziesia globularis, Vaccinium erythocarpum, Rhododendron Catawbiense*, forming very dense clumps, *Leiophyllum serpyllifolium, Sorbus Americana*, two to four feet high, *Cratægus punctata* only a foot in height, *Pyrus arbutifolia* var. *melanocarpa, Ribes rotundifolium ;* and a low and much branched species of Alder, which Mr. Curtis proposes to call *Alnus Mitchelliana*, in honor of Professor Mitchell ; but we fear it may prove to be a variety of what we deem the *A. crispa*, Ait. from the mountains of New York, New Hampshire, etc., and Newfoundland, although it has more rounded leaves with the lower surface nearly glabrous, except the primary veins ; while in the former (to which the names of *A. crispa* and *A. undulata* are not very appropriate), the leaves are often, but not always, somewhat velvety-pubescent beneath. To our list must be added any apparently undescribed species of Vaccinium, first noticed by Mr. Constable.[1] We made a hasty

[1] *Vaccinium Constablœi* (spec. nov.) : pumilum, foliis deciduis ovalibus pallidis subtus glaucis reticulato-venosisque glanduloso-mucronatis integerrimis vel obsoletissime serrulatis ciliatis, racemis brevissimis sessilibus, bracteis squamaceis parvis caducis, corollis brevissime cylindricis, antheris inclusis muticis, ovariis 10-locularibus, loculis pluri-ovulatis.
In summo jugo " Roan Mountain " dicto (Tennessee et Carolina Septentrionali), ad alt. 6000 pedes. Julio floret. — Frutex 1-3-pedalis, erectus, ramis griseo-viridibus teretis. Folia sesqui-biuncialia, lato-ovalia vel elliptica, utrinque sæpius acuta, glabra, nisi costa supra puberula

visit to the other principal summit, where we found nothing that we had not already collected, excepting *Arenaria glabra*, Michx., and descended partly by way of the contiguous Yellow Mountain.

Retracing our steps, we returned the next day to the foot of Grandfather, and reached our quarters at Jefferson the second day after. We had frequently been told of an antidote to the bite of the Rattlesnake and Copperhead (not unfrequent throughout this region), which is thought to possess wonderful efficacy, called Thurman's Snake-root after an "Indian Doctor," who first employed it; the plant was brought to us by a man who was ready to attest its virtues from his personal knowledge, and proved to be the *Silene stellata!* Its use was suggested by the markings of the root beneath the bark, in which these people find a fancied resemblance to the skin of the Rattlesnake. Nearly all the reputed antidotes are equally inert; such herbs as *Impatiens pallida*, etc., being sometimes employed; so that we are led to conclude that the bite of these reptiles is seldom fatal, or even very dangerous, in these cooler portions of the country.

About the foot of the Roan and Grandfather we obtained et margines ciliati, subsessilia, infra saturate glauca. Racemi 5-10-flori, sæpe corymbosi, ad apicem ramulorum anni præcedentis solitarii vel aggregati. Baccæ immaturæ cæruleæ, glaucæ, limbo calycis majusculo coronatæ, decem- (nunc abortu quinque?) loculares ; loculis pleio (3-6?) spermis.

Professor Dunal (in DC. Prodr. 7, p. 566) notices as an extraordinary exception to the character of Vaccinium, a species with an 8-10-celled fruit and a single (?) seed in each cell. The first-named character is not unfrequent in the genus : several of the more common species which I have cursorily examined, exhibit a more or less completely 8-10-celled ovary, but with many ovules in each cell. There is a small group, however [*Decachœna*, Torr. & Gray ined.], presenting a different structure, which is best exemplified in *V. resinosum*, Ait. The 10 carpels of this species, inclosed in the baccate calyx, are very slightly coherent with each other, and become crustaceous or bony nuts, each containing a single ascending seed. The same is the case in what I take to be *V. dumosum* and *V. hirtellum ;* and probably in some other species which have the leaves sprinkled with resinous dots. *V. frondosum*, Willd. (which is the *V. decamerocarpon* of Dunal), is similar in structure, except that the carpels appear to be more coherent and less indurated.

a few specimens of *Pycnanthemum montanum*, Michx. (Monardella, Benth.) just coming into blossom. Our plant accords with Michaux's description, except that there are frequently two or even three axillary heads besides the terminal one. The flowers have altogether the structure of Pycnanthemum, and the upper lip of the corolla is entire; so that it cannot belong to Monardella, although placed as the leading species of that genus. As to the species from which Mr. Bentham derived the generic name (*Pycnanthemum Monardella*, Michx.), I am by no means certain that it belongs either to Pycnanthemum or Monardella. The specimen in the Michauxian herbarium is not out of flower, as has been thought, but the inflorescence is undeveloped, and perhaps in an abnormal state. In examining a small portion taken from the head, I found nothing but striate-nerved bracts, obtuse and villous at the apex, and abruptly awned; the exterior involucrate and often lobed; the innermost linear, and tipped with a single awn. The aspect of the plant, also, is so like *Monarda fistulosa*, that I am strongly inclined to think it a somewhat monstrous state of that, or some nearly allied species; in which case, the genus Monardella should be restricted to the Californian species. Pursh's *P. Monardella*, I may observe, was collected beneath the Natural Bridge in Virginia, where we also obtained the plant, and subsequently met with it throughout the mountains. It is certainly a form of *Monarda fistulosa*, according to Bentham's characters, but the taste is much less pungent, the throat of the calyx less strongly bearded than is usual in that species, and the corolla nearly white. We thought it probably a distinct species; but these differences may be owing to the deep shade in which it commonly occurs. The *P. Monardella* of Elliott, according to his herbarium, is identical with that of Pursh. We collected in Ashe County several other species of Pycnanthemum, and in the endeavor to discriminate them, we encountered so many difficulties that I am induced to give a revision of the whole genus.

Some additional plants were obtained around Jefferson which were not previously in blossom, such as *Campanula*

divaricata ; Cacalia reniformis ; Sliphium perfoliatum ; the
larger form of *Coreopsis auriculata,* with nearly all the leaves
undivided; the glabrous and narrow-leaved variety of *C. seni-
folia (C. stellata,* Nutt.) which alone occurs in this region;
Melanthium Virginicum, which is a very handsome plant, with
the flowers cream-colored when they first expand ; and *Ste-
nanthium angustifolium,* Gray, which is doubtless the *Helonias
graminea* of the " Botanical Magazine." We also made an
excursion to the White Top in Virginia, twenty miles northwest from Jefferson ; a mountain of the same character as the
Roan, but on a smaller scale, and with the pasturage of its
summit more closely fed. We were not rewarded, however,
with any new plants, and the cloudy weather obscured the
prospect, which is said to be very extensive. On our return,
we found *Cedronella cordata,* Benth., nearly out of flower,
with runners often two or three feet in length. Mr. Bentham
has omitted to mention the agreeable balsamic odor of the
genus, which in our plant is much less powerful than in *C.
triphylla.* We saw plenty of *Cimicifuga Americana,* but
the flowers were still unexpanded. Our endeavors to obtain
the fruit of *Cimicifuga cordifolia* (common in this region)
were likewise unsuccessful; without which it is not always
easy to distinguish this species from *C. racemosa.* The leaf-
lets of the former are frequently very large, the terminal ones
resembling the leaves of the Vine in size and shape, as remarked by De Candolle ; in one instance we found them ten
inches in diameter; but they are generally much smaller and
more divided, apparently passing into the former species.
The number of the ovaries does not afford marked characters,
since the lowest flowers of *C. racemosa* sometimes present
two, while the upper ones of *C. cordifolia* are almost always
monogynous.

We were too early in the season for several interesting
plants, especially *Compositæ,* and did not extend our re-
searches far enough south to obtain many others ; such as
Hudsonia montana, which appears to be confined to Table
Mountain, *Rhododendron punctatum, Stuartia pentagyna,
Philadelphus hirsutus, Silene ovata* (which Mr. Curtis found

in Buncombe and Haywood counties), *Berberis Canadensis*
(which, however, Pursh collected on the mountains of Greenbrier in Virginia), *Parnassia asarifolia* (which according
to Mr. Curtis first appears in Yancey County, but Pursh
procured it from "mountain runs on the Salt Pond Mountain,
Virginia, and on the top of the Alleghanies near Christian-
burg "), and, above all, the new Thermopsis; (*T. Caroliniana*,
M. A. Curtis, MSS.) recently discovered by our friend Mr.
Curtis, in Haywood and Cherokee counties. We were likewise unsuccessful in our search for a remarkable undescribed
plant, with a habit of Pyrola and the foliage of Galax, which
was obtained by Michaux in the high mountains of Carolina.
The only specimen extant is among the " Plantæ incognitæ "
of the Michauxian herbarium, in fruit only; and we were
anxious to obtain flowering specimens, that we might complete
its history; as I have long wished to dedicate the plant to
Professor Short of Kentucky, whose attainments and eminent services in North American botany are well known and
appreciated both at home and abroad.[1]

We left this interesting region near the end of July, returning to New York by way of Raleigh, Richmond, etc. ; and
found a marked change in the vegetation immediately on
crossing the Blue Ridge. I cannot extend these remarks to
the plants observed in our homeward journey, except to mention that the Schrankia of this part of the country, which extends to the eastern slope of the Blue Ridge, is the *S. angu-*
stata, Torr. & Gray ; at least we observed no other species.

[1] *Shortia*, Torrey & Gray. Calyx quinquesepalus ; sepala imbricata,
squamacea, striata, persistentia, exteriora ovata, interiora oblonga.
Corolla — Stamina— Capsula calyce brevior, subglobosa, stylo filiformi
(subpersistente) superata, trilocularis, loculicide trivalvis, valvis medio
septiferis placenta centrali magna persistente. Semina multa, parva ;
testa nucleo conformis. Embryo teres, rectiusculus, albumine brevior.
Herba cæspitosa? subaculis, perenni, glabra ; foliis longe petiolatis,
rotundatis, subcordatis, apice nunc retusis, crenato-serratis, crenaturis
mucronatis ; scapis unifloris, nudis, apicem versus squamoso-bracteatis.
S. galacifolia, Torrey & Gray. (V. spec. sicc. in herb. Michx., cum
schedula, " Hautes montagnes de Carolinie. An Pyrola spec. ? an genus
novum ? ")

This is doubtless the *S. uncinata* of De Candolle ; but not, I think, of Willdenow. I may here remark, that the reticulate-leaved species (*S. uncinata*, Torr. & Gr.) is the Leptoglottis of De Candolle (" Mem Legum "), as I have ascertained from a fragment of the original specimen in the rich herbarium of Mr. Webb, which that gentleman obligingly sent me ; but I find no neutral flowers or sterile filaments in the numerous specimens of this plant, from different localities, which I have from time to time examined.

THE LONGEVITY OF TREES.

THE "Histoire des Arbres Forestières de l'Amérique Septentrionale "[1] of the younger Michaux is chiefly known in this country through the English translation made by Mr. Hillhouse, under the superintendence of the author, who added some new plates, and information not contained in the French edition. It was published in Paris in the year 1819. We have no intention of formally reviewing, at this late day, a work of such long-established reputation as the Sylva of Michaux. It has been the standard treatise upon the subject ever since its publication; and it well deserves the rank it holds. We wish rather to offer our grateful acknowledgments to the memory of the late Mr. William Maclure for his liberal endeavors to render this important and quite expensive work more generally accessible in the country, the noble forests of which it is designed to illustrate. In furtherance of this object, Mr. Maclure, if we are rightly informed, purchased in Paris the copies which remained unsold the year after its publication, and sold them in the United States at a very reduced price. With liberal forethought, he bought also the original copperplates of this book, and of several other expensive works of science and art; intending to have them reprinted in this country in a cheaper form, so as to insure them a wider circulation.[2] During the last twelve years of his life, or from the time of the total failure of the "great social experiment" made at New Harmony, Indiana, by the celebrated Robert Owen and himself, down to his decease in 1840, Mr. Maclure resided in the city of Mexico. The

[1] "North American Review," July, 1844.

[2] Among these are the complete copperplates of Veillot's "Histoire Naturelle des Oiseaux de l'Amérique Septentrionale "; and Audebert's "Histoire Naturelle des Oiseaux," etc.

great interest which he still felt in the advancement and diffusion of scientific knowledge in this country, was manifested
by his continued and munificent benefactions to the Academy
of Natural Sciences at Philadelphia.[1] It was his desire also
that some competent person, at his expense, should collect
additional information respecting the forest trees of North
America, and prepare a new and augmented edition of
Michaux's Sylva. Although this generous plan was not carried into execution in Mr. Maclure's lifetime, we understand
that he left testamentary instructions for its accomplishment,
or at least for the reprint of Michaux's original work.

Mr. Maclure's munificent intentions are — fulfilled, shall
we say? by the edition now before us, printed at New Harmony, Indiana, upon wretched, flimsy, whity-brown paper, of
texture scarcely firm enough to receive the impression of the
worn-out type, which seems to have done long service in the
columns of some country newspaper. The engravings, to do
the work full justice, are very good, at least in the colored
copies ; being impressions from the original plates imported
by Mr. Maclure, and colored after the French edition. The
subscription price was high enough to pay for the best typography and paper; and so popular a work would surely have
found a ready sale at any reasonable price, as the Paris edition had been long out of print. We are bound to suppose
that the testamentary instructions of the late Mr. Maclure
have been literally fulfilled. How far his generous intentions have been answered, or to what extent defeated, by the
wretched character of the reprint, are questions which we
shall not attempt to answer, except by stating that, far from
having obtained the wide circulation which Mr. Maclure
desired, these volumes have entirely disappeared from the
shelves of our booksellers, where, so far as we can learn,

[1] Besides his invaluable library of nearly 4000 volumes, containing
the choicest works in natural history, antiquities, the fine arts, voyages
and travels, etc., and many smaller contributions to this flourishing institution, of which he was president for more than twenty years, Mr. Maclure presented to it, in the latter part of his life, the sum of twenty
thousand dollars. See Dr. Morton's "Memoir of William Maclure,"
Philadelphia, 1841.

there was no demand for them; and our diligent attempt to find a single copy in the public and private libraries of three of our largest cities has proved entirely unsuccessful.

The foundation of the North American Sylva was laid by the laborious researches of the elder Michaux; who, under the auspices of the French government, devoted ten years, from 1785 to 1796, to a thorough exploration of the country, from the sunny, sub-tropical groves of Florida to the cold and inhospitable shores of Hudson's Bay; repeatedly visiting nearly all the higher peaks and deepest recesses of the Alleghany Mountains, and extending his toilsome journeys westward to the prairies of Illinois and the banks of the Mississippi. He had formed indeed, and was only prevented by untoward circumstances from executing, a plan — more hardy than we can well conceive at this late day — for ascending the Missouri to its sources, and crossing the mountains into the then untrodden but now litigated country on the Oregon. The curious reader will find an extract from his private diary in the "American Journal of Science and Arts" for January, 1842;[1] showing that he had laid his plans and proposals upon this subject before Mr. Jefferson, who was then Secretary of State. The papers submitted by him may have suggested the scheme of the national expedition of discovery, soon afterwards ordered by Jefferson, and nobly carried into effect by Lewis and Clarke.

Soon after his return to France, and the year before he fell a victim to scientific zeal upon the coast of Madagascar, the elder Michaux published his History of North American Oaks;[2] which may be deemed the nucleus of the more comprehensive work subsequently published by his son. The younger Michaux accompanied his father in the earlier portion of his travels, through South Carolina, Georgia, and Florida; but he afterwards returned to Europe. Revisiting this country in the autumn of 1801, and passing the winter in South Carolina, he traveled, during the following season,

[1] Vol. xlii. p. 7.
[2] "Histoire des Chênes de l'Amérique Septentrionale." Par A. Michaux. Paris, 1801. 1 vol. fol.

from New York, across the mountains in Pennsylvania, to the Ohio, and carefully explored the States of Kentucky and Tennessee; thence, recrossing the mountains in North Carolina to Charleston, he again embarked for France.[1] Again returning in 1807, he journeyed along the whole extent of our Atlantic coast, and visited the principal ports to examine the timber employed in shipbuilding and in workshops of every description; besides making separate excursions into the interior: "the first, along the rivers Kennebec and Sandy, passing through Hallowell, Norridgewock, and Farmington; the second, from Boston to Lake Champlain; the third, from New York to the lakes Ontario and Erie; the fourth, from Philadelphia to the borders of the rivers Monongahela, Alleghany, and Ohio; and the fifth, from Charleston to the sources of the Savannah and Oconee." Having thus faithfully collected the requisite information, his great work upon our forest trees — the fruit of so much labor — was published at Paris in 1810–13.

But this work is not the only result of the well directed industry and zeal of the elder and the younger Michaux. To these two persons, chiefly, are the French plantations indebted for their surpassingly rich collections of American trees and shrubs; which long since gave rise to the remark, as true at this day as it was twenty years ago, that an American must visit France to see the productions of his native forests. When shall it be said that the statement is no longer true? When shall we be able to point to a complete, or even a respectable, American collection of our indigenous trees and shrubs?

A few words will suffice for the second work on our list.[2]

[1] The observations made in this tour are recorded in his "Voyages à l'Ouest de Monts Alléghanys," 8vo, Paris, 1804; and in a "Mémoire sur la Naturalisation des Arbres Forestières de l'Amérique Septentrionale," 8vo, Paris, 1805.

[2] "The North American Sylva"; or a description of the forest trees of the United States, Canada, and Nova Scotia, not described in the work of F. Andrew Michaux, and containing all the forest trees discovered in the Rocky Mountains, the Territory of Oregon down to the shores of the Pacific and into the confines of California, as well as in various parts of the United States; illustrated by 122 fine plates.

Mr. Nuttall, it appears, first arrived in this country the very year that the younger Michaux finally left it. And from that time to the present, no botanist has visited so large a portion of the United States, or made such an amount of observations in the field and forest. Probably few naturalists have ever excelled him in aptitude for such observations, in quickness of eye, tact in discrimination, and tenacity of memory. In some of these respects, perhaps, he may have been equalled by Rafinesque, — and there are obvious points of resemblance between the later writings of the two, which might tempt us to continue the parallel; — but in scientific knowledge and judgment he was always greatly superior to that eccentric individual. Mr. Nuttall has also enjoyed the best opportunities for exploring the wide regions beyond the Mississippi. In 1811, along with Mr. Bradbury, he ascended the Missouri to the Mandan villages 1600 miles above its mouth; and shortly after his return published his extended and most happily executed botanical work, the " Genera of North American Plants." In 1819, at the imminent risk of his life, he ascended the Arkansas to the Great Salt River. And, in 1834, he finally succeeded in crossing the Rocky Mountains by the now well-trodden road along the sources of the Platte, and exploring the territory of Oregon and of Upper California. Mr. Nuttall was therefore peculiarly qualified for the preparation of a supplementary North American Sylva, designed especially to comprise the forest trees of these wide regions, which are now, for obvious reasons, attracting particular attention.

The work, according to the announcement on the title-page, will consist of three volumes; and we understand that the editor committed the whole manuscript to the publisher's hands more than two years ago, when he returned to his native country, to take possession of an ample family inheritance. But from some cause, the publication has been greatly delayed; only one volume having yet appeared, and that in two portions, of which the first bears the date of 1842, while the second has but just reached us. We postpone all critical remarks until the entire publication is completed; merely ob-

serving, lest the interests of the publisher should suffer from
nominal connection between this work and the New Harmony
reprint of the original Sylva, that the paper and typography
are good, and the plates, which are colored lithographs, are
respectable. Meanwhile the interest of the subject, and the
well-known scientific character of the author, will commend
the work to general attention and patronage.

The plan and object of the late Mr. Loudon's greatest
work, " The Arboretum et Fruticetum Britannicum," is fully
set forth in the copious title-page. All its promises are more
than redeemed in the execution of the work, which is truly a
fine monument of industry and careful research. We have
particular reasons, which will appear in the sequel, for commending this work to the notice of any readers interested in
these subjects, who do not already possess it. By purchasing
a copy of it, or of the valuable abridgment, the " Encyclopædia of Trees and Shrubs," a work of moderate price, they
will render important aid to the embarrassed family of the
author. To give some idea of the astonishing industry of the
late Mr. Loudon in the preparation of scientific books, we extract the following account from the " Gardeners' Magazine,"
an excellent periodical, the publication of which, after it had
continued for a period of eighteen years, terminated at the
death of its indefatigable editor.

" Mr. Loudon was brought up as a landscape-gardener, and
began to practise in 1808, when he came to England with
numerous letters of introduction to some of the first landed
proprietors in the kingdom. He afterwards took a large farm
in Oxfordshire, where he resided in 1809. In the years 1813,
1814, 1815, he made the tour of northern Europe, traversing
Sweden, Russia, Poland, and Austria; in 1819, he traveled
through Italy, and in 1828, through France and Germany.
Mr. Loudon's career as an author began in 1803, when he
was only twenty years old ; and it continued, with very little
interruption, during the space of forty years, being only concluded by his death. The first works he published were the
following : ' Observations on laying out Public Squares,'
in 1803, and on ' Plantations,' in 1804; a ' Treatise on Hot-

houses,' in 1805, and on 'Country Residences,' in 1806, both
4to ; 'Hints on the Formation of Gardens,' in 1812 ; and
three works on 'Hothouses,' in 1817 and 1818. In 1822 appeared the first edition of the 'Encyclopædia of Gardening' ;
a work remarkable for the immense mass of useful matter
which it contained, and for the then unusual circumstance of
a great quantity of woodcuts being mingled with the text :
this book obtained an extraordinary sale, and fully established
his fame as an author. Soon after was published an anony-
mous work, written either partly or entirely by Mr. Loudon,
called the 'Greenhouse Companion' ; and shortly afterwards,
'Observations on laying out Farms,' in folio, with his name.
In 1824, a second edition of the 'Encyclopædia of Gardening' was published, with very great alterations and improvements ; and the following year appeared the first edition of
the 'Encyclopædia of Agriculture.' In 1826, the 'Gardeners'
Magazine' was commenced, being the first periodical ever devoted exclusively to horticultural subjects. The 'Magazine
of Natural History,' also the first of its kind, was begun in
1828. Mr. Loudon was now occupied in the preparation of
the 'Encyclopædia of Plants,' which was published early in
1829, and was speedily followed by the 'Hortus Britannicus.'
In 1830, a second and nearly rewritten edition of the 'Encyclopædia of Agriculture' was published, and this was followed
by an entirely rewritten edition of the 'Encyclopædia of
Gardening,' in 1831 ; and the 'Encyclopædia of Cottage,
Farm, and Villa Architecture,' the first he published on his
own account, in 1832. This last work was one of the most
successful because it was one of the most useful he ever wrote,
and it is likely long to continue a standard book on the subject of which it treats. Mr. Loudon now began to prepare
his great and ruinous work, the 'Arboretum Britannicum,'
the anxieties attendant on which were, undoubtedly, the
primary cause of that decay of constitution which terminated
in his death. This work was not, however, completed till
1838, and in the mean time he began the 'Architectural
Magazine,' the first periodical devoted exclusively to architecture. The labor he underwent at this time is almost in-

credible. He had four periodicals, namely, the ' Gardeners','
' Natural History,' and ' Architectural Magazines,' and the
' Arboretum Britannicum,' which was published in monthly
numbers, going on at the same time ; and, to produce these
at the proper times, he literally worked night and day. Immediately on the conclusion of the ' Arboretum Britannicum,'
he began the ' Suburban Gardener,' which was also published
in 1838, as was the ' Hortus Lignosus Londinensis '; and in
1839 appeared his edition of Repton's ' Landscape Gardening.' In 1840 he accepted the editorship of the ' Gardeners'
Gazette,' which he retained till November, 1841 ; and in 1842
he published his ' Encyclopædia of Trees and Shrubs.' In
the same year he completed his ' Suburban Horticulturist ';
and finally, in 1843, he published his work on ' Cemeteries,'
the last separate work he ever wrote. In this list, many
minor productions of Mr. Loudon's pen have necessarily been
omitted ; but it may be mentioned that he contributed to the
' Encyclopædia Britannica,' and Brande's ' Dictionary of
Science '; and that he published numerous supplements, from
time to time, to his various works."

The adverse circumstances under which all this labor was
performed make the result appear still more remarkable.
Early in life, an attack of inflammatory rheumatism permanently stiffened one of his knees and contracted his left
arm. In 1820, his right arm was broken near the shoulder,
and the bones were never properly united. But he still continued to write with his right hand until 1825, when the arm
was broken a second time, and it became necessary to am-
putate it ; after which, having about this time lost the use of
the thumb and two fingers of the left hand, he was obliged
to employ amanuenses. It was under these trying circumstances that his great " Arboretum " was prepared. The
work was undertaken against the advice of his publishers
and friends, who foresaw that it would wholly absorb his
former earnings. But it was his favorite and crowning work,
and he unflinchingly carried it onward to completion ; though
the result verified the predictions of his advisers. It cost ten
thousand pounds sterling ; and the whole proceeds of the sale,

as well as the copyrights of his other works, were pledged to the publishers for payment. It is gratifying to learn, that, since the writer's decease, an appeal has been made to the public in England, and with great success, to purchase copies of this work and of the other publications of Mr. Loudon, so that the copyright may be redeemed, and their future proceeds applied to the benefit of his surviving family. The subjects of landscape-gardening and arboriculture are attracting increased attention in the United States, and these valuable treatises are not yet so generally known as they deserve to be ; we have thought it proper, therefore, to make this statement.

While systematically treating of the botanical character, propagation, management, and economical uses of trees, Mr. Loudon has interwoven a vast amount of curious matter respecting their history, geography, and literature. Being himself a distinguished landscape-gardener, he has successfully treated of their character and adaptation as component parts of general scenery, of which they form a most important element ; for no other constituents — no lifeless objects — produce impressions at once so strong and so widely varied as trees. He has also collected interesting statistics respecting the longevity of trees ; a subject upon which we intend to task the patience of our readers to some extent.

The most interesting ideas connected with trees are those suggested by their stability and duration. They far outlast all other living things, and form the familiar and appropriate symbols of long-protracted existence. " As the days of a tree shall be the days of my people " is one of the most beautiful and striking figures under which a blessing can be conveyed. We are naturally led to inquire, whether there is any absolute limit to their existence. If not destroyed by accident, — that is, by extrinsic causes, of whatever sort, — do trees eventually perish, like ourselves, from old age ? It is commonly thought, no doubt, that trees are fully exposed to the inevitable fate of all other living things. The opposite opinion seems to involve a paradox, and to be contradicted by every one's observation. But popular opinion is an unsafe guide ;—the more so in this case, as our ordinary conceptions on the subject spring from a false analogy, which we have unconsciously established, between plants and animals. This common analogy might, perhaps, hold good, if the tree were actually formed like the animal, all the parts of which are created at once in their rudimentary state, and soon attain their fullest development, so that the functions are carried on, throughout life, in the same set of organs. If this were the case with the tree, it would likewise die, sooner or later, of old age, — would perish from causes strictly analogous to those which fix a natural limit to the life of animals. The unavoidable induration and incrustation of its cells and vessels, apart from other causes, would put an early and sure limit to the life of the tree, just as it does in fact terminate the existence of the leaf, the proper emblem of mortality, — which, although it generally lives only a single season, may yet truly be said to die of old age. But, as the leaves are necessarily renewed every year, so also are the other essential organs of the plant. The tree is gradually developed by the successive addition of new parts. It annually renews not only its buds and leaves, but its wood and its roots ; everything, indeed, that is concerned in its life and growth. Thus, like the fabled Æson, being restored from the decrepitude of age to the bloom of early youth, — the most recent branchlets being placed, by means of the latest layer of wood, in favorable communication with the newly formed roots, and these extending at a corresponding rate into fresh soil, —

" Quae quantum vertice ad auras
Ætherias, tantum radice in Tartara tendit,"

why has not the tree all the conditions of existence in the thousandth, that it possessed in the hundredth, or the tenth, year of its age? The old and central part of the trunk may, indeed, decay; but this is of little moment, so long as new layers are regularly formed at the circumference. The tree survives ; and it is difficult to show that it is liable to death from old age in any proper sense of the term. Nor do we arrive at a different conclusion when we contemplate the tree under a less familiar but more philosophical aspect, —

considering it, not as a simple individual, like man or the higher animals, but as an aggregate of many individuals, which though ordinarily connected with the parent stock, are capable of growing by themselves, and indeed often do separate spontaneously, and in a variety of ways acquire independent existence. If, then, the tree be, as it undeniably is, a complex being, an aggregate of as many individuals, united in a common trunk, as there are, or have been, buds developed on its surface ; and if the component individuals be annually renewed, why should not the aggregate, the *tree*, last indefinitely ? To establish a proper analogy, we must not compare the tree with man, but with the coral formations, in which numberless individuals, engrafted and blended on a common base, though capable of living when detached from the mass, conspire to build up those arborescent structures so puzzling to the older naturalists that they were not inappropriately named " zoöphytes," or animal-plants. The immense coral-groves, which have thus grown up in tropical seas, have, no doubt, endured for ages ; the inner and older parts consisting of the untenanted cells of individuals that have long since perished, while fresh structures are continually produced on the surface. The individuals, indeed, perish ; but the aggregate may endure as long as time itself. So with a tree, considered under this point of view. Though the wood in the centre of the trunk and large branches — the produce of buds and leaves that have long ago disappeared — may die and decay ; yet while new individuals are formed upon the surface with each successive crop of fresh buds, and placed in as favorable communication with the soil and the air as their predecessors, the aggregate tree would appear to have no necessary, no inherent, limit to its existence.[1]

[1] A beautiful confirmation of this view may be drawn from the celebrated Banyan, or India Fig-tree, and a few other tropical trees, which freely strike root, high in the open air, from their spreading branches. These aërial roots, after reaching the earth, become in time new trunks ; and the whole tree appears like a huge tent, supported by many columns. Milton's description of the Banyan (in " Paradise Lost ") is incorrect, so far as it supposes the bending branches themselves to reach the ground, and there to strike root, just as the gardener propagates shrubs

No one denies, however, that different species may have an habitual period of death; we only insist that this is not a necessary period. In the course of things, a multitude of different accidents conspire to fix a mean limit to the life of man, which, though far below the natural period of death by old age, yet occurs with such regularity, under given conditions, that it is made a matter of calculation. So a particular kind of tree may be liable to certain accidents, which habitually insure its destruction within a definite period. A tree of rapid growth generally has a soft and fragile wood, and is therefore especially subject to decay, or to be broken or overthrown by tempests; and the chances of its destruction are fearfully multiplied with the increasing spread and weight of the branches. Each species, too, being somewhat uniformly exposed to a particular class of accidents, according to its constitution and mode of growth, may consequently exhibit something like an average duration. But death can no more be said to ensue from old age, in such a case, than in that of the ordinary mortality of mankind. The whole tree does not necessarily suffer, like the animal, from the death or amputation of its limbs; those that remain may be thereby placed, perhaps, in a more favorable condition than before. A tree may certainly be conceived to survive all ordinary accidents, or to be protected against them, and thus to live indefinitely; while animals, even if shielded from all external injury, must at last succumb to internal causes of destruction, — unavoidable, because inseparable from their organization.

by layering; whereas the roots themselves descend from a great height. When a sufficient number of these collateral trunks are formed to support the whole weight, the central, original stem may decay and disappear, as it often does, without affecting the existence of the tree; which thus increases into a grove, "high over-arched, with echoing walks between," that obviously may endure for an indefinite period. Many such trees are known, of immense magnitude, and doubtless of most extraordinary age. But the vegetable physiologist well knows that these essentially differ from ordinary trees, only in that a portion of the new wood, detached as it were from the branches, forms separate trunks instead of adhering throughout to the main trunk and contributing to its increase in circumference. These collateral trunks merely represent the outer and newer layers of ordinary trees, while the main stem represents the old and often decaying centre.

The Talipot Palm, which blossoms but once, and then perishes, — or the Century plant, which continues in our conservatories even for a hundred years without flowering, but dies when it has ripened its fruit, — may be adduced as cases of death by old age. But in its native climes, where our so-called Century-plant blossoms in the fifth or sixth year of its age, it as uniformly dies immediately afterwards. The result, in all such cases, is rather analogous to death from parturition, than to death by old age.

This doctrine of the indefinite longevity of trees — that they die from injury or disease, or, in one word, from accidents, but never really from old age — was first propounded by the distinguished De Candolle in one of his earliest writings,[1] near the commencement of the present century. It is entirely a modern doctrine (unless, indeed, we may suppose that Pliny comprehended the full meaning of his words, " Vites sine fine crescunt," which is improbable), and it is by no means surprising that it should have been received with incredulity, or vehemently controverted, by those who had not taken the pains to understand it. For the *a priori* considerations, from which the young Genevan botanist deduced his novel theory, were then, in truth, more or less hypothetical, and involved some hardy assumptions. They are now, however, amply confirmed, or at least so generally admitted by all vegetable physiologists, as to give the theory a high degree of antecedent probability. But De Candolle proceeded to indicate a mode in which its correctness might almost be tested by actual observation, and, having accumulated a great number of interesting data, he published, in 1831, the memoir [2] which, having been still further augmented, now constitutes one of the most interesting chapters of his masterly "Physiologie Végétale."

If this view be well founded, it is to be expected that different individuals of the same species should perish at very irregular periods ; and that some should be found to escape all

[1] " Flore Francaise," 1, p. 223.
[2] " *Notice sur la Longévité des Arbres.*" Par Aug. Pyr. De Candolle, in the Bibliothèque Universelle, May, 1831.

the ordinary accidents that trees are heir to, and thus attain a
longevity far transcending the habitual duration of the species. Is this view sustained by observation?

Before adducing the evidence which bears upon this question, it is necessary to inquire how the actual age of a tree
may be ascertained. In most cases, — in all those trees which
increase in diameter by annual concentric layers, — that is to
say in nearly all trees except Palms and their allies, which for
the present we may leave out of the question, the age may be
directly ascertained by counting the annual rings on a cross
section of the trunk. The record is sometimes illegible or
nearly so, but it is perfectly authentic ; and when fairly de-
ciphered, we may rely on its correctness.[1] But the venerable
trunks, whose ages we are most interested in determining, are
rarely sound to the centre ; and if they were, even the paramount interests of science would seldom excuse the arboricide.
This decisive test, therefore, can seldom be practically employed, except in the case of comparatively young trees. The
most remarkable recorded instance of its application is that
of one of the old oaks at Bordza, in Samogitia (Russian Poland) ; which, having been greatly injured by a conflagration,
was felled in the year 1812, and seven hundred and ten
concentric layers were distinctly counted on the transverse

[1] The discovery, or at least the first explicit announcement of the now
familiar fact, that ordinary trees grow by annual layers, so that the record of their age is inscribed upon the section of the trunk, is generally
attributed to Malpighi. But, probably, we should understand this celebrated anatomist as merely giving a formal statement of what was already
popularly known ; for so obvious a fact could scarcely have escaped notice. Professor Adrien de Jussieu, the present representative of that illustrious family, has, moreover, lately reproduced a passage in the " Voyage de Montaigne en Italie," written in the year 1581, nearly fifty
years before Malpighi was born, which proves this to have been the case.
" *L'ouvrier, homme ingénieux et fameux à faire de beaux instruments de
Mathématique, m'enseinga que tous les arbres portent autant de cercles qu'ils
ont duré d'années, et me le fit voir dans tous ceux qu'il avoit dans sa boutique,
travaillant en bois. Et la partie qui regard le septentrion est plus étroite,
et a les cercles plus serrés et plus denses que l'autre.*"
And now it appears that Leonardo da Vinci knew, and mentioned it —
as well as phyllotaxis. (MSS. note in Dr. Gray's handwriting.)

section, from the circumference towards the centre, where the space in which the layers could not be clearly made out was estimated to have comprised three hundred more. If the injured portion was not overestimated, the tree must have been a thousand years old. We have now before us a section of a fine trunk of the American Cypress (*Taxodium distichum*), upon the radius of which, twenty-seven inches in length, six hundred and seventy annual layers may be distinctly counted. The wood of this tree is so durable, that probably the age of trunks of more than twice that size might be ascertained by direct inspection.

When such a section cannot be obtained, we are obliged to resort to other and less direct evidence, affording only approximate, or more or less probable, conclusions. Sometimes lateral incisions, not endangering the life of the tree, furnish the means of inspecting and measuring a considerable number of the outer layers, and of computing the age of the trunk from its diameter and actual rate of growth. But as young trees grow much more rapidly than old ones, we should greatly exaggerate the age of a large trunk, if we deduced its rate of growth from the outer layers alone. We must therefore ascertain, by repeated observations, the average thickness of the layers of young trees of the same species; and by the judicious combination of both these data, a highly probable estimate may often be formed.

When unable to inspect any portion of the annual layers of remarkable old trees, we may occasionally obtain other indications upon which some reliance may be placed ; such as the amount of increase in circumference between stated intervals; but as, on the one hand, we can never depend upon the entire accuracy of two measurements made at widely distant periods, while, on the other, the growth of a small number of years, however carefully ascertained, would be an unsafe criterion, this method can seldom be employed with much confidence. A more common mode is to employ the average rate of growth of the oldest trees of which complete sections have been examined, for the approximate determination of the age of remarkably large trunks of the same species, where the size

alone is known. For often repeated observation proves that
the increase is greatest, in other words, the layers are thick-
est, in young trees ; but that afterwards — after the first
century, for instance — the tree increases in diameter at a
much slower but somewhat uniform, or else still decreasing
rate, which does not greatly vary in different trees of the
same species. Such estimates would, therefore, always tend
to underrate, rather than to exaggerate, the age of a large
tree. But it is unsafe to apply this method to other than
really venerable trunks ; for the growth of a tree is liable to
great variations during the first century or two ; either from
year to year, or between different individuals of the same species. The injury of a single leading root or branch, or the
influence of a stratum of sterile soil, may affect the whole
growth of a young tree for a series of years ; while, in an
older individual, the wide distribution of the roots and multiplication of the branches render the effect of local injuries
nearly inappreciable, and the influence of any one or more
unfavorable seasons is lost in the average of a great number.
Thus the fine Elm in Cambridge, which during the last winter fell a victim to one of the most fatal and frequent accidents which in this country interfere with the longevity of
trees, — having been cut down to make room for a petty
building, just as it had reached its hundredth anniversary, —
was fourteen feet in circumference at the height of three or
four feet from the ground.[1] The girth of its more renowned
and fortunate neighbor, the " Washington Elm," is but little
over thirteen feet ; and it might accordingly be inferred that
it is some years the junior of the " Palmer Elm." But we
learn from a very authentic source, that the celebrated Whit-
field, when excluded from the pulpits of the town and college,
preached under the shade of this tree in the summer of 1744,
— just a century ago. It is, doubtless, at least one hundred

---

[1] This "Palmer Elm," as it was called, grew with more than ordinary
rapidity for the first seventy years ; when, to casual observation, it must
have appeared nearly as large as when it was felled. For, during the
last twenty-two years, it had increased only five and one-half inches in
diameter, that is, at the rate of a quarter of an inch per annum.

and twenty or one hundred and thirty years old. We wish to place its size upon record for the use of future generations; and we therefore take this opportunity to state, that the trunk of the "Washington Elm," at Cambridge, now measures thirteen feet and two and a half inches in circumference, at the height of three feet from the ground; this being the point at which the girth is smallest, being unaffected either by the expansion of the roots below, or of the branches above, and therefore the proper place to measure it for this purpose. That this size is conformable to the age assigned is apparent from a comparison with other trees; such, for instance, as the "Aspinwall Elm, in Brookline, standing near the ancient house belonging to the family of that name, and which was known to be one hundred and eighty-one years old in 1837, when it measured sixteen feet eight inches at five feet from the ground, and twenty-six feet five inches close by the surface."[1] The noted Elm upon Boston Common should be about the same age. Its present girth, at five feet from the ground, is sixteen feet and one inch; at the height of three feet it measures seventeen feet eleven inches; near the earth, twenty-three feet and six inches. We have seen a map of Boston, published in the year 1720, upon which this Elm is delineated as a large tree. Its age, therefore, is certainly as great as that assigned to it in the subjoined account, which recently appeared in the newspapers of the day; — we know not upon what authority.[2]

[1] We quote the manuscript of an esteemed friend, who has devoted much attention to the history and growth of trees, and whose long expected volume, on the trees of New England, we hope will soon be given to the public.

[2] As such data may hereafter possess some interest, we may simply state, that the large "English Elm," one of the finest trees on Boston Common, is now eleven feet two inches in girth, at five feet from the ground; and twelve feet three inches, at the height of three feet. The American Elm, near the Botanic Garden at Cambridge, in front of the house of Judge Phillips, has a girth of thirteen feet, at six feet from the earth, and of fifteen feet, three feet lower down. Its neighbor, opposite the gardener's residence, is fourteen feet three inches in circumference, at six feet from the ground. (MSS. note in Dr. Gray's handwriting. 1874, Phillip's 14-9¾, Sander's 15-7½.)

" The ' Boston Traveller' states that this noble tree was set out about the year 1670, by Captain Daniel Henchman, and is therefore one hundred and seventy-five years old. Captain Henchman was a schoolmaster in Boston from 1666 to 1671. He joined the Ancient and Honorable Artillery Company in 1675. He was a distinguished captain, in King Philip's war, of a company of infantry. Forty-five years ago, the Great Elm had a large hollow in it, and was rapidly decaying, but was treated in a mode recommended by Forsyth, by clearing the cavity of rotten wood, and filling it with a composition composed principally of lime, rubbish from old buildings, and clay, and thus restored. It is now apparently as flourishing as ever, and without any appearance of the hollow, which was once large enough for a boy to hide himself in. The tree is a native Elm, which is the most hardy kind. Many of the old Elm-trees are what are called English Elms, with less extended roots and branches than the American."

But more commonly, perhaps, our estimates rest, either wholly or in part, upon historical evidence or tradition ; and the most numerous and best authenticated cases of this kind may be expected to occur in Europe, where many trees, especially Chestnuts, Lindens, Oaks, and Yews, may be satisfactorily traced by records through several centuries.

Having thus briefly indicated the kinds or sources of evidence which are brought to bear with more or less directness and force upon this interesting question, we proceed to offer a condensed account of some of the more remarkable or curious cases of longevity in trees ; which may show to what extent, and with what results, this various testimony has been actually applied. The evidence is cumulative. Individual cases would be little worth, if unsupported by others. But mutually strengthening each other, the obvious conclusion becomes almost irresistible, even when the testimony in particular cases is very imperfect.

We leave entirely out of view the numerous allusions to old trees that may be gathered from classical writers. Nor are the more circumstantial accounts by Pausanias, Josephus, or the Elder Pliny, available for our present purpose. The two

latter, indeed, speak of trees as old as the creation; [1] but they
have unfortunately neglected to mention the evidence upon
which their opinions were founded. Restricting ourselves,
therefore, to trees which still survive, or which have existed
within recent times, we commence our enumeration with one
which is rather remarkable for its historical associations than
for any extraordinary longevity; namely, the celebrated Sycamore Maple (*Acer Pseudo-Platanus*), which stands near
the entrance of the village of Trons, in the Grisons, the cra-
dle of liberty among the Rhetian Alps. Under the once
spreading branches of this now hollow and cloven trunk, the
Gray League — so called, either from the gray beards, or the
home-spun clothing, of the peasants who there met the nobles
favorable to their cause — was solemnly ratified in March,
1424. Upon the supposition that it was only a century old
when the meeting, to which its celebrity is owing, took place,
— and a younger tree would hardly have been selected for the
purpose, — it has now attained the age of five hundred and
twenty years. It can scarcely be younger, it may be much
older than this. In some of the earlier accounts, this tree is
said to be a Linden. Indeed, it is so called in the inscription
upon the walls of the adjacent little chapel. They were better patriots than botanists in those days; for the investigations of Colonel Bontemps leave no doubt as to the identity
of the tree.[2]

The Linden itself, however, is associated with some interesting points of Swiss history; it also affords some instances
of remarkable longevity, which the lightness and softness of
its wood would by no means lead us to expect. The Linden
in the town of Freiburg, which was planted in 1476, to commemorate the bloody battle of Morât, though now beginning
to decay, has already proved a more durable memorial than
the famous ossuary on the battlefield,

---

[1] Josephus relates, that he saw near Hebron a Terebinthus which had
existed ever since the creation (Lib. V., c. 31); and Pliny speaks of
Oaks in the Hercynian forest, which he deems coeval with the world.
(Hist. Nat., Lib. xvi., c. 2.)

[2] " Bibliothèque Univ. de Genéve," Août, 1831.

" Where Burgundy bequeathed his tombless host,
  A bony heap, through ages to remain
  Themselves their monument ; " —

and may even outlast the obelisk recently erected upon its
site. The age of this tree and the girth of its trunk being
well known, — having attained the circumference of fourteen
English feet in 364 years, — it has been employed as a standard of comparison, in computing the age of larger and more
venerable trunks of the same species.

Such a tree is still standing at the village of Villars-en-
Moing, near the town of Morât, in full health and vigor,
although portions of the bark are known to have been stripped
off about the time of the battle in 1476, when it was already
a noted tree. At four feet above the ground, the trunk has
a circumference of thirty-eight English feet, and consequently
a diameter of about twelve feet. Supposing it to have grown,
on the whole, even a little more rapidly than the Freiburg
Linden, which may be deemed a safe estimate, when we
recollect that old trees grow much more slowly than younger
ones, — supposing it to have increased in diameter at the average rate of one sixth of an inch in a year, it must have been
864 years old at the time the measurement was made, in the
year 1831. It is not probable that this estimate materially
exaggerates the age of the tree, even supposing the Linden at
Freiburg to have grown at less than the average rate for the
species. It is nearly corroborated, indeed, by the more celebrated Linden of Neustadt on the Kocher, in Würtemberg,
whose age rests wholly upon historic evidence. The readers
of Evelyn will surely remember his interesting account of this
tree ; and in recent times, some further particulars in its history have been rescued from oblivion by M. Jules Trembley,
who visited it in 1831, at the instance of the illustrious De
Candolle. It must have been already remarkable early in the
thirteenth century ; for, as is proved by documents still extant
in the registers of the town, the village of Helmbundt,·having
been destroyed in the year 1226, was rebuilt three years afterwards, at some distance from its former site, in the vicinity
of this tree, and took the name of " Neustadt an der grossen

Linden." An old poem, which bears the date of 1408, informs us, that " before the gate rises a Linden, whose branches are sustained by sixty-seven columns." The number of these columns, or pillars of stone, raised to support the heavy and widely spreading branches, one of which extends horizontally for more than a hundred feet, had increased to eighty-two when the tree was visited by Evelyn, and to one hundred and six when it was examined by Trembley. To these supports, doubtless, its preservation is chiefly owing; as the tender wood of the Linden could never sustain the enormous weight of the limbs, or resist the force of the winds. These pillars are nearly covered with inscriptions; of which the most ancient that was extant in Evelyn's time bore the date of 1551; but the oldest now legible bears the arms of Christopher, Duke of Würtemberg, with the date of 1558. At five or six feet from the ground, the trunk is thirty-five and a half English feet in circumference. If, therefore, it has grown at the actual rate of the Freiburg Linden, it must nearly have reached its thousandth anniversary. Or if, as in the case of the tree near Morât, we allow a sixth of an inch per annum for the average increase in diameter, its computed age would be a little over 800 years; surely, a moderate estimate for a tree which was called the Great Linden more than six centuries ago.

No tree of temperate climates so frequently attains an extraordinary size as the Plane, or Sycamore (Platanus); trunks of forty or fifty feet in circumference being by no means uncommon in this country. The Oriental Plane offers many equally striking instances in the south of Europe, particularly in the Levant. The celebrated tree on the island of Cos, so conspicuously seen from the channel on the Asiatic side, has recently been beautifully figured in Allen's " Pictorial Tour in the Mediterranean."

But old trunks, both of Oriental and our own very similar species, are always hollow, — mere shells; hence, in the absence of historical data, their age is only to be computed by their rate of growth; which is so rapid for the first century or two, and, at the same time, the wood is so liable to decay, that the Plane-tree is not likely to afford any instances of extreme longevity. A different conclusion might, indeed, be drawn from the account of an enormous Plane in the valley of Bou-youdereh, near Constantinople, described by Olivier, Dr. Walsh, and others; the trunk of which is one hundred and fifty feet in girth, with a central hollow of eighty feet in circumference. But the recent observations of an excellent scientific observer, Mr. Webb, leave no doubt that this monster-trunk is formed by the junction of several original trees, planted in close proximity.[1] Along the shores of the Bosphorus there are many groups of younger Planes, which, for their shade, have been designedly planted in a narrow circle, and their trunks will in time become similarly incorporated. Pliny's Lycian Plane, with a cavity of eighty-one feet in circumference, in which the consul Licinius Mutianus used to lodge with a suite of eighteen persons, may have had such an origin.

We next notice the Chestnuts, for the purpose of disposing of an analogous case of pseudo-longevity; that of the famous " Castagno di cento cavalli "; so named from the somewhat apocryphal tradition, that Jeanne of Aragon, and a hundred cavaliers of her suite, took refuge under its branches during a heavy shower, and were completely sheltered from the rain. According to Brydone, who visited it in the year 1770, the trunk, or rather trunks, — for it then had the appearance of five distinct trees, — measured two hundred and four feet in circumference; but later and more trustworthy observers reduce these dimensions to one hundred and eighty or one hundred and ninety feet. A hut has been erected in the hollow space, with an oven, in which the inhabitants dry the chestnuts and other fruits which they wish to preserve for winter, using at times, for fuel, pieces cut with a hatchet from the interior of the tree. The separation of a large hollow

---

[1] Moquin-Tandon, "Teratologie Végétale," p. 290. — By the way, although De Candolle was not at the time apprised of the real nature of this Plane, yet he was far too cautious a reasoner to estimate its age at 2000 years, as Mr. Nuttall has inadvertently stated (Sylva, i. p. 50). Whoever will read the whole paragraph in the " Physiologie Végétale," ii. p. 994, will perceive that it will by no means bear that construction.

trunk into independent portions, appearing like the remains
of as many distinct trees, is not in itself improbable. The
ancient Yew in Fortingal churchyard, Scotland, presents
a striking instance of the kind. Indeed, Brydone's guide
assured him "that, by the universal tradition and even testimony of the country, all these were once united in one stem ;
that their grandfathers remembered this, when it was looked
upon as the glory of the forest, and visited from all quarters ;
though, for many years past, it had been reduced to the ruin
we beheld. We began to examine it with more attention,
and found that there is an appearance that these five trees
were once really united in one. The opening in the middle
is at present prodigious, and it does indeed require faith to
believe that so vast a space was once occupied by solid timber. But there is no appearance of bark on the inside of
any of the stumps, nor on the sides that are opposite to one
another. . . . I have since been told by the Canon Ricupero,
an ingenious ecclesiastic of this place, that he was at the expense of carrying up peasants with tools to dig round the
'Castagno di cento cavalli' ; and he assures me, upon his
honor, that he found all these stems united below ground into
one root." [1]
It appears, however, that Brydone has not fairly represented
the worthy Canon Ricupero's opinion ; for he thought it probable that these present trunks were offshoots from the persistent base of a more ancient stem ; a conclusion which is
fully sustained by the observations of several competent naturalists, such as Duby,[2] Brunner,[3] and Philippi.[4] Every one
knows how readily the Chestnut will throw up shoots from
the root; and Philippi says it is a general custom in Sicily
to cut them down after they have attained a considerable
size, when the new stems that are thrown out from the base
shortly become trees again. Other considerations would pre-

---

[1] "Tour through Sicily and Malta."
[2] De Candolle, "Phys. Vég.," ii. p. 992.
[3] "Excursion through the East of Liguria, Sicily, and Malta."
[4] "Ueber die Vegetation am Ætna " ; in "Linnæa,"vii. p. 727 ; and in
"Comp. to Bot. Mag.," i. p. 90.

vent our assigning the highest antiquity to a tree not originally indigenous to Sicily, but doubtless introduced from the
East.

There are, however, some colossal Chestnuts upon Mount
Etna, with undoubtedly single trunks; three of which, recently measured, are found to have a circumference respectively of fifty-seven, sixty-four, and seventy feet. Some general idea of their age may perhaps be formed by a comparison
with other individuals, whose history is better known, such as
that at Sancerre, described by Bosc, which, although only
thirty-three feet in girth at six feet from the ground, has been
called the "Great Chestnut of Sancerre" for six hundred
years; or the celebrated "Tortworth Chestnut," which Strutt,
who in his "Sylva Britannica" has given a fine illustration
of its massive bole, considers as probably the largest as well
as the oldest tree standing in England, and which in the
reign of Stephen, who ascended the throne in 1135, was
already remarkable for its size, and well known as a signal
boundary to the manor of Tamworth, now Tortworth, in,
Gloucestershire. But even this tree, although it has probably
long since celebrated its thousandth anniversary, does not
equal the smallest of the three Sicilian Chestnuts, being only
fifty-two feet in circumference at five feet from the ground.

In the ascending scale of longevity, we pass from the Chestnut to the Oak, the emblem of embodied strength, one of the
longest-lived, as it is the slowest-growing, of deciduous-leaved
forest trees. The light and soft wood of the Linden, and
even of the Chestnut, seems incompatible with great longevity.
Such trees of eight hundred or a thousand years old are
extraordinary phenomena, owing their prolonged existence to
a rare conjunction of favorable circumstances, — the more
important, as they are unexpected witnesses to the truth of
our leading proposition. But this is no very uncommon age
for that

"Lord of the woods, the long-surviving Oak."

The briefest biographical notice of Oaks remarkable for
their age or size, or for historic memorials attesting their
antiquity, would alone fill our pages. We can only refer the

curious reader to the pages of Evelyn, of Gilpin, and of Strutt; to the learned, but over-labored, " Amœnitates Quercineæ " of the late Professor Burnet, in Burgess's " Eidodendron," and especially to the more accessible and standard Arboretum of Loudon, whose condensed statistical account of celebrated British Oaks, occupying thirty closely printed pages of that elaborate work, is a monument of diligence, and contains a vast amount of interesting information. Indeed, Mr. Loudon's whole account of the Oak is incomparable, and should alone suffice to immortalize his name. Among the oldest specimens now extant in England are to be enumerated, the " Parliament Oak," in Clipstone Park, supposed to be the oldest park in England, which derives its name from a Parliament having been held under it by Edward the First, in 1290; the Oak in Yardly Chase, which Cowper has immortalized; the " Winfarthing Oak," now a bleached ruin, which is said to have been called an old oak at the time of the Conquest; the Oak in Melbury Park, Dorsetshire, which Mitchell calls " as curly, surly, knotty an old monster as can be conceived "; the " Greendale Oak," in the Duke of Portland's park at Welbeck, well known from Evelyn's account, and from the series of figures which his editor, Hunter, has given of its mutilated trunk, pierced by a lofty arch through which carriages have been driven; the " Cowthorpe Oak," in Yorkshire, also figured by Hunter, the trunk of which measures seventy-eight feet in circumference near the ground, and the age is estimated as nearly coeval with the Christian era; and the " Great Oak of Salcey Forest," in Northamptonshire, " a most picturesque sylvan ruin," which is perhaps of equal antiquity.

We have already mentioned the tree at Bordza, felled some thirty years ago, which was proved, by inspection of its annual layers, to have been about a thousand years old. Its trunk was forty English feet in circumference, or twelve and a half feet in diameter. This was a goodly tree for an Oak; but it shrinks almost to insignificance when compared with one in the south of France; an account of which has quite recently been published. From a late number of the " Gardeners'

Chronicle," edited by Professor Lindley, we copy the following account, which purports to have been extracted from the Annals of the Agricultural Society of Rochelle.

"At about six miles west-southwest of Saintes (in the Lower Charente) near the road to Cozes, stands an old Oak-tree, in the large court of a modern mansion, which still promises to live many centuries, if the axe of some Vandal does not cut it down. The following are the proportions of this king of the forests of France, and probably of all Europe. The diameter of the trunk at the ground is from nine to ten yards [consequently its circumference is from eighty-five to ninety-four feet]; at the height of a man, from six and a half to seven and a half yards [from sixty to sixty-seven feet in circumference]; the diameter of the whole head, from forty to forty-three yards; the height of the trunk, eight yards; the general height of the tree, twenty-two yards. A room has been cut out of the dead wood of the interior of the trunk, measuring from nine to twelve feet in diameter, and nine feet high; and they have cut a circular seat out of the solid wood. They put a round table in the middle, when it is wanted, around which twelve guests can sit. A door and a window admit daylight into this new sort of dining-room, which is adorned by a living carpet of Ferns, Fungi, Lichens, etc. Upon a plate of wood taken from the trunk about the height of the door, two hundred annual rings have been counted, whence it results, in taking a horizontal radius from the exterior circumference to the centre of the oak, that there must have been from 1800 to 2000 of these rings; which makes its age nearly two thousand years."

We should have been told, however, from what portion of the radius this block was taken. If near the circumference, where the rings are the narrowest, the age of the tree has been over-estimated; perhaps not materially so, as it must have been growing at a slow and nearly equable rate for many centuries; if towards the centre, the computed age is within the truth. To this tree, therefore, as being probably the patriarch of the species in Europe, may well be applied the lines addressed by Cowper to the Yardley Oak: —

" O, couldst thou speak,
As in Dodona once thy kindred trees,
Oracular, I would not curious ask
The future, best unknown ; but, at thy mouth,
Inquisitive, the less ambiguous past !
By thee I might correct, erroneous oft,
The clock of history ; facts and events
Timing more punctual ; unrecorded facts
Recovering ; and misstated, setting right."

Rich although this country is, above all other parts of the
world, in different species of the Oak, it would not be difficult to explain why we cannot boast of such venerable trees,

" Whose boughs are mossed with age,
And high top bald with dry antiquity."

It is not merely, or chiefly, that, in clearing away the forest
which so recently covered the soil, " men were famous according as they had lifted up axes upon the thick trees." The
close, stifling growth of our primeval forests, like the demo-
cratic institutions which they seem to foreshadow, although
favorable to mediocrity, forbids preëminence. " A chilly,
cheerless, everlasting shade" prevents the fullest individual
development ; and even if the woodman's axe had spared the
older trees, their high-drawn trunks, no longer shielded by
the dense array of their brethren, were sure to be overthrown
by the winds. Had the aboriginal inhabitants been tillers of
the ground, our White Oaks had long since spread their
broad brawny arms, and emulated their more renowned
brethren in the parks of England. The " Charter Oak " at
Hartford, so conspicuous in the colonial history of Connecticut, and a few others of equal size, but less note, were probably mere saplings at the first settlement of the country.
" The Wadsworth Oak," in Geneseo, N. Y., however, may
claim a higher antiquity. It stands in an old " Indian clearing," on the bank of the Genesee River, which, we are sorry
to say, is gradually undermining its roots and threatening its
destruction ; — a catastrophe which we beseech the worthy
proprietor of that princely estate to avert, by a seasonable
embankment. A note in an earlier volume of this Review [1]

[1] Vol. xliv. p. 345, note.

assigns to this noble tree the age of at least five hundred years; — a credible estimate, notwithstanding the girth of the tree is somewhat overstated in that account. Its circumference at the smallest part of the trunk (four feet above the ground), — which is always the proper point for measurement, — instead of from twenty-four to twenty-seven, is only twenty-two feet four inches; although near the base, owing to the influence of the spreading roots, its girth is considerably greater.

But of all American species, the invaluable Live Oak of our southern coasts will probably be found to attain the greatest longevity; although it seldom becomes a very large, or, at any rate, a very tall tree. Like the finest European Oaks, its branches spread very widely, and contain a prodigious quantity of timber. "The trunk of the Live Oak," says Mr. Bartram, in his delightful "Travels in Florida," "is generally [on the St. John's River] from twelve to eighteen feet in girth, and rises ten or twelve feet erect from the earth; some I have seen eighteen or twenty; then divides itself into three, four, or five great limbs which continue to grow in nearly a horizontal direction, each limb forming a gentle curve, or arch, from its base to its extremity. I have stepped above fifty paces, on a straight line, from the trunk of one of these trees to the extremity of the limbs."

The younger Michaux mentions a tree felled near Charleston, whose trunk was twenty-four feet in circumference; and we learn that another individual of still greater size is still flourishing on the plantation of Mr. Middleton, near that city. According to Mr. Nuttall,[1] the tree sometimes acquires the diameter of eight or nine feet in west Florida. All these trees must have attained a great age; for this heavy and almost incorruptible wood is of extremely slow growth. May we not hope that some competent observer will collect the requisite information upon this subject, before all the larger trunks have yielded to their impending fate?

The Olive grows much more slowly than the Oak, and as its wood is very compact and durable, it is not surprising

[1] "N. Am. Sylva," Supplement, i. p. 16.

that it should furnish instances of extraordinary longevity.
In comparative youth, the stem increases in diameter only at
the rate of an eighth of an inch in a year. Therefore the
Olive at Pescio, mentioned by De Candolle, having a trunk
of twenty-four feet in girth, should be seven hundred years
old ; even supposing it to have grown, throughout, at the
ordinary rate for younger trees ; while the still larger tree at
Beaulieu, near Nice, described by Risso, and recently measured by Berthelot, doubtless the oldest of the race in Europe,
should be more than a thousand years old. Although now in
a state of decrepitude, it still bears an abundant crop of fruit,
or at least did so, as late as the year 1828.[1] It is not improbable, therefore, that those eight venerable trees, which
yet survive upon the Mount of Olives, may have been in
existence, as tradition asserts, at the time of our Saviour's
passion.

Let us now direct our attention to the class of coniferous
trees, among which, on account of the resinous matters that
commonly pervade their wood and tend to preserve it from
decay, as well as for other reasons which we will not stop to
explain, instances of longevity may be expected to occur not
inferior to those already noticed.

We begin with the classical cypress (*Cupressus sempervirens*), so celebrated in all antiquity for the incorruptibility
of its wood and its funeral uses ; doubtless, one of the longest
lived trees of southern Europe and of the East. Hunter,
in his edition of Evelyn, about a century ago, mentions the
fine avenue of Cypresses "Los Cupressos de la Reyna Sul-
tana," which adorns the garden of the Generaliffe at Granada. Under their shade, according to the well-known legend,
the last Moorish king of Granada surprised his wife with one
of the Abencerrages, which led to the massacre of thirty-six
princes of that race. This was, of course, before the year
1492, the date of the final expulsion of the Moors. These
enduring memorials of frailty and revenge were still flourishing in perennial vigor in 1831, when they were examined by

[1] Risso, "Hist. Nat. Europ. Merid.," ex Moquin-Tandon, "Teratol.
Veg.," p. 105.

Mr. Webb. Supposing them to have been only forty or fifty years old at the occurrence of the event to which they owe their celebrity, — surely a reasonable supposition, as they were then large trees, according to the legend, — they have now reached the age of about four hundred years. They are probably much older than this.

But these and all other Cypresses known in Europe are striplings in comparison with the tree at Somma, in Lombardy, which Loudon has figured in his Arboretum (p. 2470), from an original drawing furnished by Signor Manetti of Monza. The tree is greatly reverenced by the inhabitants of that part of Lombardy, who have a tradition that it was planted in the year of our Saviour's birth. Even Napoleon is said to have treated it with some deference, and to have deviated from a direct line to avoid injuring it, when laying down the plan for the great road over the Simplon. Its trunk was twenty feet in girth, according to the Abbé Belèze's measurement, in 1832, or twenty-three feet at the height of a foot from the ground, as Signor Manetti states. Since the Cypress only attains the circumference of fourteen or fifteen feet in four hundred years or more, and after that must increase with extreme slowness, we may, perhaps, place some credit in the popular tradition respecting the age of this tree, or in the testimony of the Abbé Belèze, who informed Mr. Loudon that his brother assured him, that there is an ancient chronicle extant at Milan, which proves this tree to have been in existence in the time of Julius Cæsar !

To the same class, also, belongs the goodly Cedar of Lebanon (*Cedrus Libani*), from which the sacred writers have derived so many forcible and noble images. It is generally employed as an emblem of perennial vigor and longevity. The most plausible derivation of the name is from the Arabic "kedroum" or "kèdre," signifying "power" ; and the most characteristic description of the tree, with its widespread horizontal branches and close-woven leafy canopy, is that given by the prophet Ezekiel, where it is assumed as a type of the grandeur and strength of the Assyrian empire.

" Behold, the Assyrian was a Cedar in Lebanon, with fair

branches, and with a shadowing shroud, and of an high stature; and his top was among the thick boughs. . . . Thus was he fair in his greatness, in the length of his branches; for his root was by great waters. The Cedars in the garden of God could not hide him; the Fir-trees were not like his boughs, and the Chestnut-trees were not like his branches; nor any tree in the garden of God like unto him in beauty." (Ezekiel, xxxi. 3, 7, 8.)

The celebrated grove near the summit of Mount Lebanon, to which there are particular allusions in Holy Writ, was first described in modern times by Belon, who visited it about the year 1550. The majestic old Cedars of this grove – at that time the sole, as they are still the finest, known representatives of the species — were then, as now, venerated by the Maronite Christians, who firmly believed them to have been coeval with Solomon, if not planted by his own hands, and made an annual pilgrimage to the spot, at the festival of the Transfiguration; the Patriarch celebrating high mass under one of the oldest Cedars, and very properly anathematizing all who should presume to injure them. The larger trees were described and measured by Rauwolf, an early German traveler, in 1574; by Thevenot, in 1655; and more particularly by Maundrell, in 1696; by La Roque, in 1722; by Dr. Pococke, in 1744, and by Labillardière, in 1787; since which time, De Candolle states, that all the older trees have been destroyed. But we have not been able to find the authority for this statement, and have reason to doubt its correctness. Although the number of large trees has diminished in every succeeding age, yet several recent visitors mention a few large trunks of equal size with those described by the earlier travelers. Indeed, M. Laure, an officer of the French Marine, who with the Prince de Joinville visited Mount Lebanon in the autumn of 1836, says, that all but one of the sixteen old Cedars mentioned by Maundrell are still alive, although in a decaying state; and that one of the healthiest, but perhaps the smallest trunks, measured thirty-three French feet, or about thirty-six English feet, in circumference, which, by the way, is nearly the girth of the largest that Maundrell

measured. We have little faith, however, in this particular
identification; nor do we place confidence in the rate of growth
of old Cedars, as deduced from the measurement of these trees
at different periods. For, could we be sure that any two of
these measurements were actually taken from the same trunk,
it is still very unlikely that they were made at the same height
from the ground, — a matter of great consequence, but which
is left out of view in the records of the early travelers. But
the girth of the larger trees being known by various measurements, and the average rate of growth of young Cedars being
approximately determined from individuals that have grown
in Europe, of well ascertained age and size, — such, for instance, as those in the Chelsea Botanic Garden, near London,
planted in 1683, and the fine tree which adorns the hill in the
Jardin des Plantes at Paris, and which was brought from
England in 1734 by Bernard de Jussieu, — carried, it is
said, in the crown of his hat for greater security, whose trunk,
at its centennial anniversary, had just attained the circumference of ten feet, — we only need to know the thickness of the
outer layers of these remarkable old trunks, or, in other
words, their actual and recent rate of increase, in order to
form a highly probable estimate of their age. By a few careful incisions into these trunks, the next traveler into the now
frequented East, who feels interested in such questions, might
supply this remaining desideratum, without real injury to
these renowned natural monuments, or just exposure to the
Patriarch's anathema.

From such very imperfect data as we now possess, De
Candolle deems the trees measured by Rauwolf to have been
at least six hundred years old; which would give the age of
nearly nine hundred years to any of the number that may
still survive. This estimate may fall considerably below the
truth; but our present knowledge will not warrant the assumption of a higher one. Doubtless, this remarkable forest
has existed from primeval times, while the oldest individuals,
from age to age, have decayed and disappeared. But venerable as are the present representatives, which La Martine so
grandiloquently apostrophizes, and conceives to have existed

in the days of Solomon, "yet few comparatively have the days of the years of their life been, and have not attained unto the days of the years of the life of their fathers," the real patriarchs of the world-renowned grove.

The Yew has, probably, a well founded claim to its reputation as the longest-lived tree of northern Europe; and its longevity appears the less surprising, when the closeness and incorruptibility of the wood are considered, as well as its extreme slowness of growth. A Yew

"Of vast circumference and gloom profound"

is truly, as Wordsworth has it,

"a living thing,
Produced too slowly ever to decay;
Of form and aspect too magnificent
To be destroyed."

The frequent occurrence of ancient Yews in English church-yards is simply and beautifully explained by Mr. Bowman;[1] — the Yew, being indisputably indigenous to Great Britain, and being, from its perennial verdure, its longevity, and the durability of its wood, at once an emblem and an example of immortality, its branches would be employed by our Pagan ancestors, on their first arrival, as the best substitute for the Cypress, to deck the graves of the dead, and for other sacred purposes; and the innocent custom, like others of heathen origin, would naturally be retained and engrafted upon Christianity at its first introduction.

From the inspection of various trunks of two or three hundred years old, De Candolle drew the conclusion that the trunk of the Yew increases in diameter at the rate of a little more than a line — the twelfth of an inch — in a year for the first one hundred and fifty years, and at a little less than this rate during the next century or two. De Candolle proposed, therefore, to estimate the age of ancient Yews by assuming a line per annum as their average growth in diameter. Their age would in this way be readily computed by measuring their circumference, and thence obtaining the radius in lines; the tree being reckoned as many years old as there are lines in

[1] In "Mag. Nat. Hist.," 2d ser., i. p. 86.

its diameter. Since all trees grow the more slowly as they advance in years, this method would seem to be a safe one, if we were well assured that the average rate of growth has been correctly assumed. But extended observation upon Yews in England has shown that young trees often grow much more rapidly than De Candolle supposed ; so that, from the application of his rule to Yews not more than four or five hundred years old, we should be liable greatly to exaggerate their age. But it is also found that still older trees grow so much more slowly, that the rule may be applied to very ancient Yews with reasonable probability that the estimate will fall beneath the truth, and make them appear younger than they really are. The greater the circumference of the tree, the less the danger that its more rapid early growth will falsify the estimate. The adoption of this rule leads, however, to rather startling conclusions.

The computed age of the famous Yews of Fountains' Abbey, near Ripon, in Yorkshire, is to a great extent sustained by the history of the abbey itself, as chronicled by Hugh, a monk of Kirkstall, whose narrative — still preserved, it is said, in the library of the Royal Society — forms the basis of the well known account in Burton's " Monasticon." This monastery, the noble ruins of which are now overlooked by the venerable trees that watched its erection, was founded in the year 1132, by Thurstan, Archbishop of York, for certain monks, whose consciences, being too tender to allow them to indulge in the relaxed habits of their own order, made them desirous of adopting the more rigid rule of the Cistercians, then recently introduced into England..

" At Christmas," therefore, says Burton, " the Archbishop, being at Ripon, assigned to these monks some land in the patrimony of St. Peter, about three miles west of that place, for the erecting of a monastery. The spot of ground had never been inhabited, unless by wild beasts, being overgrown with woods and brambles, lying between two steep hills and rocks, covered with wood on all sides, more proper for a retreat for wild beasts than for the human species. . . . Richard, the Prior of St. Mary's at York, was chosen Abbot by the monks, being the first of this Monastery of Fountains; with whom they withdrew into this uncouth desert, without any house to shelter them in that winter season, or provisions to subsist on, but entirely depending on Divine Providence. There stood a large Elm in the midst of the vale, on which they put some thatch or straw, and under that they lay, ate, and prayed; the Bishop for a time supplying them with bread, and the rivulet with drink. But it is supposed that they soon changed the shelter of their Elm for that of seven Yew-trees growing on the declivity of the hill on the south side of the abbey; all standing at this present time [1658], except the largest, which was blown down about the middle of the last century. They are of an extraordinary size; the trunk of one of them is twenty-six feet six inches in circumference at the height of three feet from the ground; and they stand so near each other as to form a cover almost equal to a thatched roof. Under these trees, we are told by tradition, the monks resided till they built the monastery; which seems to be very probable, if we consider how little a Yew-tree increases in a year, and to what a bulk these are grown." (Burton, Monast., fol. 141.)

We have Pennant's measurements of one of these trees, taken in 1770, giving it a diameter of eight feet five inches, or 1212 lines. Hence, according to De Candolle's rule, it was then 1200 years old.

The fine Yew at Dryburgh Abbey, which is supposed to have been planted when the abbey was founded, in 1136, and which is in full health and vigor, has a trunk only twelve feet in circumference; its estimated age would, therefore, be less than six hundred years.

The " Ankernyke Yew," near Staines, a witness of the conference between the English barons and King John, and in sight of which Magna Charta was signed (between Runnymede and Ankernyke House), and beneath whose shade the brutal Henry the Eighth first saw gospel light in Anna Boleyn's eyes, measures twenty-seven feet eight inches in circumference, and should therefore be 1100 years old, which is about the age that tradition assigns to it. The trunk of the

" Darley Yew " in Derbyshire, having a mean diameter of nine feet five inches, would, by this rule, be 1356 years old. The Yew in Tisbury churchyard, Dorsetshire, the trunk of which measures thirty-seven feet in circumference, would now be almost 1600 years old. The same computation, applied to the " superannuated Yew-tree of Braburne churchyard, Kent," which, by the measurements of Evelyn himself and of Sir George Carteret, was fifty-eight feet eleven inches in circumference in the year 1660, would give it the respectable age of 2540 years at that time! This tree has long ago disappeared. But it did not greatly exceed in size the Yew still extant in Fortingal churchyard, in Perthshire, Scotland, situated in a wild district among the Grampian Mountains, which forms a good collateral witness to the credibility of Evelyn's account. The trunk of the " Fortingal Yew " was fifty-two feet in circumference, when measured by the Hon. Daines Barrington in 1769 ; [1] or fifty-six feet six inches, according to Pennant's somewhat later measurement; [2] the discrepancy being, no doubt, attributed to the fact that the two measurements were taken at different heights. In Barrington's time, the surface was nearly entire at the base, although upon one side all the interior had decayed. Afterwards, the cavity reached the opposite surface ; and the trunk at length separated into two distinct semicircular portions, dead and decaying within, but alive and growing at the circumference, between which the rustic funeral processions were long accustomed to pass on their way to the grave. In this condition it is figured by Strutt, as the first illustration of his " Sylva Scotica " ; but he has omitted to inform us when the sketch was taken. We suspect that it represents the tree as it appeared more than fifty years ago ; for, if we rightly apprehend the account given by the excellent Dr. Neill of Edinburgh, who visited the place in the summer of 1833, one of these half-trunks has now disappeared, with the exception of some decayed portions that scarcely rise above the soil ; but the other, which still shoots forth branches from the summit,

---

[1] "Phil. Trans.," lix. p. 37.

[2] " Tours in Scotland," in Pinkerton's Gen. Coll., vol. iii.

" gives a diameter of more than fifteen feet ; so that it is easy
to conceive that the circumference of the bole, when entire,
should have exceeded fifty feet." " Considerable spoliations,"
Dr. Neill further observes, " have evidently been committed
on the tree since 1769 ; large arms have been removed, and
masses of the trunk itself carried off by the country people,
with the view of forming ' quechs ' or drinking - cups, and
other relics, which visitors were in the habit of purchasing.
Happily, further depredations have been prevented by means
of an iron rail, which now surrounds the sacred spot ; and
this venerable Yew, which in all probability was a flourishing
tree at the commencement of the Christian era, may yet survive for centuries to come." [1]

But we must not forget the typical representatives of the
class of coniferous trees, the stately Pines and Firs ; several
species of which attain a great size, and especially an unexampled height. Indeed, their mode of growth — their
straight, regularly tapering trunks, carried steadily upwards
by the continued prolongation of the leading shoot, as well as
the small lateral extension of their branches — is extremely
favorable to loftiness of stature, and to full development in the
midst of the forest. In such trees our own country abounds.
We need not dwell upon so familiar an object as our own
White Pine, which, like Saul, " from his shoulders upwards,
higher than any of the people," lifts its kingly form above its
forest brethren, to the altitude of from one hundred and fifty
to at least one hundred and eighty feet.

" Not a prince,
In all that proud old world beyond the deep,
E'er wore his crown as loftily, as he
Wears his green coronal of leaves."

The White Pine is, par excellence, a New England tree,
and has ever been identified with our commercial prosperity.
The colonists of Massachusetts Bay, at a very early period,
selected it as their cognizance, and when they first assumed
the rights of a free people, they stamped its image on their
coins. It does not seem to flourish on foreign soil ; as we

[1] Jameson's " Edinb. Phil. Jour." (1833), xv. p. 343.

infer from Loudon's description, and the ill-favored figure
which he gives as an illustration of its general appearance in
English parks and pleasure-grounds ; [1] no less than from Gilpin's complaint of its "meagreness in foliage." [2]

Yet even the White Pine is overtopped by the Douglas
Spruce (*Pinus Douglasi*), which forms the principal part of
the gloomy forests of Oregon. The extraordinary height
which this species attains was first recorded by Lewis and
Clarke ; who state that the trunk is very commonly twenty-
seven, and often thirty-six feet, in circumference, at six feet
above the earth's surface ; and rises to the height of two
hundred and thirty feet — one hundred and twenty of that
height without a limb. One which was measured by a member of their party is said to have been forty-two feet in girth,
at a height beyond the reach of an ordinary man, and was
estimated to reach the altitude of three hundred feet ! [3] This
account, so far as respects the general height of the tree, has
been amply confirmed by succeeding travelers, and especially
by that enterprising botanist, the late Mr. Douglas, whose
name the species bears, and to whom its discovery is generally
attributed.[4] Mr. Douglas was really the first to make known
the Lambert Pine (*Pinus Lambertiana*) to the scientific
world ; a species which grows on the southern frontiers of
Oregon Territory and in northern California ; the height of
which is the more extraordinary, as the trees do not form a
thick forest, but are rather sparsely scattered over the plains.

---

[1] "Arb. Brit.," iv. p. 2881 f., 2196.

[2] "Forest Scenery," i. p. 87. — The natural and economical history of
this important tree has already been fully recorded on the pages of this
Journal. Vol. xliv. p. 339, and vol. lviii. p. 300.

[3] "History of the Expedition of Lewis and Clarke," ii. p. 155. — More
surprising still, and, as to the height compared with the diameter of
the trunk, to us nearly incredible, is their account of a fallen tree of the
same species on Wappatoo Island, which, they state. "measured 318 feet
in length, although its diameter was only three feet !" (Op. cit. ii.
p. 225.)

[4] We have not found Lewis and Clarke's account anywhere cited or
alluded to, except by the accurate (former) editor of the "American Almanac," in the volume for 1838, p. 108.

To give our readers some idea of the hardships which this indefatigable collector endured, and the risks at which our nurseries have been stocked with the trees, and our gardens with the now familiar flowers of Oregon and California, we extract from the journal of Douglas a portion of the account of his visit to a group of these Lambert Pines; merely remarking that it seems to afford a fair specimen of the perils which he continually incurred. Poor fellow! to have the life at last stamped out of him by a mad bullock in a pit, while pursuing his researches upon one of the Sandwich Islands!

"Thursday, the 25th. Weather dull, cold, and cloudy. When my friends in England are made acquainted with my travels, I fear they will think that I have told nothing but my miseries. This may be true; but I now know, as they may do also, if they choose to come here on such an expedition, that the objects of which I am in quest cannot be obtained without labor, anxiety of mind, and no small risk of personal safety, of which latter statement my this day's adventures are an instance. I quitted my camp early in the morning, to survey the neighboring country, leaving my guide to take charge of the horses until my return in the evening, when I found that he had done as I wished, and in the interval dried some wet paper which I had desired him to put in order. About an hour's walk from my camp I met an Indian, who, on perceiving me, instantly strung his bow, placed on his left arm a sleeve of raccoon skin, and stood on the defensive. Being quite satisfied that this conduct was prompted by fear, and not by hostile intentions, the poor fellow having probably never seen such a being as myself before, I laid my gun at my feet on the ground, and waved my hand for him to come to me, which he did, slowly and with great caution. I then made him place his bow and quiver of arrows beside my gun, and, striking a light, gave him a smoke out of my own pipe, and a present of a few beads. With my pencil I made a rough sketch of the cone and Pine-tree which I wanted to obtain, and drew his attention to it, when he instantly pointed with his hand to the hills fifteen or twenty miles distant towards the south; and when I expressed my

intention of going thither, he cheerfully set about accompanying me. At mid-day I reached my long-wished-for Pines, and
lost no time in examining them, and endeavoring to collect
specimens and seeds. New and strange things seldom fail to
make strong impressions, and are therefore frequently over-
rated; so that, lest I should never again see my friends in
England to inform them verbally of this most beautiful and
immensely grand tree, I shall here state the dimensions of
the largest I could find among several that had been blown
down by the wind. At three feet from the ground, its circumference is fifty-seven feet nine inches; at one hundred
and thirty-four feet, seventeen feet five inches; the extreme
length two hundred and forty-five feet.[1] The trunks are
commonly straight, and the bark remarkably smooth for such
large timber, of a whitish or light-brown color, and yielding
a great quantity of bright amber gum. The tallest stems are
generally unbranched for two thirds of the height of the tree;
the branches rather pendulous, with cones hanging from their
points, like sugar-loves in a grocer's shop. These cones are,
however, only seen on the loftiest trees, and the putting myself in possession of three of these (all I could obtain) nearly
brought my life to a close. As it was impossible either to
climb the tree or hew it down, I endeavored to knock off the
cones by firing at them with ball, when the report of my gun
brought eight Indians, all of them painted with red earth,
armed with bows, arrows, bone-tipped spears, and flint knives.
They appeared anything but friendly. I endeavored to explain to them what I wanted, and they seemed satisfied, and
sat down to smoke; but presently I perceived one of them

[1] We take this to be the correct account. But, by an error in copying,
as we suppose, the length of this same tree is given at only two hundred
and fifteen feet, in the memoir inserted in the 16th volume of the " Transactions of the Linnæan Society "; whence it has been copied into Lambert's
great work on Pines, Loudon's Arboretum, the " American Almanac "
for 1838, and Hooker's " Flora Boreali-Americana.'' There is another
apparent discrepancy between the two accounts. In the Linnæan Transactions, the timber is said to be " white, soft, and light." In his journal,
Douglas says, the wood of the large tree he examined was " remarkably
fine-grained and heavy."

string his bow, and another sharpen his flint-knife with a pair
of wooden pincers, and suspend it on the wrist of the right
hand. Further testimony of their intention was unnecessary.
To save myself by flight was impossible ; so, without hesitation, I stepped back about•five paces, cocked my gun, drew
one of the pistols out of my belt, and holding it in my left
hand and the gun in my right, showed myself determined to
fight for my life. As much as possible I endeavored to preserve my coolness ; and thus we stood looking at one another
without making any movement or uttering a word for perhaps
ten minutes, when one, at last, who seemed the leader, gave
a sign that they wished for some tobacco : this I signified
that they should have, if they fetched me a quantity of cones.
They went off immediately in search of them, and no sooner
were they all out of sight, than I picked up my three cones
and some twigs of the trees, and made the quickest possible
retreat, hurrying back to my camp, which I reached before
dusk. The Indian who last undertook to be my guide to
the trees, I sent off before gaining my encampment, lest he
should betray me. How irksome is the darkness of night
to one under my present circumstances ! I cannot speak a
word to my guide, nor have I a book to divert my thoughts,
which are continually occupied with the dread lest the hostile
Indians should trace me hither and make an attack. I now
write lying on the grass, with my gun cocked beside me, and
penning these lines by the light of my Columbian candle,
namely, an ignited piece of rosiny wood." (" Companion to
Botanical Magazine," ii. pp. 130, 131.)

It is to be regretted, although, under the circumstances, it
is by no means surprising, that Mr. Douglas did not secure,
at the time, complete data for ascertaining the age of the
prostrate trunk he measured, which, as he states, was certainly
not the largest he saw. But in a block from a smaller trunk,
of the same species, he sent to England, " there are fifty-six
annual layers in a space of four and a half inches next the
outside." If we suppose the large tree to have grown at an
equivalent rate throughout, it must have been 1400 years old
when overthrown. But if it grew during the first century at

the average rate of our White Pine for the same period, the
estimate would be reduced to 1100 years; which is probably
much beneath the truth.

But the most stately tree in North America — apparently
an evergreen species of Taxodium or American Cypress —
was subsequently observed by Douglas in Upper California.
" This tree," he says, "gives the mountains a most peculiar, I
was almost going to say, awful appearance — something which
plainly tells that we are not in Europe. I have repeatedly
measured specimens of this tree, two hundred and seventy
feet long, and thirty-two feet round at three feet above the
ground. Some few I saw upwards of three hundred feet
high." [1] Truly these are trees,

> " to equal which the tallest pine,
> Hewn on Norwegian hills, to be the mast
> Of some great ammiral, were but a wand."

This naturally brings us to the proper North American Cypress (*Taxodium distichum*); one of the largest and most
remarkable trees of our southern States, but which appears
to attain its most ample development in the *tierras templadas*
of Mexico. Bartram gives a characteristic description of the
tree.

" It generally grows in the water, or in low flat lands, near
the banks of great rivers and lakes, that are covered a great
part of the year with two or three feet depth of water ; and
that part of the trunk which is subject to be under water,
and four or five feet higher up, is greatly enlarged by prodigious buttresses, or pilasters, which in full grown trees project out on every side to such a distance that several men
might easily hide themselves in the hollows between. Each
pilaster terminates under ground in a very large, strong,
serpentine root, which strikes off and branches every way
just under the surface of the earth ; and from these roots
grow woody cones, called Cypress-knees, four, five, and six
feet high, and from six to eighteen inches and two feet in
diameter at their bases. The larger ones are hollow, and

[1] Journal of Douglas's second visit to the Columbia, etc., in Hooker,
"Compan. to Bot. Mag.," ii. p. 150.

serve very well for bee-hives! A small space of the tree itself is hollow, nearly as high as the buttresses already mentioned. From this place the tree, as it were, takes another
beginning, forming a great, straight column, eighty or ninety
feet high ; when it divides every way around into an extensive, flat, horizontal top, like an umbrella, where eagles have
their secure nests, and cranes and storks their temporary rest-
ing-places. And what adds to the magnificence of their appearance is the streamers of long moss that hang from the
lofty limbs, and float in the winds." (Bartram's "Travels
through Carolina, Georgia, etc.," p. 91.)

In favorable situations, the tree sometimes attains the
height of one hundred and twenty, or one hundred and forty
feet, and a circumference of from twenty to forty feet, when
measured quite above the singular dilated base. This is
scarcely exceeded by the largest of the celebrated Cypresses
in the gardens of Chapultepec, at Mexico, called the "Cypress
of Montezuma," and which was already a remarkable tree in
the palmy days of that unfortunate monarch, three and a
half centuries ago. The girth of its trunk is forty-one feet,
according to Mr. Ward,[1] or about forty-five, according to Mr.
Exter; but its height is so great in proportion, that the
whole mass appears light and graceful.

But this tree is greatly surpassed by the famous "Ahue-
huete " (the Mexican name for the species) of the village of
Atlisco, in the intendancy of Puebla, which was first described
by Lorenzana from personal observation. The worthy Archbishop says that "the cavity of the trunk " —for the tree is
hollow— " might contain twelve or thirteen men on horseback ; and that, in the presence of the most illustrious Archbishop of Guatemala and the Bishop of Puebla, more than a
hundred boys entered it." [2] The girth of the trunk, according
to Humboldt, is a little over twenty-three metres, or seventy-
six English feet, and the diameter of the cavity about sixteen
feet.[3]

[1] "Travels in Mexico," ii. p. 230.
[2] Note to the Third Despatch of Cortes. This note is not found in
Mr. Folsom's translation.
[3] "Essai Polit. Nouv. Esp.," ed. 2, ii. p. 54.

Still more gigantic — the Nestor of the race, if not of the
whole vegetable kingdom — is the Cypress which stands in
the churchyard of the village of Santa Maria del Tule, in the
intendancy of Oaxaca, two and a half leagues east of that
city, on the road to Guatemala by the way of Tehuantepec.
In its neighborhood there are five or six other trees of the
kind, which are nearly as large as the "Cypress of Montezuma," but which this one as much surpasses as that does
the ordinary denizens of the forest. We possess three independent measurements of this enormous trunk. The first is
that given by Humboldt, who states, probably on the authority of his informant, M. Anza, that the trunk is thirty-six
metres (one hundred and eighteen English feet) in circumference. In the year 1827, Mr. Poinsett, then our minister
at the court of Mexico, transmitted to the American Philosophical Society at Philadelphia a cord which represented the
exact circumference of this tree. Its extraordinary length
naturally excited some doubts as to the correctness of the
measurement; and immediate application was made to Mr.
Poinsett for further particulars. He accordingly transmitted
a communication from Mr. Exter, an English traveler who had
just returned from Oaxaca, and who had carefully examined
the tree in question. Mr. Exter's letter was afterwards published in Loudon's "Magazine of Natural History"; and a
French translation, accompanied by some interesting com-
ments by the younger De Candolle, appeared in the "Bibliothèque Universelle" for 1831.[1] According to Mr. Exter's
measurement, the trunk is forty-six *varas* — one hundred
and twenty-two English feet — in circumference; which is
nearly in accordance with Humboldt's account. In neither
case is the height at which the trunk was measured expressly
mentioned. But this point has been duly attended to by a
recent scientific observer, M. Galeotti, who visited this celebrated tree in 1839 and in 1840, and whose careful measurement gives to the trunk the circumference of one hundred
and five French (equal to one hundred and twelve English)

---

[1] Tom. xlvi., p. 387.

feet, at the height of four feet above the surface of the soil.[1]
The previous measurements, therefore, were taken somewhat
nearer the base. The tree as yet shows no signs of decay,
although it bears less foliage in proportion to its size than
its younger fellows. But we find no authority for Mr.
Exter's statement, that this tree was mentioned by Cortès, and that
its shade once afforded shelter to his whole European army.
Perhaps he had in some way confounded it in his memory
with a Cypress which the Conquistador passed on the march
to Mexico, and which is still traditionally associated with
his name.[2]

Mr. Exter reports, and the observations of recent travelers
to some extent confirm the statement, that there are Cypresses
near the ruins of Palenque, equal in size to the tree at Santa
Maria del Tule. If this be so, they may claim a much higher
antiquity than the ruins they overshadow. They must have
witnessed the rise, the flourishing existence, the decline, and
the final extinction of a race whose whole history has sunk
into oblivion; while they are still alive.

By what means can we ascertain the age of large Cypress-
trees? Some years since, when Professor Alphonse De Candolle — the son and worthy successor of the botanist who has
rendered that name illustrious — attempted to answer this
question, the only evidence within his reach was drawn from
the rate at which trees of the kind had grown in France during
half a century. He inferred that the American Cypress, in
its early days, increases at the rate of about a foot in diameter
every fifty years; and the estimate, although surely much too
low for trees planted in favorable open situations (which have
even been known to add annually an inch to their diameter
for a series of years, both in Europe and in the United States),
is yet quite as high as our own observations will allow for
those which grow in their native forests. This rate would
give to the Cypress of Montezuma the age of about seven centuries, and would render that at Oaxaca scarcely coeval with

[1] "Bulletin de l'Acad. Roy. des Sciences de Bruxelles," 1843, tom. x.
p. 123.
[2] See Prescott's "History of the Conquest," i. p. 404.

the Christian era. Perhaps this is as great an age as we are
warranted in assuming for the Cypress of Montezuma; but
old trunks increase so much the more slowly as they advance
in age, that we must certainly assign a vastly higher antiquity
to the trees of Atlisco and Santa Maria del Tule. Yet far
the most important element in the calculation is wanting;
namely, the actual present rate of growth of these monstrous
trunks, or of other old trees of the same species. In default
of this essential evidence, De Candolle has instituted a comparison between these trees and the famous Baobabs of Senegal, upon which we place no great reliance, but from which
he infers that the great Cypress of Santa Maria del Tule, if
really the growth of a single trunk, is from four to six thousand years old, and perhaps dates its existence as far back as
the actual creation of the world.[1]

We trust that the next intelligent traveler who visits this
most ancient living monument, or any other Cypress of remarkable size, will not fail to complete the evidence that is
needed, as the full solution of this curious problem may throw
light upon some interesting questions respecting the physical
history of the world. One or more lateral incisions, not at
all endangering the existence of the tree, would at once reveal
its actual growth for the last few centuries. And if made at
proper points, and carried to a sufficient depth, they might
enable the judicious operator to disprove or confirm the surmise, that this huge bole may consist of the trunks of two or
three original trees, long since united and blended into one.
This conjecture is by no means very improbable, although
there is nothing in the external appearance of the trunk to
confirm it.[2]

Meanwhile, the Cypresses of our southern States, although
of more moderate dimensions, afford important assistance in
this inquiry. It is generally known that old trees of the

[1] Alphonse De Candolle, in "Bibl. Univ.," xlvi. p. 393. Aug. Pyr. De
Candolle, "Phys. Veg.," ii. p. 1000.

[2] In opposition to the remark of M. Anza, cited by Humboldt (Essai
Polit., the Engl. transl., ii. p. 190), we may adduce the account of Mr.
Exter, and the negative testimony of M. Galeotti.

kind grow very slowly; but there are no accounts on record, so far as we can learn, respecting their rate of growth. Our own observations, though not so extended as could be wished, incline us to adopt the standard which De Candolle assumed for the Yew; namely, the twelfth of an inch for the annual increase of old Cypresses in diameter, when growing in their native forests. But we would only apply this rule to trunks of large size, and with all the precautions that have already been mentioned; for the Cypress grows, or at least may grow, quite rapidly for the first century or two; but when old, it appears to increase quite as slowly as the Yew. We have counted sixty layers of the wood in the space of an inch. A fine section of a Cypress-trunk, which grew near Wilmington, in North Carolina, now lies before us, which, on an average radius of twenty-seven inches, or diameter of fifty-four inches, exhibits six hundred and seventy annual layers. It has, therefore, grown throughout at the average rate of less than the twenty-fourth of an inch a year, measured on the radius, or the twelfth of an inch on the diameter. The trunk was thirteen inches in diameter at the expiration of its first century, and twenty-seven inches about the close of the second; it added seven inches to its diameter during the third century, and a nearly equal amount during the fourth; and for the remaining three hundred and seventy years, it grew at a still slower, but, on the whole, nearly equable rate.

Now it is deemed a safe mode, as we have already shown, to employ the rate of growth deduced from comparatively young trees for the determination of the age of larger and older trunks of the same species. Not only is our estimate, in all such cases, likely to fall below the truth, but the larger the trunk in question the less the danger of exaggeration. Let us apply to the Mexican Cypresses the data furnished by our Wilmington tree. If the Cypress of Montezuma has grown, on the average, even a little more rapidly than the trunk before us, — has increased in diameter at the mean rate of an inch in twelve years, — it must now be fully two thousand years old. But if we suppose it to have grown at twelve times this rate (which is the maximum for young Cypresses

under the most favorable circumstances) during the whole
of the first century, we should thereby reduce the estimated
age to a thousand years. By the same computation, the Cypress at Atlisco would be 3480 years old; or 2390 years, if
we allow it the maximum rate of growth for the first century.
So, likewise, the great Cypress at Santa Maria del Tule would
be 5124 years of age, or 4024 years, with the aforesaid deduction. The latter accords perfectly with De Candolle's
minimum estimate; and it is the lowest age that, in the present state of our knowledge, can possibly be assigned to this
prodigious tree, upon the supposition that its trunk is really
single.

We are obliged to pass unnoticed those trees of unknown
species, but of surprising size, which the learned and enthusiastic Professor Martius visited in the interminable woods
that border the Amazon, and of which he has recently published such a spirited account.[1] Their trunks were so huge
that the outstretched arms of fifteen men were required to
grasp them; and so lofty, as to mock every effort for obtaining even a leaf or flower, by which the species might be determined. As to their age, Martius offers only a conjectural
estimate.

The Baobab, or Monkey-Bread (*Adansonia digitata*), of
Senegal and the Cape de Verde Islands, has long afforded
the most celebrated instances of vegetable longevity. The
tree is remarkable for the small height which it attains, compared with the diameter of the trunk or the length of its
branches. Trunks which are seventy or eighty feet in circumference rise to the height of only ten or twelve feet, when
they divide into a great number of extremely large branches,
fifty or sixty feet in length, which, spreading widely in every
direction, form a hemisphere or hillock of verdure, perhaps
one hundred and fifty feet in diameter, and only seventy in
elevation. To this peculiarity, rather than to the nature of
the wood, which is light and soft, the great longevity of the
tree is probably owing, its form opposing an effectual resist-

[1] "Flora Brasiliensis," Tab. Physiog., ix. ; "Arbores ante Christum
natum enatæ."

ance to the tempests which would overthrow ordinary trees. Its roots spread in a similar manner beneath the soil. When laid bare by a torrent that has washed away the earth, they have been traced to a distance of more than a hundred feet without reaching their extremity. The history of these Baobabs, possibly of the very trees which Adanson's account has rendered famous, reaches back to the discovery of that part of the African coast, and of the Cape de Verde Islands, by Cadamosto, in 1455 ; who, in his narrative, mentions the singular disproportion between the height and the girth of these trees.[1] But they were first fully described by the French naturalist Adanson, who examined them about a century ago. The largest trunks that Adanson measured were eighty-five feet in circumference, or twenty-seven in diameter. Golberry is said to have measured one that was over a hundred feet in girth. Quite recently, M. Perrottet has met with many Baobabs in Senegambia, varying from sixty to ninety feet in circumference, yet still in a green old age, and showing no signs of decrepitude. There can be no doubt, therefore, respecting the prodigious size which these trees attain ; and there is great reason to believe that they are among the oldest denizens of our planet. Indeed, their age is plausibly estimated at five or six thousand years. And the younger De Candolle has placed so much confidence in this estimate that he has employed it as a standard of comparison in the case of the Mexican Cypresses which we have just considered. If the evidence were really as direct as is generally thought, we could interpose no serious objection to such a conclusion. But a critical examination proves that the whole account given by recent writers, upon Adanson's authority, is strangely at variance with his own statements.

The current narrative is substantially and briefly as follows : — that Adanson observed, at the Madelaine Islands, near Cape de Verde, some Baobab-trees of thirty feet in

[1] "Arbores vero ibi sunt tantæ magnitudinis, ut earum ambitus sit pedum xvii, licet eminentia altitudinis non quadret magnitudini ; non enim altius tolluntur quam pedes xx," etc. (A. Cadamusti, Navig., c. xliii., in Grynæus, Nov. Orb., p. 45.)

diameter, upon the trunks of which he found inscriptions that had been made by former visitors three centuries before; that, by cutting through three hundred annual layers, he discovered the vestiges of these inscriptions upon the wood, thus proving that they were actually made at the date assigned; that, by measuring the thickness of these layers, he ascertained the actual increase of the trunk during the last three centuries; that, having thus obtained the rate of growth in old age, and having, by actual inspection of young trunks, learned the rate of growth during the first hundred years, he deduced from these combined data the almost inevitable conclusion, that the trees in question were five or six thousand years old.[1]

Let us compare this with Adanson's own statements, from which it purports to have been taken. His first account, which comprises all the principal facts in the case, is given in the " Voyage au Sénégal," prefixed to his volume on natural history of that country, which was published soon after his return to France, in 1753. Adanson simply relates, that, on his visit to the Madelaine Islands, he found Baobab-trees of five or six feet in diameter, which bore European names and dates, deeply engraven upon the bark. Two of these he took the trouble to renew, one of which was dated in the fifteenth, the other in the sixteenth century. The characters were about six inches in length, and as in breadth they occupied but a small part of the circumference of the trunk, Adanson reasonably inferred that they were not engraven in the early youth of these trees. He had previously seen, on the island of Senegal, trees of the kind, which were sixty-three and sixty-five feet in circumference; but he does not intimate that he inspected the layers of wood in any case. He merely remarks that these inscriptions might furnish some evidence respecting the age which Baobabs sometimes attained; " For," says he, " if we suppose that the inscriptions were engraven even in the early years of these trees, and that

---

[1] See Alphonse De Candolle, in " Bibl. Univ.," xlvi. p. 389. (Aug. Pyr. De Candolle, Phys. Veg., ii. p. 1003. Moquin-Tandon, Teratol. Veg., p. 107.)

they have grown to six feet in diameter in the course of two centuries, we may calculate how many centuries they would require to attain the full diameter of twenty-five feet." [1] Soon afterwards, Adanson communicated to the Royal Academy of Sciences of Paris a full account of the Baobab; which was published in the volume of Memoirs of that society for the year 1761; and, lastly, he wrote the article " Baobab " for the supplement to the great French Encyclopædia, published in the year 1776. These accounts, although more detailed, embody no essential additions to what has already been given. He says that the trees in question were two in number, upon the bark of which the names of Europeans were engraved, with dates, some posterior to the year 1600; and others, as far back as 1555, were probably the work of those who accompanied Thevet, who, in his voyage to antarctic lands, saw some of these trees that same year.[2] Some of the dates appeared to be anterior to 1500, but these were somewhat equivocal. Neglecting, therefore, the indistinct dates in the fourteenth century, continues Adanson, and even allowing that the inscriptions were made when the trees were very young, which is highly improbable, as they occupied less than an eighth of the entire circumference, it is evident, that, if the Baobab has attained six feet in diameter between 1555 and 1749, that is, in two hundred years, it would require more than eight centuries to attain the diameter of twenty-five feet, supposing the growth to continue at a uniform rate. But Adanson goes on to say that trees grow the more slowly as they advance in age; so that such an estimate would fall

[1] "Voyage au Sénégal," Paris, 1757, p. 66.

[2] " Aupres du promontoire Verd, y a trois petites isles prochaines de terre ferme, autres que celles, que nous appellōs Isles de Cap Verd, dont nous parlerons cy apres, assez belles, pour les beaux arbres, qu'elles produissent ; toutesfois elles ne sont habitées. . . . En l'une de ces isles se trouve un arbre, lequel porte feuilles semblables a celles de noz figuiers ; le fruit est lōg de deux pieds ou envirō, et gros en proportion," etc. (Thevet, "Singularités de la France Antarctique ; " Anvers, 1558, p. 18.) Thevet proceeds to describe the fruit, its edible character, its furnishing food for monkeys, etc., so as to leave no doubt as to its being a Baobab.

below the truth. As to its rate of growth when young, he
states that the tree acquires the diameter of an inch or an
inch and a half in the first year; the diameter of a foot in ten
years, and about a foot and a half in thirty years; but so far
from having extended these data, and employed them in the
manner which is attributed to him, he says, that, although it
might be desirable thus to employ them, a sound geometry
teaches that they are quite insufficient for the purpose.
Hence, instead of attempting any precise determination, he
merely offers the probable conjecture, that these largest
Baobabs may have been in existence several thousand years,
or nearly from the period of the universal deluge ; which
would give them a claim to be considered the most ancient
living monuments in the world.[1]

We cannot learn that Adanson ever made any further
statements upon the subject; and, as he never revisited the
African coast, he cannot have collected additional facts. His
original writings plainly show that he never pretended to
have obtained the data and made the estimates which have
so long been attributed to him. To whom belongs the credit
of falsifying his testimony we are unable to ascertain, as the
authors above mentioned do not cite their immediate authority ; — perhaps to one M. Duchesne, whose name the elder
De Candolle has casually alluded to, as having drawn up a
table, exhibiting the diameter of the Baobab at different
periods, doubtless upon the very plan that Adanson pointed
out and condemned. We are only surprised that such accurate and judicious writers as the De Candolles, father and
son, should have relied upon second-hand authorities in any
case where the originals were accessible, and especially in
what they term " the most celebrated case of extreme longevity
that has yet been observed with precision." [2]

---

[1] "Mém. Acad. Sciences," 1761, p. 231 ; and "Encycl. Suppl.," vol.
i. p. 798.

[2] A passage which has met our eye in Mirbel's "Elémens de Physiologie Végétale," i. p. 116, shows that no such data as those which have
been, as we suppose, falsely assumed, were known to that author down
to the year 1815.

We close our enumeration, already too protracted, with a case of longevity, perhaps transcending that of the oldest Baobabs, or of the Mexican Cypresses; namely, the famous Dragon-tree (*Dracœna Draco*) of the city of Orotava, in Teneriffe. This tree has been visited by many competent observers; and among others, by that prince of scientific travelers, the veteran Humboldt, who has given a good figure of it, as it appeared about seventy years ago, from a drawing made by M. Ozonne in 1776. A later and much fuller account was published about twenty years since, by M. Berthe-lot,[1] who has assiduously devoted many years to the study of the civil and natural history of the Canary Islands; and a fine figure of the mutilated trunk, as it appeared after the terrible storm of the 21st of July, 1819, forms one of the most striking pictorial illustrations of that elaborate and excellent work, the " Histoire Naturelle des Iles Canaries," by P. Barker Webb, Esq., and M. Berthelot.

The trunk is by no means equal in size to some of the trees already noticed. It is only fifty feet in girth at the base, and not more than sixty or seventy in elevation. But, at the discovery of Teneriffe in 1402, nearly four and a half centuries ago, this Dragon-tree was nearly as large as at the present day, and had been immemorially an object of veneration among the Guanches. After the conquest, at the close of the fifteenth century, the trunk was employed as a boundary in dividing the lands, and as such is mentioned in ancient documents. It had changed very little since that period, except that the centre had been hollowed by slow decay, until the summer of 1819, when a third of its spreading top was carried away by a tempest. But it still continues to vegetate; and its remaining branches are still annually crowned, — as they have been each returning autumn, perhaps for hundreds of centuries, with its beautiful clusters of white, lily-like blossoms, — emblems of " the eternal youth of nature."

The Dragon-tree, like its allies the Palms, and unlike ordinary trees, does not increase in diameter by annual concentric layers. The usual means of investigation are here of

[1] In " Nova Acta Acad. Nat. Cur.," xiii., 1827, p. 781.

no account; and, apart from historic evidence, we can only form a somewhat conjectural estimate of the age of this celebrated trunk, by a comparison with young trees of the same species, which are known to grow with extreme slowness. M. Berthelot, who has attempted the comparison under the most favorable circumstances, — having lived many years upon the island, — declares that the calculations which he has made, upon the supposition that the trunk has increased in size even at the rate of young Dragon-trees up to within the last eight hundred or one thousand years, have more than once confounded his imagination. We cannot but assign the very highest antiquity to a tree like this, which the storms and casualties of four centuries have scarcely changed.

Upon the whole, we cannot resist the conclusion, that many trees have far survived what we are accustomed to consider their habitual duration; that even in Europe, where man has so often and so extensively changed the face of the soil, as his wants or caprices have dictated, some trees, favored by fortune, have escaped destruction for at least one or two thousand years; while in other, and particularly in some tropical countries, either on account of a more favorable climate, or because they have been more respected, or haply more neglected, by the inhabitants, a few may with strong probability be traced back to twice that period; and, perhaps, almost to that epoch which the monuments both of history and geology seem to indicate as that of the last great revolution of the earth's surface. After making every reasonable allowance for errors of observation and too sanguine inference, and assuming, in the more extraordinary cases, those estimates which give minimum results, we must still regard some of these trees, not only as the oldest inhabitants of the globe, but as more ancient than any human monument, — as exhibiting a living antiquity, compared with which the mouldering relics of the earliest Egyptian civilization, the pyramids themselves, are but structures of yesterday.

THE FLORA OF JAPAN.[1]

IT is interesting to notice that, notwithstanding the comparative proximity of Japan to western North America, fewer of its species are represented there than in far distant Europe. Also, — showing that this difference is not owing to the separation by an ocean, — that far more Japanese plants are represented in eastern North America than in either. It is, indeed, possible that my much better knowledge of American botany than of European may have somewhat exaggerated this result in favor of Atlantic North America as against Europe, but it could not as against western North America.

If we regard the identical species only, in the several floras, the preponderance is equally against western as compared with eastern North America, but is more in favor of Europe. For the number of species in the Japanese column [2] which likewise occur in western North America is about 120 ; in eastern North America, 134 ; in Europe, 157.

Of the 580 Japanese entries, there are which have corresponding

European representatives, a little above 0.48 per cent. ; of identical species, 0.27.

Western North American representatives, about 0.37 per cent. ; of identical species, 0.20.

[1] Extract from the concluding part of a " Memoir on the Botany of Japan, in its relations to that of North America, and of other parts of the Northern Temperate Zone." (Memoirs of the American Academy of Arts and Science, new series, vi. 1859.)

It is this paper which fixed the attention of the scientific world upon Professor Gray and established his reputation as a philosophical naturalist. — C. S. S.

[2] The column in a tabular view of the distribution of Japanese plants and their nearest allies in the northern temperate zone.

126   *ESSAYS.*

Eastern North American representatives, about 0.61 per
cent.; of identical species, 0.23.

So geographical continuity favors the extension of identical
species; but still eastern North America has more in common
with Japan than western North America has.

The relations of this kind between the floras of Japan and
of Europe are obvious enough; and the identical species are
mostly such as extend continuously — as they readily may —
throughout Russian Asia, some few only to the eastern confines of Europe, but most of them to its western borders. To
exhibit more distinctly the features of identity between the
floras of Japan and of North America, and also the manner
in which these are distributed between the eastern and western portions of our continent, — after excluding those species which range around the world in the northern hemisphere,
or the greater part of it, or (which is nearly the same thing
in the present view) which are unknown in Europe, — I will
enumerate the remaining peculiar species which Japan possesses in common with America: —

In Japan.	In W. N. America.	In E. N. America.
Anemone Pennsylvanica		A. Pennsylvanica
(Coptis asplenifolia?)	C. asplenifolia	
(Trautvetteria palmata)	T. palmata	T. palmata
Caulophyllum thalic-		
troides		C. thalictroides
Diphylleia cymosa		D. cymosa
Brasenia peltata	(B. peltata)	B. peltata
Geranium erianthum	G. erianthum	
Rhus Toxicodendron	R. Toxicodendron, var.	R. Toxicodendron
Vitis Labrusca (Thunb.)		V. Labrusca
Thermopsis fabacea	T. fabacea	
Prunus Virginiana?		P. Virginiana
Spiræa betulæfolia	S. betulæfolia	S. betulæfolia
Photinia arbutifolia, in		
Bonin.	P. arbutifolia	
Pyrus rivularis?	P. rivularis	
Ribes laxiflorum	R. laxiflorum	
(Penthorum sedoides,		
China)		P. sedoides
Cryptotænia Canadensis		C. Canadensis
Heracleum lanatum	H. lanatum	H. lanatum

(Archemora rigida ?)		A. rigida
(Archangelica Gmelini)	A. Gmelini	A. Gmelini
Cymopterus littoralis ?	C. littoralis	
Osmorrhiza longistylis	O. longistylis	O. longistylis
Echinopanax horridus	E. horridus	
Aralia quinquefolia		A. quinquefolia
Cornus Canadensis	C. Canadensis	C. Canadensis
Viburnam plicatum		V. plicatum (lantanoides)
*Achillea Sibirica	*A. Sibirica	
*Artemisia borealis	*A. borealis	*A. borealis
Vaccinium macrocarpon	V. macrocarpon	V. macrocarpon
Menziesia ferruginea	M. ferruginea	M. ferruginea
(Boschniakia glabra ?)	B. glabra	
*Pleurogyne rotata	*P. rotata	*P. rotata
(Asarum Canadense ?)		A. Canadense
*Polygonum Bistorta	P. Bistorta	
Rumex persicarioides	R. persicarioides	R. persicarioides
Liparis liliifolia		L. liliifolia
Pogonia ophioglossoides		P. ophioglossoides
Iris setosa	*I. setosa	
Trillium erectum, var.		T. erectum
(Smilacina trifolia)		S. trifolia
Polygonatum giganteum		P. giganteum
(Streptopus roseus)	S. roseus	S. roseus
Veratum viride	V. viride	V. viride
Juncus xiphioides	J. xiphioides	
(Cyperus Iria)		C. Iria
Carex rostrata		C. rostrata
Carex stipata	C. stipata	C. stipata
Carex macrocephala	C. macrocephala	
Sporobolus elongatus	S. elongatus	S. elongatus
Agrostis scabra	A. scabra	A. scabra
Festuca pauciflora	F. pauciflora	
Adiantum pedatum	A. pedatum	A. pedatum
Onoclea sensibilis		O. sensibilis
Osmunda cinnamomea		O. cinnamomea
Lycopodium lucidulum		L. lucidulum
(Lycopodium dendroi- deum)	L. dendroideum	L. dendroideum

The names inclosed in parentheses are of species which I
have not seen from Japan : some of them inhabit the adjacent
mainland ; some are imperfectly identified. Those marked *
are high northern species in America.

Of these fifty-six extra-European species, thirty five inhabit

western, and forty-one eastern North America. And fifteen
are western and not eastern ; twenty-one eastern and not
western ; and twenty common to both sides of the continent.
Eight or ten of these fifty-six species extend eastward into
the interior of Asia.

On the other hand, the only species which I can mention
as truly indigenous both to Japan and to Europe, but not
recorded as ranging through Asia, are : —

*Euonymus latifolius, Valeriana dioica, Pyrola media,
Fagus sylvatica, Streptopus amplexifolius, Blechnum spi-
cant, Athyrium fontanum.*

Two of these species extend across the northern part of the
American continent and on to the Asiatic ; another occurs on
the northwest coast of America ; and another, the Fagus, is
represented in eastern America by a too closely related species. It is noteworthy that not one of these seven plants is
of a peculiarly European genus, or even a Europæo-Siberian
genus ; while of the fifty-six species of the Americo-Japanese
region wanting in Europe, twenty are of the extra-European
genera, seventeen are of genera restricted to the North
American, east Asian, and Himalayan regions (except that
Brasenia has wandered to Australia) ; fourteen of the genera
(most of them monotypic) are peculiar to America and Japan
or the districts immediately adjacent ; one is peculiar to our
northwest coast and Japan ; and eight are monotypic genera
wholly peculiar (Brasenia excepted) to the Atlantic United
States and Japan.   Add to these the similar cases of other
American species (nearly all of them particularly Atlantic-
American) which have been detected in the Himalayas or in
northern Asia, — such as *Menispermum Canadense (Dau-
ricum,* DC.), *Amphicarpœa monoica? Clitoria Mariana,
Osmorrhiza brevistylis, Monotropa uniflora, Phryma leptostachya, Tipularia discolor?* etc., — and it will be almost impossible to avoid the conclusion that there has been a peculiar
intermingling of the eastern American and eastern Asian
floras, which demands explanation.

The case might be made yet stronger by reckoning some
subgeneric types as equivalent to generic in the present view.

and by distinguishing those species or genera which barely
enter the eastern borders of Europe; *e. g.*, *Cimicifuga fœ-
tida*, *Mœnringia lateriflora*, *Geum strictum*, *Spirœa salici-
folia*, etc.

It will be yet more strengthened, and the obvious conclusion will become irresistible, when we take the nearly allied,
as well as the identical, species into account. And also when
we consider that, after excluding the identical species, only
fifteen per cent. of the entries in the European column of the
detailed tabular view are in italic type (*i. e.* are closely representative of Japanese species); while there are twenty-two
per cent. of this character in the American column.

For the latter, I need only advert to some instances of such
close representation, as of

Trollius patulus	by	T. Americanus,
Aguilegia Burgeriana	"	A. Canadensis,
Rhus vernicifera	"	R. venenata,
Celastras scandens	"	C. articulatus,
Negundo cissifolium	"	N. aceroides,
Sophora Japonica	"	S. affinis,
Sanguisorba tenuifolia	"	S. Canadensis,
Astilbe Thunbergii and Japonica	"	A. decandra,
Mitchella undulata	"	M. repens,
Hamamelis Japonica	"	H. Virginica,
Clethra barbinervis	"	C. acuminata,
Rhododendron brachycarpum	"	R. Catawbiense,
Amsonia elliptica	"	Tabernæmontana,
Saururus Loureiri	"	S. cernuus,

and many others of the same sort, — several of which, when
better known, may yet prove to be conspecific; while an
equally large number could be indicated of species which,
altogether more positively different, are yet no less striking
counterparts.

To demonstrate the former proposition, I have only to
contrast the extra-American genera common to Europe and
Japan with the extra-European genera common to North
America and Japan. The principal European genera of this
category are Adonis, Epimedium, Chelidonium, Malachium,
Lotus, Anthriscus, Hedera, Asperula, Rubia, Carpesium, Ligu-
laria, Lampsana, Picris, Pæderota, Ajuga, Thymus, Nepeta,

Lamium, Ligustrum, Kochia? Daphne, Thesium, Buxus, Mercurialis, Cephalanthera, Paris, Asparagus, — to which may as well be added Pæonia and Bupleurum, the former having a representative on the mountains, and the latter in the arctic regions, of western America, but both absent from the rest of our continent. Excepting Pæderota and Buxus (the latter a rather doubtful native of eastern Asia), none of these genera are peculiar to Europe, but all extend throughout Asia and elsewhere over large parts of the world.

The following incomplete list of North American genera or peculiar subgeneric types represented in Japan and its vicinity, but unknown in Europe, presents a very different appearance. Those which are absent from the flora of western North America are italicized.

Trautvetteria	Philadelphus	*Asarum,* § *Heterotropa*
Cimicifuga (barely	*Penthorum*	*Phytolacca*
reaches Europe)	*Hamamelis*	*Benzoin* and *Sassafras?*
Illicium	*Liquidambar*	*Tatranthera*
Magnolia	*Cryptotœnia*	*Saururus*
Cocculus and *Meni-*	Cymopterus?	*Pachysandra*
spermum?	*Archemora*	*Laportea*
Mahonia	Osmorrhiza	*Pilea*
Caulophyllum	Aralia and § Ginseng	*Bœhmeria*
Diphylleia	Echinopanax	*Microptelea*
Brasenia	*Diervilla*	*Maclura*
Nelumbium	*Mitchella*	*Juglans*
Dicentra	*Oldenlandia*	Abies, § Tsuga
Stuartia (& *Gordonia?*)	(Siegesbeckia, in	Chamæcyparis
Zanthoxylum	Mexico)	Torrea
Cissus	*Cacalia* (reaches	*Arisæma*
Ampelopsis	E. Europe)	Arctiodracon
Berchemia	Gaultheria	*Pogonia*
Æsculus	*Leucothoë*	*Arethusa*
Sapindus	*Pieris*	Dioscorea
Negundo	*Clethra*	*Aletris*
Thermopsis	Menziesia	*Coprosmanthus*
Wistaria	*Symplocos*	Trillium
Desmodium	*Ardisia*	Clintonia
Lespedeza	Boschniakia	*Streptopus,* § *Hekori-*
Rhynchosia	*Catalpa*	*ma*
Sophora	*Tecoma*	*Chamælirium?*
Photinia	Dicliptera	Sporobolus

Astilbe	Leptandra	Arundinaria
Mitella	Callicarpa	Adiantum
Hydrangea	Cedronella	Onoclea
Itea	Amsonia	

Here are about ninety extra-European genera or forms, sixty-four of which are absent from western North America out of the tropics (the latter comprising a very large part of the most striking representative species), and almost as many more are divided between North America and extra-tropical (chiefly northern and eastern) Asia. About forty of the latter genera are groups of single, or of two or few closely related species, peculiar, or nearly peculiar, to the regions just mentioned.

This list should be supplemented by those additional North American genera which have one or more closely representative species in the Himalayan region only, such as Podophyllum, Pyrularia, etc. ; and also by the numerous cases in which eastern American plants are represented in the Himalayo-Japanese region by strikingly cognate, although not congeneric species; such as our Macrotys by Pityrosperma; Schizandra by Kadsura and Sphærostema ; Neviusia by Kerria and Rhodotypus; Calycanthus by Chimonanthus; *Cornus florida* by Benthamia ; Prosartes by Disporum ; Helonias by Heloniopsis ; and so of others, which have been mentioned in the former part of this memoir, and exhibited in the accompanying tabular view.

I had long ago, in Silliman's Journal, presented some data illustrative of this remarkable parallelism, and also more recently in my " Statistics of the Flora of the Northern United States " (vol. xxii., second series) ; where I had noticed the facts, — (1) that a large percentage of our extra-European types are shared with eastern Asia ; and (2) that no small part of these are unknown in western North America. But Mr. Bentham was first to state the natural conclusion from all these data, — though I know not if he has even yet published the remark, — namely, that the interchange between the temperate floras even of the western part of the Old World and of the New has mainly taken place via Asia. Notwithstand-

ing the few cases which point in the opposite direction (*e. g.*
Eriocaulon septangulare, Spartina, Subularia, *Betula alba*),
the general statement will be seen to be well sustained. Also,
in the "Journal of the Proceedings of the Linnæan Society,"
ii. p. 34, Mr. Bentham "calls to mind how frequently large
American genera (such as Eupatorium, Aster, Solidago, Solanum, etc.) are represented in eastern Asia by a small number
of species, which gradually diminish or altogether disappear
as we proceed westward toward the Atlantic limits of Europe ;
whilst the types peculiar to the extreme west of Europe (excluding of course the arctic flora) are wholly deficient in
America. These are among the considerations which suggest
an ancient continuity of territory between America and Asia,
under a latitude, or at any rate with a climate, more meridional than would be effected by a junction through the chains
of the Aleutian and the Kurile Islands."

I shall presently state why connection in a more meridional
latitude need not be supposed.

The deficiency in the temperate American flora of forms at
all peculiar to western Europe is almost complete, and is most
strikingly in contrast with the large number of eastern
American forms repeated or represented in eastern Asia. Of
genera divided between eastern North America and Europe,
I can mention only Ostrya, Narthecium, Psamma, the maritime Cakile, and perhaps Scolopendrium. Hottonia might
have been added, but for a species accredited to Java. And
if we extend the range across our continent, we add only
Cercis and Lœflingia. Of the ampler genera at all characteristic of the European flora, I can enumerate from the flora
of the northern United States nothing more important than
Helianthemum and Valerianella, two or three species of each
(but those of the former hardly congeners of the European
ones), adding that Hieracia and perhaps Cirsia are somewhat
more plentiful in eastern than in western America. Let it
also be noted, that there are even fewer western European
types in the Pacific than in the Atlantic United States, notwithstanding the similarity of the climate !

That representation by allied species of genera peculiar or

nearly peculiar to two regions furnishes evidence of similar nature and of equal pertinency with representation by identical species, will hardly be doubted. Whether or not susceptible of scientific explanation, it is certain that related species of phænogamous plants are commonly associated in the same region, or are found in comparatively approximate (however large) areas of similar climate.[1] Remarkable exceptions may indeed be adduced, but the fact that they are remarkable goes to confirm the proposition. Indeed, the general expectation of botanists in this regard sufficiently indicates the common, implicit opinion. The discovery of a new Sarracenia or of a new Halesia in the Atlantic United States, or of a new Eschscholtzia, Platystemon, or Calais west of the Rocky Mountains, would excite no surprise. A converse discovery, or the detection of any of these genera in a remote region, would excite great surprise. The discovery of numerous closely related species thus divided between two widely separated districts might not, in the present state of our knowledge, suggest former continuity, migration, or interchange ;

[1] The fundamental and most difficult question remaining in natural history is here presented — the question whether this actual geographical association of congeneric or other nearly related species is primordial, and therefore beyond all scientific explanation, or whether even this may be to a certain extent a natural result. The only noteworthy attempt at a scientific solution of the problem, aiming to bring the variety as well as the geographical association of existing species more within the domain of cause and effect, is that of Mr. Darwin and (later) of Mr. Wallace, — partially sketched in their short papers " On the Tendency of Species to form Varieties, and on the Perpetuation of Varieties and Species by natural Means of Selection," in the "Journal of the Proceedings of the Linnæan Society," vol. iii. (Zoölogy), p. 45. The views there suggested must bear a prominent part in future investigations into the distribution and probable origin of species. It will hardly be doubted that the tendencies and causes indicated are really operative ; the question is as to the extent of their operation. But I am already disposed, on these and other grounds, to admit that what are termed closely related species may in many cases be lineal descendants from a pristine stock, just as domesticated races are ; or, in other words, that the limits of occasional variation in species (if by them we mean primordial forms) are wider than is generally supposed, and that derivative forms when segregated may be as constantly reproduced as their originals.

but that of identical species peculiar to the two inevitably would.

Why should it? Evidently because the natural supposition is that individuals of the same kind are descendants from a common stock, or have spread from a common centre; and because the progress of investigation, instead of eliminating this preconception from the minds of botanists, has rather confirmed it. Every other hypothesis has derived its principal support from difficulties in the application of this. A review of what has been published upon the subject of late years makes it clear that the doctrine of the local origin of vegetable species has been more and more accepted, although, during the same period, species have been shown to be much more widely dispersed than was formerly supposed. Facts of the latter kind, and the conclusions to which they point, have been most largely and cogently brought out by Dr. Hooker, and are among the very important general results of his extensive investigations. And the best evidence of the preponderance of the theory of the local origin of species — notwithstanding the great increase of facts which at first would seem to tell the other way — is furnished by the works of the present De Candolle upon geographical botany. This careful and conscientious investigator formerly adopted and strenuously maintained Schouw's hypothesis of the double or multiple origin of species. But in his great work, the "Géographie Botanique Raisonnée," published in the year 1855, he has in effect discarded it, and this not from any theoretical objections to that view, but because he found it no longer needed to account for the general facts of distribution. This appears from his qualified though dubious adherence to the hypothesis of a double origin, as a *dernier ressort*, in the few and extraordinary cases which he could hardly explain in any other way. His decisive instance, indeed, is the occurrence of the eastern American *Phryma leptostachya* in the Himalaya Mountains.

The facts presented in the present memoir effectually dispose of this subsidiary hypothesis, by showing that the supposed single exception belongs to a not uncommon case.

Indeed, so many species are now known to be common to eastern and northern Asia and eastern North America, — some of them occurring also in northwestern America and some not, — and so many genera are divided between these two regions, that the antecedent improbability of such occurrence is done away, and more cases of the kind may be confidently expected. However others may regard them, it is clear that De Candolle would now explain these cases in accordance with the general views of distribution adopted by him, under which they naturally fall, — so abandoning the notion of a separate creation.

I know not whether any botanist continues to maintain Schouw's hypothesis. But its elements have been developed into a different and more comprehensive doctrine, that of Agassiz, which should now be contemplated. It may be denominated the autochthonal hypothesis.

In place of the ordinary conception, that each species originated in a local area, whence it has been diffused, according to circumstances, over more or less broad tracts, — in some cases becoming widely discontinuous in area through climatic or other physical changes operating during a long period of time, — Professor Agassiz maintains, substantially, that each species originated where it now occurs, probably in as great a number of individuals occupying as large an area, and generally the same area, or the same discontinuous areas as at the present time.

This hypothesis is more difficult to test, because more ideal than any other. It might suffice for the present purpose to remark, that, in referring the actual distribution, no less than the origin, of existing species to the Divine will, it would remove the whole question out of the field of inductive science. Regarded as a philosophical question, Maupertius's well-known "principle of least action" might be legitimately urged against it, namely, "that it is inconsistent with our idea of Divine wisdom that the Creator should use more power than was necessary to accomplish a given end." This philosophical principle holds so strictly true in all the mechanical adaptations of the universe, as Professor Pierce has

shown, that we cannot think it inapplicable to the organic
world also, and especially to the creation of beings endowed
with such enormous multiplying power, and such means and
facilities for dissemination, as most plants and animals.
Why then should we suppose the Creator to do that super-
naturally which would be naturally effected by the very in-
strumentalities which he has set in operation?

Viewed, however, simply in its scientific applications to the
question under consideration (the distribution of plants in
the temperate zone of the northern hemisphere), the autochthonal hypothesis might be tested by inquiring whether the
primitive or earliest range of our species could possibly have
remained unaffected by the serious and prolonged climatic
vicissitudes to which they must needs have been subject; and
whether these vicissitudes, and their natural consequences,
may not suffice to explain the partial intermingling of the
floras of North America and northern Asia, upon the supposition of the local origin of each species.   Let us bring to the
inquiry the considerations which Mr. Darwin first brought to
bear upon such questions, and which have been systematically
developed and applied by the late Edward Forbes, by Dr.
Hooker, and by Alphonse De Candolle.

No one now supposes that the existing species of plants are
of recent creation, or that their present distribution is the result of a few thousand years.   Various lines of evidence conspire to show that the time which has elapsed since the close
of the tertiary period covers an immense number of years;
and that our existing flora may in part date from the tertiary
period itself.   It is now generally admitted that about twenty
per cent. of the Mollusca of the middle tertiary (miocene
epoch), and forty per cent. of the pliocene species on the Atlantic coast still exist; and it is altogether probable that as
large a portion of the vegetation may be of equal antiquity.
From the nature of the case, the direct evidence as respects
the flora could not be expected to be equally abundant.   Still,
although the fossil plants of the tertiary and the post-tertiary
of North America have only now begun to be studied, the
needful evidence is not wanting.

On our northwestern coast, in the miocene of Vancouver's Island, among a singular mixture of species referable to Salix, Populus, Quercus, Planera, Diospyros, Salisburia, Ficus, Cinnamomum, Personia, or other *Proteaceæ*, and a Palm (the latter genera decisively indicating a tropical or subtropical climate), Mr. Lesquereux has identified one existing species, a tree characteristic of the same region ten or fifteen degrees farther south, namely, the Redwood or *Sequoia sempervirens*. In beds at Somerville referred to the lower or middle pliocene by Mr. Lesquereux, this botanist has recently identified the leaves of *Persea Carolinensis*, *Prunus Caroliniana*, and *Quercus myrtifolia*, now inhabiting the warm seacoast and islands of the southern States.[1]

The pliocene quadrupeds of Nebraska also show that the climate east of the Rocky Mountains at this epoch was much warmer than now. About the upper Missouri and Platte there were then several species of Camel (Procamelus) and allied Ruminantia and a Rhinoceros, besides a Mastodon, an Elephant, some Horses and their allies, not to mention a corresponding number of carnivorous animals. These herbivora probably fed in a good degree upon herbage and grasses of still existing species. For herbs and grasses are generally capable of enduring much greater climatic changes, and are therefore likely to be even more ancient, than trees. These animals must have had at least a warm-temperate climate to live in: so that in latitude 40°- 43° they could not have been anywhere near the northern limit of the temperate flora of those days ; indeed the temperate flora, which now in western Europe touches the Arctic Circle, must then have reached equally high latitudes in central or western North America. In other words, the temperate floras of America and Asia must then have been conterminous (with small oceanic separation), and therefore have commingled, as conterminous floras of similar climate everywhere do.

At length, as the post-tertiary opened, the glacier epoch

---

[1] These and other data, obligingly communicated by Mr. Lesquereux, have been published in the May number of the " American Journal of Science and Arts," 3 ser., xvii.

came slowly on, — an extraordinary refrigeration of the
northern hemisphere, in the course of ages carrying glacial
ice and arctic climate down nearly to the latitude of the Ohio.
The change was evidently so gradual that it did not destroy
the temperate flora, at least not those enumerated above as
existing species. These and their fellows, or such as survive,
must have been pushed on to lower latitudes as the cold advanced, just as they now would be if the temperature were to
be again lowered ; and between them and the ice there was
doubtless a band of subarctic and arctic vegetation, — portions of which, retreating up the mountains as the climate
ameliorated and the ice receded, still scantily survive upon
our highest Alleghanies, and more abundantly upon the colder
summits of the mountains of New York and New England ; —
demonstrating the existence of the present arctic-alpine vegetation during the glacial era; and that the change of climate
at its close was so gradual that it was not destructive to vegetable species.

As the temperature rose, and the ice gradually retreated, the
surviving temperate flora must have returned northward *pari
passu*, and — which is an important point — must have advanced much farther northward, and especially northwestward, than it now does; so far, indeed, that the temperate
floras of North America and of eastern Asia, after having been
for long ages most widely separated, must have become a second time conterminous. Whatever doubts may be entertained
respecting the existence of our present vegetation generally
before the glacial era, its existence immediately after that
period will hardly be questioned. Here, therefore, may be
adduced the direct evidence recently brought to light by Mr.
Lesquereux, who has identified our Live Oak (*Quercus virens*),
Pecan (*Carya olivæformis*), Chinquapin (*Castanea pumila*),
Planer-tree (*Planera aquatica*), Honey-Locust (*Gleditschia
triacanthos*), *Prinos coriaceus*, and *Acorus Calamus*, — besides an Elm and a Ceanothus doubtfully referable to existing
species, — on the Mississippi, near Columbus, Kentucky, in
beds which Mr. Lesquereux regards as anterior to the drift.
Professor D. D. Owen has indicated their position " as about

one hundred and twenty feet lower than the ferrugineous sand in which the bones of the *Megalonyx Jeffersonii* were found." So that they belong to the period immediately succeeding the drift, if not to that immediately preceding it. All the vegetable remains of this deposit, which have been obtained in a determinable condition, have been referred, either positively or probably, to existing species of the United States flora, most of them now inhabiting the region a few degrees farther south.

If, then, our present temperate flora existed at the close of the glacial epoch, the evidence that it soon attained a high northern range is ready to our hand. For then followed the second epoch of the post-tertiary, called the fluvial by Dana, when the region of the St. Lawrence and Lake Champlain was submerged, and the sea there stood five hundred feet above its present level; when the higher temperate latitudes of North America, and probably the arctic generally, were less elevated than now, and the rivers vastly larger, as shown by the immense upper alluvial plains, from fifty to three hundred feet above their present beds; and when the diminished breadth and lessened height of northern land must have given a much milder climate than the present.

Whatever the cause, the milder climate of the fluvial epoch is undoubted. Its character, and therefore that of the vegetation, is decisively shown, as geologists have remarked, by the quadrupeds. While the Megatherium, Mylodon, Dicotyles, etc., demonstrate a warmer climate than at present in the southern and middle United States, the *Elephas primigenius*, ranging from Canada to the very shores of the Arctic Ocean, equally proves a temperate climate and a temperate flora in these northern regions. This is still more apparent in the species of the other continent, where, in Siberia, not only the *Elephas primigenius*, but also a Rhinoceros roamed northward to the arctic sea-coast. The quadrupeds that inhabited Europe in the same epoch are well known to indicate a warm temperate climate as far north as Britain, in the middle, if not the later post-tertiary. North America then had its herds of Mastodons, Elephants, Buffaloes or Bisons of dif-

ferent species, Elks, Horses, Megalonyx, the Lion, etc. ; and,
from the relations between this fauna and that of Europe,
there is little doubt that the climate was as much milder
than the present on this as on the other side of the ocean.
All the facts known to us in the tertiary and post-tertiary,
even to the limiting line of the drift, conspire to show that
the difference between the two continents as to temperature
was very nearly the same then as now, and that the isothermal lines of the northern hemisphere curved in the directions
they now do.

A climate such as these facts demonstrate for the fluvial
epoch would again commingle the temperate floras of the
two continents at Behring's Straits, and earlier — probably
through more land than now — by way of the Aleutian and
Kurile Islands. I cannot imagine a state of circumstances
under which the Siberian Elephant could migrate, and temperate plants could not.

The fluvial was succeeded by the "terrace epoch," as Dana
names it, " a time of transition towards the present condition,
bringing the northern part of the continent up to its present
level and down to its present cool temperature," [1] — giving
the arctic flora its present range, and again separating the
temperate floras of the New and of the Old World to the
extent they are now separated.

Under the light which these geological considerations throw
upon the question, I cannot resist the conclusion, that the extant vegetable kingdom has a long and eventful history, and
that the explanation of apparent anomalies in the geographical distribution of species may be found in the various and
prolonged climatic or other physical vicissitudes to which they
have been subject in earlier times ; that the occurrence of certain species, formerly supposed to be peculiar to North America, in a remote or antipodal region affords itself no presumption that they were originated there, and that the interchange
of plants between eastern North America and eastern Asia is
explicable upon the most natural and generally received hy-

[1] For the collocation and communication of the geological data here
presented, I am indebted to the kindness of my friend, Professor Dana.

pothesis (or at least offers no greater difficulty than does the arctic flora, the general homogeneousness of which round the world has always been thought compatible with local origin of the species), and is perhaps not more extensive than might be expected under the circumstances. That the interchange has mainly taken place in high northern latitudes, and that the isothermal lines have in earlier times turned northward on our eastern, and southward on our northwest coast, as they do now, are points which go far towards explaining why eastern North America, rather than Oregon and California, has been mainly concerned in this interchange, and why the temperate interchange, even with Europe, has principally taken place through Asia.

Brasenia peltata. — To the remarks upon the known range of this species, I have now to add the interesting fact, that it exists upon the northwestern coast of America, having been gathered by Dr. Pickering, in Wilkes's South Sea Exploring Expedition, in a stream which falls into Gray's Harbor, lat. 47°. It must be local on the western side of the continent, or it would have been met with before. When this remarkable plant was known to occur only in eastern North America and eastern Australia, it made the strongest case in favor of double creation that perhaps has ever been adduced. But since it has been found to occur throughout the eastern Himalayas and in Japan, and has now been detected in northwestern America also, the case seems to crown the conclusions to which this memoir arrives. (Note to reprint in " American Journal of Science and Arts," 3 ser., xviii. 199.)

SEQUOIA AND ITS HISTORY.[1]

THE session being now happily inaugurated, your presiding officer of the last year has only one duty to perform before he surrenders his chair to his successor. If allowed to borrow a simile from the language of my own profession, I might liken the President of this Association to a biennial plant. He flourishes for the year in which he comes into existence, and performs his appropriate functions as presiding officer. When the second year comes round, he is expected to blossom out in an address and disappear. Each President, as he retires, is naturally expected to contribute something from his own investigations or his own line of study, usually to discuss some particular scientific topic.

Now, although I have cultivated the field of North American botany, with some assiduity, for more than forty years, have reviewed our vegetable hosts, and assigned to no small number of them their names and their place in the ranks, yet, so far as our own wide country is concerned, I have been to a great extent a closet botanist. Until this summer I had not seen the Mississippi, nor set foot upon a prairie.

To gratify a natural interest, and to gain some title for addressing a body of practical naturalists and explorers, I have made a pilgrimage across the continent. I have sought and viewed in their native haunts many a plant and flower which for me had long bloomed unseen, or only in the *hortus siccus*. I have been able to see for myself what species and what forms constitute the main features of the vegetation of each successive region, and record — as the vegetation unerringly does — the permanent characteristics of its climate.

---

[1] The address of the retiring President of the American Association for the Advancement of Science. Delivered at Dubuque, Iowa, August, 1872. (Proceedings American Association, xxi. 1.)

Passing on from the eastern district, marked by its equably distributed rainfall, and therefore naturally forest-clad, I have seen the trees diminish in number, give place to wide prairies, restrict their growth to the borders of streams, and then disappear from the boundless drier plains; have seen grassy plains change into a brown and sere desert, — desert in the common sense, but hardly anywhere botanically so; have seen a fair growth of coniferous trees adorning the more favored slopes of a mountain range high enough to compel summer showers; have traversed that broad and bare elevated region shut off on both sides by high mountains from the moisture supplied by either ocean, and longitudinally intersected by sierras which seemingly remain as naked as they were born; and have reached at length the westward slopes of the high mountain barrier which, refreshed by the Pacific, bear the noble forests of the Sierra Nevada and the Coast Range, and among them trees which are the wonder of the world. As I stood in their shade, in the groves of Mariposa and Calaveras, and again under the canopy of the commoner Redwood, raised on columns of such majestic height and ample girth, it occurred to me that I could not do better than to share with you, upon this occasion, some of the thoughts which possessed my mind. In their development they may, perhaps, lead us up to questions of considerable scientific interest.

I shall not detain you with any remarks — which would now be trite — upon the size or longevity of these far-famed Sequoia trees, or of the Sugar Pines, Incense-Cedar, and Firs associated with them, of which even the prodigious bulk of the dominating Sequoia does not sensibly diminish the grandeur. Although no account and no photographic representation of either species of the far-famed Sequoia trees gives any adequate impression of their singular majesty — still less of their beauty, — yet my interest in them did not culminate merely or mainly in considerations of their size and age. Other trees, in other parts of the world, may claim to be older. Certain Australian Gum-trees (Eucalypti) are said to be taller. Some, we are told, rise so high that they might even

cast a flicker of shadow upon the summit of the pyramid of
Cheops. Yet the oldest of them doubtless grew from seed
which was shed long after the names of the pyramid-builders
had been forgotten. So far as we can judge from the actual
counting of the layers of several trees, no Sequoia now alive
can sensibly antedate the Christian era.

Nor was I much impressed with an attraction of man's
adding. That the more remarkable of these trees should bear
distinguishing appellations seems proper enough; but the
tablets of personal names which are affixed to many of them
in the most visited groves — as if the memory of more or less
notable people of our day might be made more enduring by
the juxtaposition — do suggest some incongruity. When we
consider that a hand's breadth at the circumference of any one
of the venerable trunks so placarded has recorded in annual
lines the lifetime of the individual thus associated with it, one
may question whether the next hand's breadth may not measure the fame of some of the names thus ticketed for adventitious immortality. Whether it be the man or the tree that is
honored in the connection, probably either would live as long,
in fact and in memory, without it.

One notable thing about these Sequoia trees is their isola-
tion. Most of the trees associated with them are of peculiar
species, and some of them are nearly as local. Yet every
Pine, Fir, and Cypress in California is in some sort familiar,
because it has near relatives in other parts of the world. But
the Redwoods have none. The Redwood — including in that
name the two species of "Big-trees" — belongs to the general
Cypress family, but is *sui generis*. Thus isolated systematically, and extremely isolated geographically, and so wonderful
in size and port, they more than other trees suggest questions.

Were they created thus local and lonely, denizens of California only; one in limited numbers in a few choice spots on
the Sierra Nevada, the other along the Coast Range from the
Bay of Monterey to the frontiers of Oregon? Are they veritable Melchizedeks, without pedigree or early relationship,
and possibly fated to be without descent?

Or are they now coming upon the stage — or rather were

they coming but for man's interference — to play a part in the future?

Or are they remnants, sole and scanty survivors of a race that has played a grander part in the past, but is now verging to extinction? Have they had a career, and can that career be ascertained or surmised, so that we may at least guess whence they came, and how, and when?

Time was, and not long ago, when such questions as these were regarded as useless and vain, — when students of natural history, unmindful of what the name denotes, were content with a knowledge of things as they now are, but gave little heed as to how they came to be so. Now, such questions are held to be legitimate, and perhaps not wholly unanswerable. It cannot now be said that these trees inhabit their present restricted areas simply because they are there placed in the climate and soil of all the world most congenial to them. These must indeed be congenial, or they would not survive. But when we see how Australian Eucalyptus trees thrive upon the Californian coast, and how these very Redwoods flourish upon another continent; how the so-called Wild Oat (*Avena sterilis* of the Old World) has taken full possession of California; how that cattle and horses, introduced by the Spaniard, have spread as widely and made themselves as much at home on the plains of La Plata as on those of Tartary; and that the Cardoon-thistle seeds, and others they brought with them, have multiplied there into numbers probably much exceeding those extant in their native lands; indeed, when we contemplate our own race, and our own particular stock, taking such recent but dominating possession of this New World; when we consider how the indigenous flora of islands generally succumbs to the foreigners which come in the train of man; and that most weeds (*i. e.*, the prepotent plants in open soil) of all temperate climates are not "to the manner born," but are self-invited intruders, — we must needs abandon the notion of any primordial and absolute adaptation of plants and animals to their habitats, which may stand in lieu of explanation, and so preclude our inquiring any further. The harmony of Nature and its admirable perfection need not be regarded as inflexible and changeless. Nor need Nature be likened to a statue, or a cast in rigid bronze, but rather to an organism, with play and adaptability of parts, and life and even soul informing the whole. Under the former view, Nature would be " the faultless monster which the world ne'er saw," but inscrutable as the Sphinx, whom it were vain, or worse, to question of the whence and whither. Under the other, the perfection of Nature, if relative, is multifarious and ever renewed; and much that is enigmatical now may find explanation in some record of the past.

That the two species of Redwood we are contemplating originated as they are and where they are, and for the part they are now playing, is, to say the least, not a scientific supposition, nor in any sense a probable one. Nor is it more likely that they are destined to play a conspicuous part in the future, or that they would have done so, even if the Indian's fires and the white man's axe had spared them. The Redwood of the coast (*Sequoia sempervirens*) had the stronger hold upon existence, forming as it did large forests throughout a narrow belt about three hundred miles in length, and being so tenacious of life that every large stump sprouts into a copse. But it does not pass the Bay of Monterey, nor cross the line of Oregon, although so grandly developed not far below it. The more remarkable *Sequoia gigantea* of the Sierra exists in numbers so limited that the separate groves may be reckoned upon the fingers, and the trees of most of them have been counted, except near their southern limit, where they are said to be more copious. A species limited in individuals holds its existence by a precarious tenure; and this has a foothold only in a few sheltered spots, of a happy mean in temperature, and locally favored with moisture in summer. Even there, for some reason or other, the Pines with which they are associated (*Pinus Lambertiana* and *P. ponderosa*), the Firs (*Abies grandis* and *A. magnifica*), and even the Incense-Cedar (*Libocedrus decurrens*) possess a great advantage, and, though they strive in vain to emulate their size, wholly overpower the Sequoias in numbers. " To him that hath shall be given." The force of numbers eventually wins. At least in the commonly visited groves *Sequoia gigantea* is invested in its last stronghold, can neither advance into more exposed positions above, nor fall back into drier and barer ground below, nor hold its own in the long-run where it is, under present conditions; and a little further drying of the climate, which must once have been much moister than now, would precipitate its doom. Whatever the individual longevity, certain if not speedy is the decline of a race in which a high death-rate afflicts the young. Seedlings of the big trees occur not rarely, indeed, but in meagre proportion to those of associated trees; and small indeed is the chance that any of these will attain to " the days of the years of their fathers." " Few and evil " are the days of all the forest likely to be, while man, both barbarian and civilized, torments them with fires, fatal at once to seedlings, and at length to the aged also. The forests of California, proud as the State may be of them, are already too scanty and insufficient for her uses. Two lines, such as may be drawn with one sweep of a brush over the map, would cover them all. The coast Redwood — the most important tree in California, although a million times more numerous than its relative of the Sierra — is too good to live long. Such is its value for lumber and its accessibility, that, judging the future by the past, it is not likely, in its primeval growth, to outlast its rarer fellow-species.

Happily man preserves and disseminates as well as destroys. The species will doubtless be preserved to science, and for ornamental and other uses, in its own and other lands; and the more remarkable individuals of the present day are likely to be sedulously cared for, all the more so as they become scarce.

Our third question remains to be answered : Have these famous Sequoias played in former times and upon a larger stage a more imposing part, of which the present is but the epilogue? We cannot gaze high up the huge and venerable trunks, which one crosses the continent to behold, without wishing that these patriarchs of the grove were able, like the long-lived antediluvians of Scripture, to hand down to us, through a few generations, the traditions of centuries, and so tell us something of the history of their race. Fifteen hundred annual layers have been counted, or satisfactorily made out, upon one or two fallen trunks. It is probable that close to the heart of some of the living trees may be found the circle that records the year of our Saviour's nativity. A few generations of such trees might carry the history a long way back. But the ground they stand upon, and the marks of very recent geological change and vicissitude in the region around, testify that not very many such generations can have flourished just there, at least in an unbroken series. When their site was covered by glaciers, these Sequoias must have occupied other stations, if, as there is reason to believe, they then existed in the land.

I have said that the Redwoods have no near relatives in the country of their abode, and none of their genus anywhere else. Perhaps something may be learned of their genealogy by inquiring of such relatives as they have. There are only two of any particular nearness of kin; and they are far away. One is the Bald Cypress, our southern Cypress (Taxodium), inhabiting the swamps of the Atlantic coast from Maryland to Texas, thence extending — with, probably, a specific difference — into Mexico. It is well known as one of the largest trees of our Atlantic forest-district, and although it never — except perhaps in Mexico, and in rare instances — attains the portliness of its western relatives, yet it may equal them in longevity. The other relative is Glyptostrobus, a sort of modified Taxodium, being about as much like our Bald Cypress as one species of Redwood is like the other.

Now species of the same type, especially when few, and the type peculiar, are, in a general way, associated geographically, *i. e.,* inhabit the same country, or (in a large sense) the same region. Where it is not so, where near relatives are separated, there is usually something to be explained. Here is an instance. These four trees, sole representatives of their tribe, dwell almost in three separate quarters of the world : the two Redwoods in California, the Bald Cypress in Atlantic North America, its near relative, Glyptostrobus, in China.

It was not always so.   In the tertiary period, the geological botanists assure us, our own very Taxodium or Bald Cypress, and a Glyptostrobus, exceedingly like the present Chinese tree, and more than one Sequoia, coexisted in a fourth quarter of the globe, namely, in Europe!   This brings up the question: Is it possible to bridge over these four wide intervals of space and the much vaster interval of time, so as to bring these extraordinarily separated relatives into connection?   The evidence which may be brought to bear upon this question is various and widely scattered.   I bespeak your patience while I endeavor to bring together, in an abstract, the most important points of it.

Some interesting facts may come out by comparing generally the botany of the three remote regions, each of which is the sole home of one of these genera, i. e., Sequoia in California, Taxodium in the Atlantic United States,[1] and Glyptostrobus in China, which compose the whole of the peculiar tribe under consideration.

Note then, first, that there is another set of three or four peculiar trees, in this case of the Yew family, which has just the same peculiar distribution, and which therefore may have the same explanation, whatever that explanation be.   The genus Torreya, which commemorates our botanical Nestor and a former president of this association, Dr. Torrey, was founded upon a tree rather lately discovered (that is, about thirty-five years ago) in northern Florida.   It is a noble, Yew-like tree, and very local, being, so far as known, nearly confined to a few miles along the shores of a single river.   It seems as if it had somehow been crowded down out of the Alleghanies into its present limited southern quarters; for in cultivation it evinces a northern hardiness.   Now another species of Torreya is a characteristic tree of Japan ; and one very like it, if not the same, inhabits the mountains of north-

[1] The phrase "Atlantic United States" is here used throughout in contradistinction to Pacific United States.   To the former of course belongs, botanically and geographically, the valley of the Mississippi and its tributaries up to the eastern border of the great woodless plains, which constitute an intermediate region.

ern China, — belongs, therefore, to the eastern Asiatic temperate region, of which northern China is a part, and Japan, as we shall see, the portion most interesting to us. There is only one more species of Torreya, and that is a companion of the Redwoods in California. It is the tree locally known under the name of the California Nutmeg. Here are three or four near brethren, species of the same genus, known nowhere else than in these three habitats.

Moreover, the Torreya of Florida is associated with a Yew; and the trees of this grove are the only Yew-trees of eastern North America; for the Yew of our northern woods is a decumbent shrub. A Yew-tree, perhaps the same, is found with Taxodium in the temperate parts of Mexico. The only other Yews in America grow with the Redwoods and the other Torreya in California, and extend northward into Oregon. Yews are also associated with Torreya in Japan; and they extend westward through Mandchuria and the Himalayas to western Europe, and even to the Azores Islands, where occurs the common Yew of the Old World.

So we have three groups of coniferous trees which agree in this peculiar geographical distribution, with, however, a notable extension of range in the case of the Yew: first, the Redwoods, and their relatives, Taxodium and Glyptostrobus, which differ so as to constitute a genus for each of the three regions; second, the Torreyas, more nearly akin, merely a different species in each region; third, the Yews, still more closely related while more widely disseminated, of which it is yet uncertain whether they constitute seven, five, three, or only one species. Opinions differ, and can hardly be brought to any decisive test. However it be determined, it may still be said that the extreme differences among the Yews do not surpass those of the recognized variations of the European Yew, the cultivated races included.

It appears to me that these several instances all raise the very same question, only with different degrees of emphasis, and, if to be explained at all, will have the same kind of explanation.

Continuing the comparison between the three regions with

which we are concerned, we note that each has its own species of Pines, Firs, Larches, etc., and of a few deciduous-leaved trees such as Oaks and Maples; all of which have no peculiar significance for the present purpose, because they are of genera which are common all round the northern hemisphere. Leaving these out of view, the noticeable point is that the vegetation of California is most strikingly unlike that of the Atlantic United States. They possess some plants, and some peculiarly American plants in common, — enough to show, as I imagine, that the difficulty was not in the getting from the one district to the other, or into both from a common source, but in abiding there. The primordially unbroken forest of Atlantic North America, nourished by rainfall distributed throughout the year, is widely separated from the western region of sparse and discontinuous tree-belts of the same latitude on the western side of the continent, where summer rain is wanting, or nearly so, by immense treeless plains and plateaux of more or less aridity, traversed by longitudinal mountain ranges of similar character. Their nearest approach is at the north, in the latitude of Lake Superior, where, on a more rainy line, trees of the Atlantic forest and that of Oregon may be said to interchange. The change of species and of the aspect of vegetation in crossing, say on the forty-seventh parallel, is slight in comparison with that on the thirty-seventh or near it. Confiding our attention to the lower latitude, and under the exceptions already specially noted, we may say that almost every characteristic form in the vegetation of the Atlantic States is wanting in California, and the characteristic plants and trees of California are wanting here.

California has no Magnolia nor Tulip trees, nor Star-anise-tree; no so-called Papaw (Asimina); no Barberry of the common single-leaved sort; no Podophyllum or other of the peculiar associated genera; no Nelumbo nor White Water-lily; no Prickly Ash nor Sumach; no Loblolly-bay nor Stuartia; no Basswood nor Linden-trees; neither Locust, Honey-locust, Coffee-trees (Gymnocladus), nor Yellow-wood (Cladrastis); nothing answering to Hydrangea or Witch-

hazel, to Gum-trees (Nyssa and Liquidambar), Viburnum or
Diervilla; it has few Asters and Golden-rods; no Lobelias;
no Huckleberries and hardly any Blueberries; no Epigæa,
the charm of our earliest eastern spring, tempering an icy April
wind with a delicious wild fragrance; no Kalmia nor Clethra,
nor Holly, nor Persimmon; no Catalpa-tree, nor Trumpet-
creeper (Tecoma); nothing answering to Sassafras, nor to
Benzoin-tree, nor to Hickory; neither Mulberry nor Elm; no
Beech, true Chestnut, Hornbeam, nor Ironwood, nor a proper
Birch-tree; and the enumeration might be continued very
much further by naming herbaceous plants and others familiar
only to botanists.

In their place California is filled with plants of other types,
— trees, shrubs, and herbs, of which I will only remark that
they are, with one or two exceptions, as different from the
plants of the eastern Asiatic region with which we are concerned (Japan, China, and Mandchuria), as they are from
those of Atlantic North America. Their near relatives, when
they have any in other lands, are mostly southward, on the
Mexican plateau, or many as far south as Chili. The same
may be said of the plants of the intervening great plains, except that northward and in the subsaline vegetation there are
some close alliances with the flora of the steppes of Siberia.
And along the crests of high mountain ranges the arctic-alpine
flora has sent southward more or less numerous representatives through the whole length of the country.

If we now compare, as to their flora generally, the Atlantic
United States with Japan, Mandchuria, and northern China,
— i. e., eastern North America with eastern north Asia, half
the earth's circumference apart, — we find an astonishing
similarity. The larger part of the genera of our own region,
which I have enumerated as wanting in California, are present
in Japan or Mandchuria, along with many other peculiar
plants, divided between the two. There are plants enough of
the one region which have no representatives in the other.
There are types which appear to have reached the Atlantic
States from the south; and there is a larger infusion of sub-
tropical Asiatic types into temperate China and Japan;

among these there is no relationship between the two countries to speak of. There are also, as I have already said, no small number of genera and some species which, being common all round or partly round the northern temperate zone, have no special significance because of their occurrence in these two antipodal floras, although they have testimony to bear upon the general question of geographical distribution. The point to be remarked is, that many, or even most, of the genera and species which are peculiar to North America as compared with Europe, and largely peculiar to Atlantic North America as compared with the Californian region, are also represented in Japan and Mandchuria, either by identical or by closely similar forms. The same rule holds on a more northward line, although not so strikingly. If we compare the plants, say of New England and Pennsylvania (lat. 45°– 47°), with those of Oregon, and then with those of northeastern Asia, we shall find many of our own curiously repeated in the latter, while only a small number of them can be traced along the route even so far as the western slope of the Rocky Mountains. And these repetitions of east American types in Japan and neighboring districts are in all degrees of likeness. Sometimes the one is undistinguishable from the other; sometimes there is a difference of aspect, but hardly of tangible character; sometimes the two would be termed marked varieties if they grew naturally in the same forest or in the same region; sometimes they are what the botanist calls representative species, the one answering closely to the other, but with some differences regarded as specific; sometimes the two are merely of the same genus, or not quite that, but of a single or very few species in each country; when the point which interests us is, that this peculiar limited type should occur in two antipodal places, and nowhere else.

It would be tedious, and, except to botanists, abstruse, to enumerate instances; yet the whole strength of the case depends upon the number of such instances. I propose therefore, if the Association does me the honor to print this discourse, to append in a note a list of the more remarkable

ones.[1] But I would here mention certain cases as specimens.

Our *Rhus Toxicodendron*, or Poison Ivy, is very exactly
repeated in Japan, but is found in no other part of the world,
although a species much like it abounds in California. Our
other poisonous Rhus (*R. venenata*), commonly called Poison
Dogwood, is in no way represented in western America, but
has so close an analogue in Japan that the two were taken
for the same by Thunberg and Linnæus, who called them both
R. Vernix.

Our northern Fox-grape, *Vitis Labrusca*, is wholly confined to the Atlantic States, except that it reappears in Japan
and that region.

The original Wistaria is a woody leguminous climber with
showy blossoms, native to the middle Atlantic States; the
other species, which we so much prize in cultivation, *W.
Sinensis*, is from China, as its name denotes, or perhaps only
from Japan, where it is certainly indigenous.

Our Yellow-wood (Cladrastis) inhabits a very limited district on the western slope of the Alleghanies. Its only and
very near relative, Maackia, is in Mandchuria.

The Hydrangeas have some species in our Alleghany region; all the rest belong to the Chino-Japanese region and
its continuation westward. The same may be said of Philadelphus, except that there are one or two mostly very similar
species in California and Oregon.

Our Blue Cohosh (Caulophyllum) is confined to the woods
of the Atlantic States, but has lately been discovered in
Japan.[2] A peculiar relative of it, Diphylleia, confined to the
higher Alleghanies, is also repeated in Japan, with a slight
difference, so that it may barely be distinguished as another
species. Another relative is our Twin-leaf (Jeffersonia) of
the Alleghany region alone; a second species has lately turned
up in Mandchuria. A relative of this is Podophyllum, our
Mandrake, a common inhabitant of the Atlantic United
States, but found nowhere else. There is one other species
of it, and that is in the Himalayas. Here are four most

---

[1] See Appendix, I.   [2] Appendix, II.

peculiar genera of one family, each of a single species in the
Atlantic United States, which are duplicated on the other
side of the world, either in identical species, or in an analogous species, while nothing else of the kind is known in any
other part of the world.

I ought not to omit Ginseng, the root so prized by the
Chinese, which they obtained from their northern provinces
and Mandchuria, and which is now known to inhabit Corea
and northern Japan. The Jesuit Fathers identified the plant
in Canada and the Atlantic States, brought over the Chinese
name by which we know it, and established the trade in it,
which was for many years most profitable. The exportation
of Ginseng to China probably has not yet entirely ceased.
Whether the Asiatic and the Atlantic American Ginsengs
are to be regarded as of the same species or not is somewhat
uncertain, but they are hardly, if at all, distinguishable.

There is a shrub, Elliottia, which is so rare and local that
it is known only at two stations on the Savannah River, in
Georgia. It is of peculiar structure, and was without near
relative until one was lately discovered in Japan (Tripeta-
leia), so like it as hardly to be distinguishable except by having the parts of the blossom in threes instead of fours, — a
difference which is not uncommon in the same genus, or even
in the same species.

Suppose Elliottia had happened to be collected only once,
a good while ago, and all knowledge of the limited and obscure
locality were lost; and meanwhile the Japanese form came to
be known. Such a case would be parallel with an actual one.
A specimen of a peculiar plant (*Shortia galacifolia*) was
detected in the herbarium of the elder Michaux, who collected
it (as his autograph ticket shows) somewhere in the high
Alleghany Mountains, more than eighty years ago. No one
has seen the living plant since or knows where to find it, if
haply it still flourishes in some secluded spot. At length it
is found in Japan; and I had the satisfaction of making the
identification.[1] One other relative is also known in Japan;
and another, still unpublished, has just been detected in
Thibet.

[1] "Amer. Jour. Science," 1867, p. 402; "Proc. Amer. Acad.," viii. p. 244.

Whether the Japanese and the Alleghanian plants are exactly the same or not, it needs complete specimens of the two to settle. So far as we know, they are just alike; and even if some difference were discerned between them, it would not appreciably alter the question as to how such a result came to pass. Each and every one of the analogous cases I have been detailing — and very many more could be mentioned — raises the same question, and would be satisfied with the same answer.

These singular relations attracted my curiosity early in the course of my botanical studies, when comparatively few of them were known, and my serious attention in later years, when I had numerous and new Japanese plants to study in the collections made by Messrs. Williams and Morrow, during Commodore Perry's visit in 1853, and especially by Mr. Charles Wright, in Commodore Rodgers's expedition in 1855. I then discussed this subject somewhat fully, and tabulated the facts within my reach.[1]

This was before Heer had developed the rich fossil botany of the arctic zone, before the immense antiquity of existing species of plants was recognized, and before the publication of Darwin's now famous volume on the " Origin of Species " had introduced and familiarized the scientific world with those now current ideas respecting the history and vicissitudes of species with which I attempted to deal in a moderate and feeble way.

My speculation was based upon the former glaciation of the northern temperate zone, and the inference of a warmer period preceding and perhaps following. I considered that our own present vegetation, or its proximate ancestry, must have occupied the arctic and subarctic regions in pliocene times, and that it had been gradually pushed southward as the temperature lowered and the glaciation advanced, even beyond its present habitation; that plants of the same stock and kindred, probably ranging round the arctic zone as the present arctic species do, made their forced migration southward upon widely different longitudes, and receded more or less as the climate

[1] " Mem. Amer. Acad.," vol. vi. pp. 377–458 (1859).

grew warmer; that the general difference of climate which
marks the eastern and the western sides of the continents —
the one extreme, the other mean — was doubtless even then
established, so that the same species and the same sorts of
species would be likely to secure and retain foothold in the
similar climates of Japan and the Atlantic United States, but
not in intermediate regions of different distribution of heat
and moisture; so that different species of the same genus, as
in Torreya, or different genera of the same group, as Redwood, Taxodium, and Glyptostrobus, or different associations
of forest trees, might establish themselves each in the region
best suited to their particular requirements, while they would
fail to do so in any other. These views implied that the
sources of our actual vegetation and the explanation of these
peculiarities were to be sought in, and presupposed, an ancestry in pliocene or still earlier times, occupying the higher
northern regions. And it was thought that the occurrence of
peculiarly North American genera in Europe in the tertiary
period (such as Taxodium, Carya, Liquidambar, Sassafras,
Negundo, etc.), might be best explained on the assumption of
early interchange and diffusion through north Asia, rather
than by that of the fabled Atlantis.

The hypothesis supposed a gradual modification of species
in different directions under altering conditions, at least to
the extent of producing varieties, sub-species, and representative species, as they may be variously regarded; likewise the
single and local origination of each type, which is now almost
universally taken for granted.

The remarkable facts in regard to the eastern American
and Asiatic floras which these speculations were to explain
have since increased in number, more especially through the
admirable collections of Dr. Maximowicz in Japan and adjacent countries, and the critical comparisons he has made and
is still engaged upon.

I am bound to state that, in a recent general work [1] by
a distinguished European botanist, Professor Grisebach, of

[1] "Die Vegetation der Erde nach ihrer klimatischen Anordnung."
1871.

Göttingen, these facts have been emptied of all special significance, and the relations between the Japanese and the Atlantic United States flora declared to be no more intimate than might be expected from the situation, climate, and present opportunity of interchange. This extraordinary conclusion is reached by regarding as distinct species all the plants common to both countries between which any differences have been discerned, although such differences would probably count for little if the two inhabited the same country, thus transferring many of my list of identical to that of representative species; and then by simply eliminating from consideration the whole array of representative species, *i. e.*, all cases in which the Japanese and the American plant are not exactly alike. As if, by pronouncing the cabalistic word species, the question were settled, or rather the greater part of it remanded out of the domain of science; as if, while complete identity of forms implied community of origin, anything short of it carried no presumption of the kind; so leaving all these singular duplicates to be wondered at, indeed, but wholly beyond the reach of inquiry.[1]

Now the only known cause of such likeness is inheritance; and as all transmission of likeness is with some difference in individuals, and as changed conditions have resulted, as is well known, in very considerable differences, it seems to me that, if the high antiquity of our actual vegetation could be rendered probable, not to say certain, and the former habitation of any of our species or of very near relatives of them in high northern regions could be ascertained, my whole case would be made out. The needful facts, of which I was ignorant when my essay was published, have now been for some years made known, — thanks, mainly, to the researches of Heer upon ample collections of arctic fossil plants. These are confirmed and extended by new investigations, by Heer and Lesquereux, the results of which have been indicated to me by the latter.[2]

---

[1] See Appendix, II.

[2] Reference should also be made to the extensive researches of Newberry upon the tertiary and cretaceous floras of the western United

The Taxodium, which everywhere abounds in the miocene formations in Europe, has been specifically identified, first by Gœppert, then by Heer, with our common Cypress of the southern States. It has been found fossil in Spitzbergen, Greenland, and Alaska, — in the latter country along with the remains of another form, distinguishable, but very like the common species ; and this has been identified by Lesquereux in the miocene of the Rocky Mountains. So there is one species of tree which has come down essentially unchanged from the tertiary period, which for a long while inhabited both Europe and North America, and also, at some part of the period, the region which geographically connects the two (once doubtless much more closely than now), but which has survived only in the Atlantic United States and Mexico.

The same Sequoia which abounds in the same miocene formations in northern Europe has been abundantly found in those of Iceland, Spitzbergen, Greenland, Mackenzie River, and Alaska. It is named *S. Longsdorfii*, but is pronounced to be very much like *S. sempervirens*, our living Redwood of the Californian coast, and to be the ancient representative of it. Fossil specimens of a similar, if not the same, species have recently been detected in the Rocky Mountains by Hayden, and determined by our eminent palæontological botanist Lesquereux ; and he assures me that he has the common Redwood itself from Oregon in a deposit of tertiary age. Another Sequoia (*S. Sternbergii*), discovered in miocene deposits in Greenland, is pronounced to be the representative of *S. gigantea*, the Big Tree of the Californian Sierra. If the

States. See especially Professor Newberry's Paper in the "Boston Journal of Natural History," vol. vii. No. 4, describing fossil plants of Vancouver's Island, etc. ; his "Notes on the Later Extinct Floras of North America," etc., in "Annals of the Lyceum of Natural History," vol. ix., April, 1868 ; "Report on the Cretaceous and Tertiary Plants collected in Raynolds and Hayden's Yellowstone and Missouri Exploring Expedition, 1859-1860," published in 1869 ; and an interesting article entitled "The Ancient Lakes of Western America, their Deposits and Drainage," published in "The American Naturalist," January, 1871.

The only document I was able to consult was Lesquereux's Report on the Fossil Plants, in Hayden's Report of 1872.

Taxodium of the tertiary time in Europe and throughout the
arctic regions is the ancestor of our present Bald Cypress, —
which is assumed in regarding them as specifically identical,
— then I think we may, with our present light, fairly assume
that the two Redwoods of California are the direct or collateral descendants of the two ancient species which so closely
resemble them.

The forests of the arctic zone in tertiary times contained at
least three other species of Sequoia, as determined by their
remains, one of which, from Spitzbergen, also much resembles
the common Redwood of California. Another, " which appears to have been the commonest coniferous tree on Disco,"
was common in England and some other parts of Europe. So
the Sequoias, now remarkable for their restricted station and
numbers, as well as for their extraordinary size, are of an
ancient stock : their ancestors and kindred formed a large
part of the forests, which flourished throughout the polar regions, now desolate and ice-clad, and which extended into low
latitudes in Europe. On this continent one species, at least,
had reached to the vicinity of its present habitat before the
glaciation of the region. Among the fossil specimens already
found in California, but which our trustworthy palæontological botanist has not yet had time to examine, we may expect
to find evidence of the early arrival of these two Redwoods
upon the ground which they now, after much vicissitude,
scantily occupy.

Differences of climate, or circumstances of migration, or
both, must have determined the survival of Sequoia upon the
Pacific, and of Taxodium upon the Atlantic coast. And still
the Redwoods will not stand in the east, nor could our Taxodium find a congenial station in California. Both have probably had their opportunity in the olden time, and failed.

As to the remaining near relative of Sequoia, the Chinese
Glyptostrobus, a species of it, and its veritable representative,
was contemporaneous with Sequoia and Taxodium, not only
in temperate Europe, but throughout the arctic regions from
Greenland to Alaska. According to Newberry, it was abundantly represented in the miocene flora of the temperate zone
of our own continent, from Nebraska to the Pacific.

Very similar would seem to have been the fate of a more familiar gymnospermous tree, the Gingko or Salisburia. It is now indigenous to Japan only. Its ancestor, as we may fairly call it, — since, according to Heer, " it corresponds so entirely with the living species that it can scarcely be separated from it," — once inhabited northern Europe and the whole arctic region round to Alaska, and had even a representative farther south, in our Rocky Mountain district. For some reason, this and Glyptostrobus survive only on the shores of eastern Asia.

Libocedrus, on the other hand, appears to have cast in its lot with the Sequoias. Two species, according to Heer, were with them in Spitzbergen. *L. decurrens*, the Incense Cedar, is one of the noblest associates of the present Redwoods. But all the rest are in the southern hemisphere, two at the southern extremity of the Andes, two in the South Sea Islands. It is only by bold and far-reaching suppositions that they can be geographically associated.

The genealogy of the Torreyas is still wholly obscure; yet it is not unlikely that the Yew-like trees, named Taxites, which flourished with the Sequoias in the tertiary arctic forests, are the remote ancestors of the three species of Torreya, now severally in Florida, in California, and in Japan.

As to the Pines and Firs, these were more numerously associated with the ancient Sequoias of the polar forests than with their present representatives, but in different species, apparently more like those of eastern than of western North America. They must have encircled the polar zone then, as they encircle the present temperate zone now.

I must refrain from all enumeration of the angiospermous or ordinary deciduous trees and shrubs, which are now known, by their fossil remains, to have flourished throughout the polar regions when Greenland better deserved its name and enjoyed the present climate of New England and New Jersey. Then Greenland and the rest of the north abounded with Oaks, representing the several groups of species which now inhabit both our eastern and western forest districts; several Poplars, one very like our Balsam Poplar, or Balm of

Gilead-tree; more Beeches than there are now, a Hornbeam, and a Hop-Hornbeam, some Birches, a Persimmon, and a Planer-tree, near representatives of those of the Old World, at least of Asia, as well as of Atlantic North America, but all wanting in California; one Juglans like the Walnut of the Old World, and another like our Black Walnut; two or three Grapevines, one near our southern Fox Grape or Muscadine, another near our northern Frost Grape; a Tilia, very like our Basswood of the Atlantic States only; a Liquidambar; a Magnolia, which recalls our *M. grandiflora;* a Liriodendron, sole representative of our Tulip-tree; and a Sassafras, very like the living tree.

Most of these, it will be noticed, have their nearest or their only living representatives in the Atlantic States, and when elsewhere, mainly in eastern Asia. Several of them, or of species like them, have been detected in our tertiary deposits, west of the Mississippi, by Newberry and Lesquereux. Herbaceous plants, as it happens, are rarely preserved in a fossil state, else they would probably supply additional testimony to the antiquity of our existing vegetation, its wide diffusion over the northern and now frigid zone, and its enforced migration under changes of climate.[1]

Concluding, then, as we must, that our existing vegetation is a continuation of that of the tertiary period, may we suppose that it absolutely originated then? Evidently not. The preceding cretaceous period has furnished to Carruthers in Europe a fossil fruit like that of the *Sequoia gigantea* of the famous groves, associated with Pines of the same character as those that accompany the present tree; has furnished to Heer,

---

[1] There is at least one instance so opportune to the present argument that it should not pass unnoticed, although I had overlooked the record until now. *Onoclea sensibilis* is a Fern peculiar to the Atlantic United States (where it is common and widespread) and to Japan. Professor Newberry identified it several years ago in a collection obtained by Dr. Hayden of miocene fossil plants of Dacota Territory, which is far beyond its present habitat. He moreover regards it as probably identical with a fossil specimen " described by the late Professor E. Forbes, under the name of *Filicites Hebridicus,* and obtained by the Duke of Argyll from the Island of Mull."

from Greenland, two more Sequoias, one of them identical
with a tertiary species, and one nearly allied to *Sequoia
Langsdorfii*, which in turn is a probable ancestor of the
common Californian Redwood ; has furnished to Newberry
and Lesquereux in North America the remains of another
ancient Sequoia, a Glyptostrobus, a Liquidambar which well
represents our Sweet Gum, Oaks analogous to living ones,
leaves of a Plane-tree, which are also in the tertiary and are
scarcely distinguishable from our own *Platanus occidentalis*,
of a Magnolia and a Tulip-tree, and " of a Sassafras undistinguishable from our living species." I need not continue
the enumeration. Suffice it to say that the facts justify the
conclusion which Lesquereux — a scrupulous investigator —
has already announced : " that the essential types of our
actual flora are marked in the cretaceous period, and have
come to us after passing, without notable changes, through
the tertiary formations of our continent."

According to these views, as regards plants at least, the
adaptation to successive times and changed conditions has
been maintained, not by absolute renewals, but by gradual
modifications. I, for one, cannot doubt that the present existing species are the lineal successors of those that garnished
the earth in the old time before them, and that they were as
well adapted to their surroundings then, as those which flourish and bloom around us are to their conditions now. Order
and exquisite adaptation did not wait for man's coming, nor
were they ever stereotyped. Organic nature, — by which I
mean the system and totality of living things, and their adaptation to each other and to the world, — with all its apparent
and indeed real stability, should be likened, not to the ocean,
which varies only by tidal oscillations from a fixed level to
which it is always returning, but rather to a river, so vast
that we can neither discern its shores nor reach its sources,
whose onward flow is not less actual because too .slow to be
observed by the ephemeræ which hover over its surface, or
are borne upon its bosom.

Such ideas as these, though still repugnant to some, and
not long since to many, have so possessed the minds of the

naturalists of the present day, that hardly a discourse can be
pronounced or an investigation prosecuted without reference
to them. I suppose that the views here taken are little, if at
all, in advance of the average scientific mind of the day. I
cannot regard them as less noble than those which they are
succeeding.

An able philosophical writer, Miss Frances Power Cobbe,
has recently and truthfully said : [1]

" It is a singular fact, that when we can find out how anything is done, our first conclusion seems to be that God did
not do it. No matter how wonderful, how beautiful, how intimately complex and delicate has been the machinery which
has worked, perhaps for centuries, perhaps for millions of
ages, to bring about some beneficent result, if we can but
catch a glimpse of the wheels its divine character disappears."

I agree with the writer that this first conclusion is prema-
ture and unworthy, — I will add, deplorable. Through what
faults or infirmities of dogmatism on the one hand, and skep-
ticism on the other, it came to be so thought, we need not
here consider. Let us hope, and I confidently expect, that it
is not to last ; that the religious faith which survived without
a shock the notion of the fixity of the earth itself may equally
outlast the notion of the absolute fixity of the species which
inhabit it ; that in the future even more than in the past,
faith in an order, which is the basis of science, will not — as
it cannot reasonably — be dissevered from faith in an Or-
dainer, which is the basis of religion.

APPENDIX.

I.

In the following table the names in the left-hand column are from
my " Manual of the Botany of the Northern United States," and
from Dr. Chapman's " Flora of the Southern United States," the
two together comprehending the flora of the Atlantic United States
east of the Mississippi River. Alpine plants on the one hand, and
subtropical plants on the other, are excluded.

[1] " Darwinism in Morals," in " Theological Review," April, 1871.

The entries in the middle column, when there are any, are of identical or representative species occurring in Oregon or California. Those in the right-hand column are of such species in Japan, or other parts of northeastern Asia, including the Himalayas and Siberia as far west as the Altai Mountains. When these are not identical, or so closely related to the American species that the one may be said strictly to represent the other, also when genera or parts of genera are adduced merely as representing the same type in these respective regions, the names are included in parentheses.

Species which extend through Europe into northeastern Asia, and therefore nearly round the temperate zone, are also left out of view, the object being to consider the peculiar relations of the floras of eastern North America and eastern temperate Asia. The table has been drawn up off-hand, from the means within reach. Probably the example might be considerably increased.

EXTRA-EUROPEAN (TEMPERATE) GENERA AND SPECIES OF THE ATLANTIC UNITED STATES (*i. e.*, EAST OF THE MISSISSIPPI) REPRESENTED BY IDENTICAL OR STRICTLY REPRESENTATIVE SPECIES, OR ELSE BY LESS INTIMATELY RELATED SPECIES (THE LATTER INCLUDED IN PARENTHESES).

	1. *In the Pacific United States.*	2. *In Northeastern Asia, Japan to Altai and the Himalayas.*
Anemone Pennsylvanica.		Anemone dichotoma = Pennsylvanica.
" parviflora.		Anemone parviflora?
Ranunculus alismæfolius.	Ranunculus alismæfolius.	Ranunculus alismæfolius.
" Cymbalaria.	" Cymbalaria.	" Cymbalaria.
" Gmelini.	" Gmelini.	" Gmelini.
" Pennsylvanicus.	" Pennsylvanicus.	" Pennsylvanicus.
Trautvetteria palmata.	Trautvetteria palmata.	Trautvetteria palmata.
Hydrastis Canadensis.		Hydrastis Jesoensis.
Trollius Americanus.	Trollius Americanus var.	Trollius patulus var. = Americanus, Ledeb.
Aconitum uncinatum.		Aconitum uncinatum ex Hook. f.
Actæa spicata, var. rubra.	Actæa spicata, var. arguta.	Actæa spicata, var. rubra.
" alba.		" alba ?
Cimicifuga Americana.		Cimicifuga fœtida, barely occurs in N. Europe also.
Cimicifuga racemosa and cordifolia.	(Cimicifuga elata.)	(Cimicifuga Dahurica and § Pityrosperma, 3 spp.)
Illicium Floridanum and parviflorum.		Illicium anisatum, religiosum, etc.
Schizandra coccinea.		(Schizandra nigra, etc.)
Magnolia, 7 spp.		(Magnolia, 8–12 spp.)
Menispermum Canadense.		Menispermum Dahuricum.
Caulophyllum thalictroides.[1]		Caulophyllum thalictroides.
Diphylleia cymosa.		Diphylleia Grayi.
Jeffersonia diphylla.		Jeffersonia = Plagiorhegma dubium.
Podophyllum peltatum.		(Podophyllum Emodi.)
Brasenia peltata.	Brasenia peltata.	Brasenia peltata.
Nelumbium luteum.		Nelumbium speciosum.
Stylophorum diphyllum.		(Stylophorum Japonicum and lactucoides.)

[1] See Appendix, II.

Dicentra eximia.	Dicentra eximia or formosa	(Dicentra spp.)
Corydalis aurea.	Corydalis aurea var.	Corydalis aurea var., etc.
Viola Selkirkii.		Viola Selkirkii.
" Canadensis.	Viola Canadensis var.	" Canadensis var.
Claytonia, Virginica and Ca-	Claytonia lanceolata.	(Claytonia, spp. Siberia.)
roliniana.		
Elodes Virginica.		Elodes Virginica.
" petiolata.		" petiolata.
Tilia Americana (American		Tilia sp., American type, one of
type).		which reaches Hungary.
Tilia heterophylla.		
Stuartia, 2 spp.		(Stuartia, 3 spp.)
Xanthoxylum spp.		(Xanthoxylum spp.)
Rhus venenata.		Rhus vernicifera, etc.
" Toxicodendron.	(Rhus diversiloba.)	" Toxicodendron.
Vitis Labrusca.		Vitis Labrusca.
" indivisa.		" humulifolia.
Ampelopsis quinquefolia.		(Ampelopsis tricuspidata.)
Berchemia volubilis.		(Berchemia racemosa, etc.)
Sageretia Michauxii.		(Sageretia theæsans.)
Celastrus scandens.		(Celastrus, 5 spp.)
Æsculus glabra.		Æsculus Chinensis and Hippo-
		castanum.
" flava and Pavia.		(Æsculus dissimilis.)
" parviflora.	(Æsculus Californica.)	(" Punduana Wall.)
Acer spicatum.		Acer spicatum var.
" Pennsylvanicum.		" tegmentosum.
Negundo aceroides.	Negundo aceroides Califor-	(Negundo cissifolium and spp.)
	nicum.	
Wistaria frutescens.		Wistaria Sinensis and spp.
Desmodium, many spp.		(Desmodium, several spp.)
Lespedeza spp.		(Lespedeza, spp.)
Rhynchosia spp.		(Rhynchosia, sp.)
Amphicarpæa monoica.		(Amphicarpæa, 5 spp.)
Thermopsis, 3 spp.	(Thermopsis fabacea & sp.)	Thermopsis fabacea.
Cladrastis tinctoria.		(Maackia Amurensis.)
Cassia spp.		(Cassia spp.)
Gleditschia triacanthos and		Gleditschia Chinensis, etc.
monosperma.		
Neptunia lutea.		(Neptunia spp.)
Spiræa (Neillia) opulifolia.	Spiræa opulifolia.	(Neillia spp., Himalayas.)
" corymbosa.	" betulæfolia.	Spiræa betulæfolia.
Neviusa Alabamensis.		(Stephanandra, Kerria.)
Geum macrophyllum.	Geum macrophyllum.	Geum Japonicum.
Potentilla Pennsylvanica.	Potentilla Pennsylvanica.	Potentilla Pennsylvanica.
Rubus triflorus.		Rubus triflorus, var. Japonicus.
" strigosus.	Rubus strigosus.	" strigosus.
Pyrus Americana and sambu-	Pyrus sambucifolia.	Pyrus Americana and sambuci-
cifolia.		folia.
Amelanchier Canadensis and	Amelanchier Canadensis	Amelanchier Canadensis var.
vars.	var.	
Calycanthus, 3 spp.	Calycanthus occidentalis.	(Chimonanthus fragrans.)
Ribes Cynosbati.	(Ribes spp.)	Ribes Cynosbati.
" lacustre.	" setosum.	" lacustre.
" prostratum.	" laxiflorum.	" laxiflorum.
Philadelphus, 2 spp.	Philadelphus, 2 spp.	Philadelphus spp.
Itea Virginica.		(Itea spp.)
Hydrangea, 3 spp.		Hydrangea, many spp.
Astilbe decandra.		Astilbe Thunbergii and spp.
Boykinia aconitifolia.	Boykinia occidentalis and	(Boykinia? = Saxifr. tellimioides
	elata.	Maxim.)
Mitella nuda.	Mitella nuda.	Mitella nuda, Siberia.
" diphylla.	Mitella § Mitellastra, etc.,	(Mitella § Mitellastra, sp.)
	spp.	
Tiarella cordifolia.	Tiarella unifoliata.	Tiarella polyphylla.
Penthorum sedoides.		Penthorum sedoides ? = Chi-
		nense and humile.
Hamamelis Virginica.		Hamamelis Japonica, etc.
Fothergilla alnifolia.		(Corylopsis spp., etc.)
Heracleum lanatum.	Heracleum lanatum.	Heracleum lanatum.
Archangelica Gmelini.	Archangelica Gmelini.	Archangelica Gmelini.
Sium lineare.		Sium cicutæfolium.
Cryptotænia Canadensis.		Cryptotænia Canadensis.
Archemora, 2 spp.		(Peucedanum ? Sieboldii.)

Osmorrhiza longistylis and brevistylis.	Osmorrhiza longistylis, etc.	Osmorrhiza longistylis, etc.
Aralia spinosa.		Aralia spinosa var.
" racemosa.	Aralia humilis.	" edulis, etc.
" nudicaulis.		(" cordata.)
" (Ginseng) quinquefolia.		" repens, Ginseng, etc.
Cornus Canadensis.	Cornus Canadensis.	Cornus Canadensis.
" florida.	" Nuttallii.	Benthamia spp.
" stolonifera.		" alba.
Diervilla, 2 spp.		(Diervilla § Weigela, spp.)
Triosteum, 2 spp.		Triosteum sinuatum and Himalaicum.
Viburnum lantanoides.		Viburnum lantanoides (and related species.)
" dentatum & pubescens.	Viburnum ellipticum.	Viburnum dilatatum, etc.
Mitchella repens.		Mitchella undulata.
Adenocaulon bicolor.	Adenocaulon bicolor.	Adenocaulon adhærescens.
Boltonia, spp.		(Boltonia spp.)
Aster § Biotia, corymbosus and spp.		Aster § Biotia, corymbosus and spp.
Aster § Conyzopsis, angustus.	Aster angustus.	Aster angustus = Brachyactis ciliata, Ledeb.
Artemisia Canadensis.	Artemisia Canadensis.	Artemisia Canadensis ? = comutata.
" biennis.	" biennis.	Artemisia biennis.
" frigida.	" frigida.	" frigida.
Senecio pseudo-arnica.		Senecio pseudo-arnica.
Nabalus spp.		(Nabalus ochroleucus acerifolius)
Cacalia spp.		(Cacalia spp.)
Mulgedium pulchellum.	Mulgedium pulchellum.	Mulgedium Sibiricum.
Vaccinium § Oxycoccus macrocarpum.		Vaccinium macrocarpum, forma ambigua.[1]
" " erythrocarpum.		" Japonicum (ab erythrocarpo vix differt).
" § Batodendron, 2 spp.		(" § Batodendron, spp.)
" § Cyanococcus, 15 spp.		(" § Cyanococcus, one sp. near Pennsylvanicum.
" ovalifolium.	Vaccinium ovalifolium.	" ovalifolium.
Chiogenes hispidula.		Chiogenes hispidula.
Epigæa repens.		Epigæa Asiatica.
Gaultheria procumbens.		Gaultheria pyroloides.
Leucothoe axillaris and Catesbæi.		Leucothoe Keiskei.
Leucothoe racemosa and recurva.		(" Grayana and Tschonoskii.)
Andromeda § Portuna floribunda.		(Andromeda § Portuna sp.)
Andromeda § Pieris spp.		(" § Pieris spp.)
Clethra, 2 spp.		(Clethra sp.)
Menziesia ferruginea, var. globularis.	Menziesia globularis.	Menziesia pentandra and others.
Rhododendron Catawbiense.	Rhododendron Californicum.	(Rhododendron brachycarpum.)
" maximum.		(" Metternichii.)
" punctatum.		(" Keiskei.)
Rhodora Canadensis.		(" spp.)
Azalea, 4 spp.	(Azalea occidentalis).	(Azalea spp.)
Elliottia racemosa.		Tripetaleia paniculata and bracteata.

[1] " Ob flores revera terminales, bracteolas lineari-lanceolatas scariosas et folia acuta." Dr. Maximowicz (in " Mel. Biolog. Diagn." decas 12) refers this to *V. Oxycoccus*, instead of to *V. macrocarpum*, which is "semper bene distinctum floribus axillaribus, bracteolis ovatis foliaceis et foliis obtusis." But in one of my specimens the axis of the umbel is continued into a leafy shoot, as in *V. macrocarpum ;* and the bracteolæ vary from linear to ovate, and from thin and scarious to chartaceous or coriaceous in both species ; they are never (so far as I know) "foliaceous" in *V. macrocarpum*, but the bracteæ sometimes are. The leaves are sometimes acutish in the latter, and also very obtuse in *V. Oxycoccus;* in the Japanese specimens under consideration they are often half an inch in length. I must add that in the length of the filaments they accord with the character which I assigned to *V. Oxycoccus* in the Manual. In fact, a form combining the characters of the two species survives in Japan.

Pyrola elliptica.		Pyrola elliptica.
Monotropa uniflora.		Monotropa uniflora.
Shortia galacifolia.		Shortia galacifolia = Schizocodon uniflorus.
Ilex § Prinos spp.		(Ilex § Prinos spp.)
Diospyros Virginiana.		(Diospyros spp.)
Symplocos sp.		(Symplocos spp.)
Tecoma radicans.		Tecoma grandiflora.
Catalpa bignonioides.		Catalpa Kæmpferi.
Veronica Virginica.		Veronica Virginica.
Callicarpa Americana.		(Callicarpa; 3 spp.)
Phryma Leptostachya.		Phryma Leptostachya.
Lycopus Virginicus.	Lycopus Virginicus.	Lycopus parviflorus.
Teucrium Canadense.		Teucrium Japonicum.
Hedeoma, 4 spp.		(Hedeoma sp.)
Lophanthus spp.	Lophanthus spp.	(Lophanthus sp.)
Scutellaria (nuculis alatis) nervosa.		Scutellaria (nuculis alatis) Guilielmi.[1]
Halenia deflexa.		Halenia Sibirica, and spp.
Phlox subulata.	(Phlox Douglasii and spp.)	Phlox Sibirica.
Gelsemium sempervirens.		(Gelsemium elegans.)
Mitreola, 2 spp.		Mitreola oldenlandioides.
Apocynum androsæmifolium.	Apocynum androsæmifolium.	(Apocynum venetum.)
Amsonia Tabernæmontana.		(Amsonia elliptica.)
Asarum Virginicum and arifolium.	Asarum caudatum.	Asarum variegatum and Blumei.
Asarum Canadense.		" caulescens and Sieboldii.
Phytolacca decandra.		Phytolacca Kæmpferi, etc.
Corispermum hyssopifolium.		Corispermum hyssopifolium.
Polygonum arifolium.		Polygonum perfoliatum.
" sagittatum.		" sagittatum and Sieboldii.
Sassafras officinale.		(Lindera triloba, etc.)
Lindera Benzoin, etc.		" hypoglauca, etc.
Tetranthera geniculata.	(Tetranthera Californica.)	(Tetranthera spp.)
Pyrularia oleifera.		(Pyrularia = Sphærocarpa spp.)
Saururus cernuus.		Saururus Loureiri.
Stillingia spp.		(Stillingia spp.)
Pachysandra procumbens.		Pachysandra terminalis.
Planera aquatica.		Planera Japonica (and kichardi).
Maclura aurantiaca.		Maclura gerontogæa.
Pilea pumila.		Pilea pumila.
Laportea Canadensis.		Laportea evitata, etc.
Bœhmeria cylindrica.		(Bœhmeria spp.)
Parietaria debilis.		Parietaria debilis.
Juglans nigra.	(Juglans rupestris.)	(Juglans regia.)
" cinerea.		Juglans Mandchurica, stenocarpa.
Corylus rostrata.	Corylus rostrata var.	Corylus rostrata, var. Mandchurica.
Betula glandulosa.		Betula glandulosa.
" nigra.		(" ulmifolia, etc.)
Alnus maritima.		Alnus maritima.
Myrica cerifera.	Myrica Californica.	(Myrica Nagi.)
Pinus resinosa.		(Pinus densiflora, etc.)
" Strobus.	Pinus monticola.	" excelsa.

[1] *Scutellaria Guilielmi*, n. sp. Perilomioides: slender, branched from the base, stoloniferous? leaves membranaceous, minutely pubescent, crenately dentate, the lower round-cordate and slender-petioled, the others ovate or oblong with rounded or truncate base and short-petioled, the floral similar but gradually smaller; flowers solitary in the axils; peduncles about the length of the calyx; corolla ("light purple," only three lines long) hardly more than twice the length of the calyx, its lips of nearly equal length; nutlets surrounded by an abrupt and reflexed denticulate wing, upper face of the disk muricate, the lower as if squamellate. *S. hederacea?* Gray, in Perry's "Japan Exped." iii. p. 316, and "Bot. Contrib. Proc. Amer. Acad." viii. p. 370, not of Kunth and Bouché. It appears from a note by Vatke, in "Bot. Zeit.," 1872, p. 717, that *S. hederacea* is identical with the Tasmanian *S. humilis,* and its nutlets were originally described as echinulate-tuberculate, and by implication wingless. So our plant may be named in honor of Dr. S. W. Williams, who first collected a little of it at Simoda, Japan. Better and fruiting specimens were gathered on the Loo-Choo Islands, by Charles Wright.

Abies Canadensis.	Abies Mertensiana.	Abies Tsuga and diversifolia.
Thuja occidentalis.	Thuja gigantea, etc.	Thuja Japonica.
Taxodium distichum.		(Glyptostrobus heterophyllus.)
Cupressus (Chamæcyparis) thuyoides.	C. Nutkaensis.	Cupressus pisifera, obtusa, etc.
Taxus Canadensis.	Taxus brevifolia.	Taxus cuspidata.
Torreya taxifolia.	Torreya Californica.	Torreya nucifera and grandis.
Arisæma, 3 spp.		Arisæma, 9 spp.
Symplocarpus fœtidus.	(Lysichiton Camtschatcense.)	Symplocarpus fœtidus? and Lysichiton Camtschatcense.
Listera australis.		(Listera Japonica, etc.)
Arethusa bulbosa.		(Arethusa Japonica.)
Pogonia ophioglossoides.		Pogonia ophioglossoides.
Microstylis ophioglossoides.		(Microstylis Japonica.)
Liparis liliifolia.		Liparis liliifolia?
Cypripedium acaule.		(Cypripedium Japonicum.)
Habenaria virescens.		Habenaria fucescens.
Aletris farinosa and aurea.		Aletris Japonica.
Iris cristata.		Iris tectorum = cristata, Miq.
Dioscorea villosa.		(Dioscorea spp.)
Smilax hispida.		Smilax Sieboldii.
" herbacea and peduncularis.		" herbacea = Nipponica.
Smilax tamnifolia.		(" higoensis.)
Croomia pauciflora.		Croomia pauciflora.
Trillium grandiflorum.	Trillium obovatum.	Trillium obovatum.
" erectum.		" erectum var.
Tofieldia glutinosa and pubens.		(Tofieldia Japonica and nutans.)
Helonias bullata.		Heloniopsis pauciflora, breviscapa, Japonica.
Chamælirium luteum.		Chamælirium luteum.
Zygadenus, 3 spp.	Zygadenus glaucus, etc.	(Zygadenus Japonicus.)
Streptopus roseus.	Streptopus roseus.	Streptopus roseus.
Prosartes lanuginosa.	Prosartes Hookeri, etc.	Prosartes viridescens, etc.
Clintonia borealis.	Clintonia uniflora.	Clintonia Udensis.
Polygonatum giganteum.		Polygonatum giganteum.
Smilacina trifolia.		Smilacina trifolia.
" racemosa.	Smilacina racemosa var.	" Japonica.
" stellata.	" stellata.	" Davarica.
Erythronium Americanum and albidum.	Erythronium grandiflorum.	Erythronium grandiflorum.
Narthecium Americanum.		Narthecium Asiaticum.
Scirpus Eriophorum.		Scirpus Eriophorum.
Carex rostrata.		Carex rostrata.
Carex stipata	Carex stipata.	Carex stipata.
Zizania aquatica.		Zizania=Hydropyrum latifolium.
Arundinaria macrosperma.		(Arundinaria Japonica.)
Avena striata and Smithii.		Avena collosa.
Adiantum pedatum.	Adiantum pedatum.	Adiantum pedatum.
Pellæa gracilis.		Pellæa Stelleri = gracilis.
Aspidium fragrans.		Aspidium fragrans.
Asplenium thelypteroides.		Asplenium thelypteroides.
Camptosorus rhizophyllus.		Camptosorus Sibiricus.
Onoclea sensibilis.		Onoclea sensibilis.
Osmunda cinnamonea.		Osmunda cinnamonea.
" Claytoniana.		" Claytoniana.
Lygodium palmatum.		(Lygodium Japonicum.)
Botrychium Virginicum.		Botrychium Japonicum.
Lycopodium lucidulum.		Lycopodium lucidulum.
" dendroideum.		" dendroideum.

It appears that two thirds of the middle column is blank; namely, that only a third of the species or forms which are more or less peculiar to temperate Atlantic North America (*i. e.,* east of the Mississippi and south of the Great Lakes and the St. Lawrence) and to temperate eastern Asia, are represented in Oregon and California. Moreover, eighty of the genera here treated of are peculiar to North America and temperate Asia ; and sixty-three (*i. e.*, more than three quarters) of these are not met with in western North America. This table may be compared, or rather contrasted, with the following one.

EXTRA-EUROPEAN PLANTS OF TEMPERATE EASTERN ASIA WHICH ARE REPRESENTED IDENTICALLY OR BY SOME NEAR RELATIVE IN OREGON (SOUTH OF LAT. 48°) OR CALIFORNIA (ARCTIC-ALPINE PLANTS EXCLUDED), BUT NOT IN THE ATLANTIC UNITED STATES : —

Thalictrum sparsiflorum.	Thalictrum sparsiflorum.
Ranunculus affinis.	Ranunculus affinis.
Coptis occidentalis.	Coptis occidentalis.
" brachypetala and Teeta.	" asplenifolia.
Aconitum delphinifolium.	Aconitum delphinifolium.
Pœonia spp.	(Pœonia Rossii.)
Berberis § Mahonia spp.	(Berberis § Mahonia spp.
Epimedium § Aceranthus sp.	Vancouveria hexandra.
Achlys Japonica.	Achlys triphylla.
Corydalis pæoniæfolia.	Corydalis pæoniæfolia.
Mœhringia umbrosa.	Mœhringia macrophylla.
Linum perenne.	Linum perenne.
Thermopsis fabacea.	Thermopsis fabacea.
Astragalus adsurgens.	Astragalus adsurgens.
Chamærhodos erecta.	Chamærhodos erecta.
Spiræa callosa.	Spiræa Nobleana.
Rubus spectabilis.	Rubus spectabilis.
Pyrus rivularis ?	Pyrus rivularis.
Cratægus sanguinea.	Cratægus Douglasii.
Rosa Kamtschatica.	Rosa Kamtschatica.
Photinia arbutifolia.	(Photinia serrulata.)
Saxifraga Sibirica.	Saxifraga Sibirica.
Mitella § Mitellaria spp.	(Mitella § Mitellaria spp.)
Glehnia littoralis.	Glehnia littoralis.
Oplopanax horrida.	Oplopanax horrida.
Echenais carlinoides.	Echenais carlinoides.
Lonicera Maximowiczii.	Lonicera Breweri.
Gaultheria adenothrix.	Gaultheria Myrsinites.
Rhododendron ovatum and semibarbatum.	Rhododendron albiflorum.
Pyrola subaphylla.	Pyrola aphylla.
Villarsia Crista-Galli.	Villarsia Crista-Galli.
Lycopus lucidus.	Lycopus lucidus.
Boschniakia glabra.	Boschniakia glabra.
Echinospermum patulum.	Echinospermum patulum.
" Redowskii.	" Redowskii.
Hottuynia cordata.	Anemiopsis Californica.
Quercus spp.	(Quercus densiflora.)
Castanopsis spp.	(Castanopsis chrysophylla.)
Lysichiton Camtschatcense.	Lysichiton Camtchatcense.
Erythronium grandiflorum.	Erythronium grandiflorum.
Carex macrocephala.	Carex macrocephala.
Triticum ægilopoides.	Triticum ægilopoides.
Elymus Sibiricus.	Elymus Sibiricus.
Abies Menziesii.	Abies Menziesii.
Woodwardia radicans.	Woodwardia radicans.

The entries are only forty-five ; and the representation, when at all close, is by identical or nearly identical species. Only seven of the genera here noted are peculiar to northeastern Asia and northwestern America : namely, Glehnia, Oplopanax, and Lysichiton, each of a single species common to both coasts ; Achlys, of which there is a Japanese species said to differ from the American ; Boschniakia, of a common high northern species, and a peculiar one in California ; Echinais, of one or two Asiatic species, one of them lately found in California and Colorado, but possibly of recent introduction ; and Castanopsis, a rather large and characteristic east Asian genus, represented by a single but very distinct species in Oregon and California.

Small, under the circumstances, as is the number of cognate plants or forms in these two floras, it is large in comparison with those which are peculiar to the United States and Europe, excluding, as before, all Arctic-alpine species. The following seem to be the principal : —

Anemone nemorosa, of which there is a peculiar Pacific form, perhaps reaching the eastern borders of Asia.

Myosorus minimus, which may be a recently introduced plant.

Cakile, a maritime genus.

Saxifraga aizoides.

Bellis integrifolia, which may be compared with the European *B. annua.*

Lobelia Dortmanna.

Primula Mistassinica.

Centunculus lanceolatus, a mere form of *C. minimus.*

Hottonia inflata, which represents *H. palustris.*

Utricularia minor.

Salicornia Virginica, the *S. mucronata* of Bigelow and probably of Lagasca also.

Corema Conradi, representing the Portuguese *C. alba.*

Vallisneria spiralis, which appears to be absent from northern Asia.

Spiranthes Romanzoviana, with its single station on the Irish coast. It extends across the American continent well northward, but seemingly not into the adjacent parts of Asia.

Eriocaulon septangulare, restricted in the Old World to a few stations on west British coasts.

Carex extensa, C. flacca (or *Barrattii*), and one or two others.

Cinna arundinacea, var. *pendula.*

Leersia oryzoides.

Spartina stricta and *S. juncea.*

Equisetum Telmateia.

Lycopodium inundatum.

Calluna vulgaris, which holds as small and precarious a tenure on this continent as *Spiranthes Romanzoviana* does in Europe.

Barely two dozen; and three or four of these are more or less
maritime.   Only two or three of them extend west of the Mississippi
Valley.

Narthecium is not in the list, a form or near ally of the European
and Atlantic-American species having been detected in Japan ; the
genus is unknown on the Pacific side of our continent.

II.

SINCE the foregoing tables were prepared, a letter from Mr. Dall
(who has returned from an arduous and successful exploration of the
Alaskan region, made under the authority of the United States Coast
Survey) informs me that his party met with Caulophyllum upon one
of the Shumagin Islands.   These islands lie off the southern shore
of the peninsula of Alaska, about in latitude 55°, longitude 160°.
No specimen occurs in the beautiful collection of dried plants made
in this expedition, mainly by Mr. Harrington ; nor indeed any other
plants which affect so southern a range as our Caulophyllum.   Yet
the plant may well have been rightly identified ; although it should
be seen by botanists before any conclusions are drawn from it.   But
the occurrence of an intermediate station like this would probably
lead Professor Grisebach to rank the north Asiatic Caulophyllum
no longer as a representative species, but as identical with our Atlantic plant, as Miquel and Maximowicz, as well as myself, have
already done upon evidence derived from the specimens.

Then, — upon Professor Grisebach's idea that, while identical
species are to be referred to a single origin and the disseverance
accounted for through means and causes now in operation, representative species have somehow arisen independently under similar
climates, — Caulophyllum must be explained as a case of migration,
but Diphylleia (in the same predicament, only with a perceptible
difference between the two plants) as a case of double origination.
So of the *Shortia galacifolia* and the *Schizocodon uniflorus*, of
which the corolla and stamens in both are still wanting.   If these,
when found, should prove to be exactly alike in the two, the very
difficult problem of accounting for the world-wide separation under
present circumstances is to be encountered ; if a difference appears,
the problem is to consider how, and upon what, similar climates can
have acted to have originated almost identical species upon opposite
sides of the world.   Professor Grisebach's views imply that " each
species has arisen under the influence of physical and other external
conditions," and that gradual alterations in a climate somehow pro-

duce adaptive " changes in organization " ; wherefore, as the President of the Linnæan Society has aptly remarked,[1] " We have a right to ask of him, What is the previous organization upon which he imagines climate to have worked to produce allied species in one region and representative species in distant regions ? " The difference here between Grisebach's conception and our own is, that we consider climate and other external conditions to have acted upon common ancestors in each case ; but he apparently declines to conjecture what they acted upon.

In conclusion I may advert to one instance, in which it would appear, either that widely different climates have originated the same or closely similar species, or else that one and the same species (one of those common to the United States and Japan) has been dispersed over the globe in a manner and to an extent that place it beyond the reach of explanations limited to the results of forces still in activity and means of dispersion still available. *Brasenia peltata* inhabits : 1. The Atlantic United States, from Canada to Texas ; 2. Oregon, or rather Washington Territory, a single known station at Gray's Harbor, on the Pacific, latitude 47° ; and Clear Lake, in California, latitude 39° ; 3. Japan ; 4. Khasya and Bhotan, altitude 4–6000 feet ; 5. Australia, Moreton Bay, etc. ; 6. West Africa, in a lake in Angola !

[1] Address of George Bentham, Esq , President of the Linnæan Society, etc., read May 24, 1872.

DO VARIETIES WEAR OUT OR TEND TO WEAR OUT?[1]

THIS question has been argued from time to time for more than half a century, and is far from being settled yet. Indeed, it is not to be settled either way so easily as is sometimes thought. The result of a prolonged and rather lively discussion of the topic about forty years ago in England, in which Lindley bore a leading part on the negative side, was, if we rightly remember, that the nays had the best of the argument. The deniers could fairly well explain away the facts adduced by the other side, and evade the force of the reasons then assigned to prove that varieties were bound to die out in the course of time. But if the case were fully re-argued now, it is by no means certain that the nays would win it. The most they could expect would be the Scotch verdict "not proven." And this not because much, if any, additional evidence of the actual wearing out of any variety has turned up since, but because a presumption has been raised under which the evidence would take a bias the other way. There is now in the minds of scientific men some reason to expect that certain varieties would die out in the long run, and this might have an important influence upon the interpretation of the facts that would be brought forward. Curiously enough, however, the recent discussions to which our attention has been called seem, on both sides, to have over-looked this matter.

But, first of all, the question needs to be more specifically stated if any good is to come from a discussion of it. There are varieties and varieties. They may, some of them, disappear or deteriorate, but yet not wear out — not come to an end from any inherent cause. One might even say, the younger they are the less chance of survival unless well cared

[1] New York Tribune, semi-weekly edition, December 8, 1874.

for. They may be smothered out by the adverse force of superior numbers; they are even more likely to be bred out of existence by unprevented cross-fertilization, or to disappear from mere change of fashion. The question, however, is not so much about reversion to an ancestral state, or the falling off of a high-bred stock into an inferior condition. Of such cases it is enough to say that, when a variety or strain of animal or vegetable is led up to unusual fecundity or size or product of any organ, for our good, and not for the good of the plant or animal itself, it can be kept so only by high feeding and exceptional care; and that with high feeding and artificial appliances come vastly increased liability to disease, which may practically annihilate the race. But then the race, like the bursted boiler, could not be said to wear out, while if left to ordinary conditions, and allowed to degenerate back into a more natural, if less useful state, its hold on life would evidently be increased rather than diminished.

As to natural varieties or races under normal conditions, sexually propagated, it could readily be shown that they are neither more nor less likely to disappear from any inherent cause than the species from which they originated. Whether species wear out, *i. e.*, have their rise, culmination, and decline from any inherent cause, is wholly a geological and very speculative problem, upon which, indeed, only vague conjectures can be offered. The matter actually under discussion concerns cultivated domesticated varieties only, and, as to plants, is covered by two questions.

First: Will races propagated by seed, being so fixed that they come true to seed, and purely bred (not crossed with any other sort), continue so indefinitely, or will they run out in time — not die out, perhaps, but lose their distinguishing characters? Upon this, all we are able to say is, that we know no reason why they should wear out or deteriorate from any inherent cause. The transient existence or the deterioration and disappearance of many such races are sufficiently accounted for otherwise; as in the case of extraordinarily exuberant varieties, such as mammoth fruits or roots, by increased liability to disease, already adverted to, or by the failure of the high feeding they demand. A common cause, in ordinary cases, is cross-breeding, through the agency of wind or insects, which is difficult to guard against. Or they go out of fashion and are superseded by others thought to be better, and so the old ones disappear.

Or, finally, they revert to an ancestral form. As offspring tend to resemble grandparents almost as much as parents, and as a line of close-bred ancestry is generally prepotent, so newly originated varieties have always a tendency to reversion. This is pretty sure to show itself in some of the progeny of the earlier generations, and the breeder has to guard against it by rigid selection. But the older the variety is — that is, the longer the series of generations in which it has come true from seed — the less the chance of reversion: for now, to be like the immediate parent, is also to be like a long line of ancestry; and so all the influences concerned — that is, both parental and ancestral heritability — act in one and the same direction. So, since the older a race is the more reason it has to continue true, the presumption of the unlimited permanence of old races is very strong.

Of course the race itself may give off new varieties; but that is no interference with the vitality of the original stock. If some of the new varieties supplant the old, that will not be because the unvaried stock is worn out or decrepit with age, but because in wild nature the newer forms are better adapted to the surroundings, or, under man's care, better adapted to his wants or fancies.

The second question, and one upon which the discussion about the wearing out of varieties generally turns, is : Will varieties propagated from buds, *i. e.*, by division, grafts, bulbs, tubers and the like, necessarily deteriorate and die out? First, Do they die out as a matter of fact? Upon this, the testimony has all along been conflicting. Andrew Knight was sure that they do, and there could hardly be a more trustworthy witness.

" The fact," he says, fifty years ago, " that certain varieties of some species of fruit which have been long cultivated cannot now be made to grow in the same soils and under the

same mode of management which was a century ago so perfectly successful, is placed beyond the reach of controversy.
Every experiment which seemed to afford the slightest prospect of success was tried by myself and others to propagate
the old varieties of the Apple and Pear which formerly con-
stituted the orchards of Herefordshire, without a single
healthy or efficient tree having been obtained ; and I believe
all attempts to propagate these varieties have, during some
years, wholly ceased to be made."

To this it was replied, in that and the next generation,
that cultivated vines have been transmitted by perpetual division from the time of the Romans, and that several of the
sorts, still prized and prolific, are well identified, among them
the ancient Græcula, considered to be the modern Corinth or
Currant grape, which has immemorially been seedless ; that
the old Nonpareil apple was known in the time of Queen
Elizabeth ; that the White Beurré pears of France have been
propagated from earliest times ; and that Golden pippins,
St. Michael pears, and others said to have run out, were still
to be had in good condition.

Coming down to the present year, a glance through the
proceedings of pomological societies and the debates of farmers' clubs, bring out the same difference of opinion. The
testimony is nearly equally divided. Perhaps the larger number speak of the deterioration and failure of particular old
sorts; but when the question turns on "wearing out," the
positive evidence of vigorous trees and sound fruits is most
telling. A little positive testimony outweighs a good deal of
negative. This cannot readily be explained away, while the
failure may be, by exhaustion of soil, incoming of disease, or
alteration of climate or circumstances. On the other hand
it may be urged, that, if a variety of this sort is fated to become decrepit and die out, it is not bound to die out all at
once or everywhere at the same time. It would be expected
first to give way wherever it was weakest, from whatever
cause. This consideration has an important bearing upon the
final question, Are old varieties of this kind on the way to die
out on account of their age or any inherent limit of vitality?

Here, again, Mr. Knight took an extreme view. In his essay in the "Philosophical Transactions," published in the year 1810, he propounded the theory, not merely of a natural limit to varieties from grafts and cuttings, but even that they would not survive the natural term of the life of the seedling trees from which they were originally taken. Whatever may have been his view of the natural term of the life of a tree, and of a cutting being merely a part of the individual that produced it, there is no doubt that he laid himself open to the effective replies which were made from all sides at the time, and have lost none of their force since. Weeping-Willows, Bread-fruits, Bananas, Sugar-cane, Tiger-lilies, Jerusalem Artichokes, and the like, have been propagated for a long time in this way, without evident decadence.

Moreover, the analogy upon which his hypothesis is founded will not hold. Whether or not one adopts the present writer's conception, that individuality is not actually reached or maintained in the vegetable world, it is clear enough that a common plant or tree is not an individual in the sense that a horse or man, or any one of the higher animals is — that it is an individual only in the sense that a branching zoöphyte or mass of coral is. *Solvitur crescendo:* the tree and the branch equally demonstrate that they are not individuals, by being divided with impunity and advantage, with no loss of life but much increase. It looks odd enough to see a writer like Mr. Sisley reproducing the old hypothesis in so bare a form as this: "I am prepared to maintain that varieties are individuals, and as they are born they must die, like other individuals." "We know that Oaks, Sequoias and other trees live several centuries, but how many we do not exactly know. But that they must die no one in his senses will dispute." Now what people in their senses do dispute is, not that the tree will die, but that other trees, established from cuttings of it, will die with it.

But does it follow from this that non-sexually propagated varieties are endowed with the same power of unlimited duration that are possessed by varieties and species propagated sexually — *i. e.* by seed? Those who think so jump too soon at their conclusion. For, as to the facts, it is not enough to point out the diseases or the trouble in the soil and in the atmosphere, to which certain old fruits are succumbing, nor to prove that a parasitic fungus (*Peronospora infestans*) is what is the matter with potatoes. For how else would constitutional debility, if such there be, more naturally manifest itself than in such increased liability or diminished resistance to such attacks? And if you say that, anyhow, such varieties do not die of old age, — meaning that each individual attacked does not die of old age, but of manifest disease, — it may be asked in return, What individual man ever dies of old age in any other sense than of a similar inability to resist invasions which in earlier years would have produced no noticeable effect? Aged people die of a slight cold or a slight accident, but the inevitable weakness that attends old age is what makes these slight attacks fatal.

Finally, there is a philosophical argument which tells strongly for some limitations of the duration of non-sexually-propagated forms, one that Knight probably never thought of, but which we should not have expected recent writers to overlook. When Mr. Darwin announced that the principle of cross-fertilization between the individuals of a species is the plan of nature, and is practically so universal that it fairly sustains his inference that no hermaphrodite species continually self-fertilized would continue to exist, he made it clear to all who apprehend and receive the principle, that a series of plants propagated by buds only must have a weaker hold of life than a series reproduced by seed. For the former is the closest kind of breeding. Upon this ground such varieties may be expected ultimately to die out ; but the mills of the gods grind so exceeding slow that we cannot say that any particular grist has been actually ground out under human observation.

If it be asked how the asserted principle is proved or made probable, we can here merely say that the proof is wholly inferential. But the inference is drawn from such a vast array of facts that it is wellnigh irresistible. It is the legitimate explanation of those arrangements in nature to secure cross-fertilization in the species, either constantly or occasionally, which are so general, so varied and diverse, and we may add so exquisite and wonderful, that, once propounded, we see that it must be true. What else, indeed, is the meaning and use of sexual reproduction? Not simply increase in numbers; for that is otherwise effectually provided for by budding propagation in plants and many of the lower animals. There are plants, indeed, of the lower sort, in which the whole multiplication takes place in this way, and with great rapidity. These also have sexual reproduction; but in it two old individuals are always destroyed to make a single new one! Here propagation diminishes the number of individuals fifty per cent. Who can suppose that such a costly process as this, and that all the exquisite arrangements for cross-fertilization in hermaphrodite plants, do not subserve some most important purpose? How and why the union of two organisms, or generally of two very minute portions of them, should reinforce vitality, we do not know and can hardly conjecture. But this must be the meaning of sexual reproduction.

The conclusion of the matter from the scientific point of view is, that sexually propagated varieties, or races, although liable to disappear through change, need not be expected to wear out, and there is no proof that they do; but that nonsexually propagated varieties, though not liable to change, may theoretically be expected to wear out, but to be a very long time about it.

ÆSTIVATION AND ITS TERMINOLOGY.[1]

THE term æstivation, to denote the arrangement of the parts of the calyx, corolla, etc., in the bud, as well as that of vernation for leaves in a leaf-bud, was introduced by Linnæus. He did not elaborate the former subject as he did the latter, and the few terms given to the modes he recognized are for the most part defined merely by a reference to their use in vernation. Æstivation as a botanical character is comparatively recent, and its terminology is not yet quite satisfactorily settled. I propose to consider, (1) what the leading modes are, and (2) how they are to be designated.

(1) In the first place, the modes of æstivation may be conveniently divided into two classes, those in which the parts overlap, and those in which they do not.

Of overlapping æstivation, only two principal kinds need be primarily distinguished, namely : 1. where some pieces overlap and others are overlapped, i. e., some have both margins exterior and others both margins interior or covered ; 2. where each piece of a circle is overlapped by its neighbor on one side while it overlaps its neighbor on the other. There are mixtures and subordinate modifications of these two, but no third mode.

In æstivation without overlapping, there is, first, the rare case in which the parts of the whorl or cycle never come into contact in the bud ; and, secondly, that in which they impinge by their edges only. There is also the case in which both margins of each piece are rolled or bent inward, and the rarer one in which they are turned outward ; and the apex of each piece may comport itself in any of these ways. But these dispositions are those of the pieces or leaves taken separately, and the terms applied to them are the same as in vernation or

[1] American Journal of Science and Arts, 3 ser., x. 339. (1875.)

prefoliation, are used in the same sense, and so are not at all peculiar to æstivation or prefloration. The like may be said of a remaining mode, which belongs, however, to a different category, that in which the parts being united into a tube or cup, this is bodily plaited into folds, or otherwise disposed; in which case the margin of the tube or cup, or such lobes as it may have, may exhibit any of the modes of æstivation above indicated.

Without further notice, then, of this last, the plicate or plaited æstivation, and of analogous conformations of the tube or cup of a calyx or corolla, or of the disposition of each piece individually (whether revolute, involute, reflexed, inflexed, and the like), about the terminology of which there is no question, — omitting, likewise, for the latter reason, the case of open æstivation, — there are left three types to deal with:

I. With some pieces of the set wholly exterior in the bud to others.

II. With each piece covered at one margin, and covering by the other.

III. With each piece squarely abutting against its neighbors on either side, without overlapping.

In modes II and III, the pieces are all on the same level and are to be viewed as members of a whorl. In mode I, although they may sometimes be members of a whorl, some parts of which have become external to others in the course of growth, they may, and in many cases must belong either to two or more successive whorls (as in the corolla of *Papaveraceæ*, and even the calyx of *Cruciferæ*, the upper or inner of course covered by the lower or outer), or to the spiral phyllotaxy of alternate leaves.

The type of the latter, and the common disposition when the parts are five, is with two pieces exterior, the third exterior by one edge and interior by the other, and two wholly interior. This is simply a cycle in $\frac{2}{5}$ phyllotaxy, the third piece being necessarily within and covered at one margin by the first, while it is exterior to and with its other margin covers the fifth, this and the fourth being of course wholly interior. So, likewise, when the parts are three, one exterior, one half exterior, and one interior or overlapped, the æstivation accords with ⅓ phyllotaxy. When of eight or higher numbers the spiral order is usually all the more manifest. When of four or six, the case is one of whorls (opposite leaves representing the simplest whorl), either of a pair of whorls (as in Epimedium, Berberis, etc.), or a single whorl, the parts of which have overlapped in cyclic order.

(2) As to the terminology. Linnæus in the " Philosophia Botanica " treats only of vernation, there termed " Foliatio." For this the former term was substituted, and that of æstivation for the disposition of petals in a flower-bud, introduced, as I suppose (not having the volume to consult), in the Termini Botanici, published in the sixth volume of the " Amœnitates Academicæ," 1762. I refer to it only through Giseke's edition, 1781. Here the terms are convoluta, imbricata, conduplicata, defined only by reference to the section vernatio, and valvata, unhappily explained by a reference to the glumes of Grasses, also " inæquivalvis ; si magnitudine discrepant." Imbricata is the only term besides valvata which directly relates to the arrangement of petals, etc., *inter se ;* and the reference takes us back to something " tectus, ut nudus non appareat," covered as with tiles, we may infer. In the " Philosophia Botanica," under the section Foliatio, the definition of imbricata is " quando parallele, superficie recta, sibi invicem incumbunt." This would apply either to mode I, or mode II, according as invicem is understood ; but the diagram (tab. x. 6) shows that case I is intended. Convoluta refers to the rolling of a petal or leaf by itself, as does conduplicata to its folding ; but Linnæus gives two figures, one of a single rolled-up leaf, the other of one leaf rolled up within another.

Finally, among the modes of vernation indicated by Linnæus, there is one which it is important here to notice, relating as it does to the arrangement of a pair of leaves in the bud, and evidently quite as applicable to a whorl of a larger number of parts than two ; *i. e.* :

" Obvoluta, quum margines alterni comprehendunt oppositi folii marginem rectum " (Philosophia Botanica, 105).

Or, in Termini Botanici, "pagina superiore lateribus approximatis ita ut alterum latus distinguat alterum folium." This, as the definition and the diagram in the "Philosophia Botanica" show, answers in æstivation to mode II. It was early taken up as such by Mirbel (Elem. Phys. Veg. et Bot., 1815, ii. 738, 739), where the polypetalous corolla of Hermannia and Oxalis and the gamopetalous corolla of *Apocyneæ* are cited as examples.

Valvate æstivation, our mode III, is rightly defined by Mirbel in the same place, and still earlier by Brown.

Linnæus made no use of æstivation as a character. Nor did Jussieu, except merely that, in his "Genera Plantarum," the petals of Malvaviscus are said to be convolute.

In De Candolle's "Théorié Elémentaire," 1813 — a still unsurpassed treatise, upon which, next to the "Philosophia Botanica," our botanical glossology rests — neither the word æstivation, nor its synonym, prefloration, is mentioned, and even vernation or prefoliation is equally omitted.

But the history of æstivation as a botanical character began in a work published three years earlier, namely, in R. Brown's "Prodromus Floræ Novæ Hollandiæ," 1810. The preface notes that it was first accurately observed by Grew. In it Brown defines only the valvate mode, "ubi margines foliolorum vel laciniarum integumenti invicem applicati sunt, capsulæ valvularum in modum." In the body of the work, wherever it is important, the æstivation is noted as valvate, imbricate, plicate, induplicate, etc.; and the open æstivation (aperta) is named by him in a subsequent paper.

Being the first to employ æstivation systematically, and to develop its value, Brown's terminology for its modes may well be considered authoritative. And so indeed it is, as far as it goes. But he did not make one important distinction, namely, that between our I and II. Imbricate, in his use, comprises all kinds of overlapping, that of the corolla of *Apocyneæ* and of a Gentian, as well as that of a Primrose. He must have not only noticed the difference, but also appreciated its general importance, notwithstanding the occasional passage of the one into the other. He must have also observed that in many cases, as in Asclepias for instance, the mode II passes into mode III, the valvate, and may possibly have discerned that under a phyllotaxic view these are more nearly related than either is to mode I. I find, however, only one instance in which he has indicated the distinction, namely, in the character of Burchellia, furnished to the "Botanical Register," t. 435, 1820. Of its corolla it is said: "æstivatione mutuo imbricata contorta." The phrase is interesting, as it seems to recognize the distinction between the mode of overlapping (which is that of our mode II) and the torsion, which only now and then accompanies it. Looking over the "Plantæ Javanicæ Rariores" to see if there is any later use, I find no instance in which Brown has occasion to speak of this mode II; but it occurs in the portion of his associate, Mr. Bennett, who (on p. 212) describes the petals of Sonerila as "æstivatione convoluta." Had this term been thus employed by Brown himself, and at an earlier date, I should regard the terminology of these three modes of æstivation as settled, namely: I, imbricata, II, convoluta, III, valvata. The first and the third are established beyond question, although somewhat remains to be said about the first.

But meanwhile another use has prevailed as respects the second. In De Candolle's "Prodromus," the first general or considerable work after Brown in which terms of æstivation are employed, this mode is almost uniformly characterized as contorta. I cannot at this moment trace the term to its origin. It was probably suggested by the name *Contortæ,* said to have been given by Linnæus to the Apocyneous natural order; and it seemed appropriate to the instances in which the strong convolution of rounded petals, as in Oxalis, or their lobes, as in Phlox, give an appearance like that of twisting, although there is no twist or torsion. But it is to just such cases, in which there is most of seeming twisting on account of the strong convolution, that the term convolute is now and then assigned in the "Prodromus"; as in the character of *Byttneriaceæ,* and that of Malvaviscus. The latter may perhaps be explained by the peculiarity that the petals do not uncoil in anthesis. But in *Apocynaceæ,* in the "Prodromus," the terms convoluta and contorta are seemingly employed synonymously, or nearly so (the latter most frequently); at least I see no difference between the æstivation of Allamanda, said to be contorted, and that of Vinca (rosea), said to be convolute. Endlicher in this regard follows the "Prodromus." In the new "Genera Plantarum" by Bentham and Hooker this mode is most commonly designated as contorta, sometimes as contorto-imbricata, rarely (Philadelphus, etc.) convoluta. I have myself, from a period as early as 1840, employed the term convolute, thinking it unadvisable to have two names for the same thing, and wishing to restrict, if it might be, the term contorted to cases of torsion. Adrien de Jussieu, on the other hand, used convolute (with strict Linnæan propriety) for regular imbrication with a high degree of overlapping, thus giving two names to different degrees of the same thing.

It being conceded, I presume, that the mode II should be specifically distinguished, what name, on the whole, ought it to bear? If we follow prevalent usage, contorta will be the term. But this term was unknown in this sense to the founders of æstivation, Linnæus and Brown; it correctly expresses the real state of things in only a few cases; and where there is torsion, it leads to a most awkward way of expressing it. We have to write — "lobes of the corolla contorted and twisted: corollæ lobi contorti et torti," introducing dextrorsum or sinistrorsum,[1] to express the direction of the overlapping and of the torsion, which are not always the same. So the most current name is the least appropriate. Convoluta is as good a name as can be, and its use in the present sense is not unconformable with the Linnæan use in vernation. When well carried out, three or five or more petals, as the case may be, are simply rolled up together. When the overlapping is slight, there is simply the tendency to convolution. But if, as in other nomenclature, priority gives a paramount

[1] I note with satisfaction that Bentham and Hooker use these terms to signify from left to right, or from right to left, of a person, supposed to stand outside of the closed bud, which is surely the natural position of the observer.

claim, obvoluta will be the proper term, beginning as it did with Linnæus for vernation, and taken up, as it was very early, by Mirbel for æstivation. The only objections to it are, first, that it has never come into systematic use, and, second, that *ob*, in the composition of botanical terms, commonly stands for obversely or inversely. But obvoluta is not burdened with this signification : it is classical for "wrapped round," as is convoluta for rolled together. I conclude that one or the other of these two terms ought to be used.

Finally, although there is little, if any, practical misuse, there is some mis-definition, of the term imbricate as applied to æstivation. Adrien de Jussieu defines it well (in Cours Élémentaire, 308) in the phrase "La préfloraison spirale est aussi nommé imbriquée " ; and in noting that when the number stops at five, the pieces fall into two exterior, two interior, and one (the third in the spiral) intermediate, this making what is called " æstivatio quincuncialis." [1]   This is clear and to the point. But other authors have had a fancy for distinguishing between quincuncial and imbricate (as if the former were not the typical case of the latter when the parts are five), and so have had to devise something else to answer to imbricate. Alphonse De Candolle (in his Introd. Bot., i. 154, written before phyllotaxy was well understood), after relegating imbricative to the category of a crowd of verticils, and remarking that the quincuncial is sometimes confounded with the imbricate, adds that some confound also under this latter name the case in which there is one exterior piece, one interior, and three covered at one margin but free at the other. I know not where this began ; but its latest reproduction is in Le Maout and Decaisne's " Traité Générale " and in the English translation of it. In the diagram the pieces are numbered directly round the circle from 1 to 5, the fifth coming next the first : " so they thus complete one turn of a spiral," — which shows that Le Maout had vague ideas of phyllotaxy, of

---

[1] The name quincuncial answers the purpose after definition, and has long been in use ; but this arrangement in diagram is wholly unlike the quincunx, with its four pieces or stars in the periphery, or at the angles of a square, and one in the centre.

which he seems to have invented a new ($\frac{1}{5}$) order. Moreover, this is essentially identical with the "cochlear" æstivation of the same work (not of Lindley) ; and Eichler, in his "Blüthendiagramme," adopts this name (unsuitable though it be) for this particular arrangement, whatever be the position of the inclosed or inclosing petal. A glance shows that this supposed "true imbricate æstivation" is a slight and not very uncommon deviation (by the displacement of what should be the interior margin of one of the petals during growth) of the mode II, variously termed obvolute, convolute, or contorted æstivation. But it is so intermediate between this and the quincuncially imbricate as perhaps to justify Brown in applying the name imbricate generically to all the overlapping modes. I see, since the above was written, that Eichler, in his "Blüthendiagramme," in effect does this. I find also, that Eichler uniformly employs the term convolute, or convolutive, as I have done, instead of contorted. I should hope, rather than immediately expect, that this use would become general.

A PILGRIMAGE TO TORREYA.[1]

ORDERED to go south until I should meet the tardy spring and summer, I was expected to follow the beaten track to east Florida. But I wished rather to avoid the crowd of invalids and pleasure travelers, and turned my attention in preference to western Florida, determined that, if possible, I would make a pious pilgrimage to the secluded native haunts of that rarest of trees, the *Torreya taxifolia*.

All that I knew, or could at the moment learn, was, that this peculiar evergreen Yew-like tree — prized by arboriculturists for its elegance, and dear to us botanists for the name it bears and commemorates — grew on the banks of the Apalachicola River, somewhere near the confluence of the Flint and Chattahoochee, which by their union form it. It was there discovered, nearly forty years ago, by Mr. Hardy B. Croom, and had since been seen, at two or three stations, by his surviving associate Dr. Chapman, of Apalachicola, author of the Southern Flora. Mr. Croom, upon ascertaining that he was the fortunate discoverer of an entirely new type of coniferous trees, desired that it should bear Dr. Torrey's name; and the genus Torreya was accordingly so named and characterized by the Scotch botanist Arnott. It is of the Yew family, in foliage and in male flowers much resembling the Yew itself, but more graceful than the European Yew-tree, wholly destitute of the berry-like cup which characterizes the latter genus, and with the naked seed itself fleshly-coated, and larger than an olive, which it resembles in shape and appearance. One young tree, brought or sent by Mr. Croom himself, has been kept alive at New York — showing its aptitude for a colder climate than that of which it is a native — and has been more or less multiplied by cuttings.[2] Sprigs from this tree

[1] "American Agriculturist," 1875, 262.
[2] "The American Agriculturist" for May states that the tree spoken

or its progeny were appropriately borne by the members of
the Torrey Botanical Club, at its founder's funeral, two years
ago, and laid upon his coffin. But very few botanists have
ever seen the tree growing wild and in its full development.
I was desirous to be one of the number.

Among the broad, black lines with which the railway map
is chequered, I found one which terminates at Chattahoochee.
This was the objective point, and the way to it seemed plain
enough, though long. Pilgrimages to famous shrines by railway, in the Old World, are nowadays systematized and made
easy. The untried one which I undertook appeared to offer no
privation or difficulty, except the uncertainty whether I should
be fortunate enough to find the grove which I sought. And,
indeed, there was little privation to speak of. It was, however, rather trying to us (*i. e.*, to myself and my companion
in travel and life), when, after leaving Savannah on an early
April morning, with the assured understanding that we should
reach Chattahoochee late that evening, we learned that we were
to be left for twenty hours at a small hamlet on the borders
of east Florida, named Live Oak, a manifest *lucus a non
lucendo*, as there were no Live Oak-trees in the neighborhood, but a prevalent growth of Long-leaved Pines. There
was some good botanizing to console us, and, thanks to the
railroad conductor for directing us aright, unpretending but
truly comfortable quarters for the night. Then, the next day,
resuming our journey after a twelve-o'clock dinner, which we
were to mend with a supper at Tallahassee, we were at length
informed that we were to be supperless; that the stations
both of Tallahassee and Quincy were out of town and out of
reach of all edibles; that Chattahoochee station, to be reached
after ten o'clock, was only a freight-house on the wild and
wooded bank of the river, built upon piles in the swamp,

of, or its seed, "was brought from Florida by the late distinguished Major
Le Conte." I am confident that this is a mistake, and that Le Conte knew
nothing of this tree in its native station. If my recollection is correct, at
least two seedling trees were placed in Dr. Torrey's hands by Mr. Croom,
one of which was consigned to A. J. Downing, of Newburgh, the ultimate
fate of which is unknown to me, the other to Mr. Hogg, senior, which, as
"The American Agriculturist" states, is now in Central Park.

reached at ordinary times over a mile of trestles, and now so overflowed that it probably could not be reached at all, certainly not that night; that the train would stop for the night two or three miles back in the woods, where the agent had taken up his abode in a box-car; that the town of Chattahoochee, a mile away, large as it appeared on the map, consisted mainly of a state-prison, and a couple of grocery shops, neither of which was quite proper for passing a night in, even if we could reach it; in fine, that our only course would be to sleep in the car (which made no provision for it), and crave from the agent of the road a share of his breakfast.

The kind and intelligent fellow-travelers as far as Tallahassee and Quincy, who gave us this disheartening information, finding that we were not disposed to stop short of our object, remarked that they had set us down as eminently philosophical people, since we had passed a night at Live Oak and still possessed our souls in patience (a view which a couple who had stopped at the hotel there practically confirmed), and so left us with their good wishes, but evidently faint hopes. The weekly steamboat, which was to call at the landing next day, would eventually relieve us; and so we resolved to make the best of it. The worthy young conductor, who was to sleep in the car also, kindly proffered a share of his supper; but we fortunately had a bottle of cold tea, some crusts of bread ten days old, and wafer-biscuits, upon which we scantily supped, and then, folding around us such drapery and wraps as we had, lay down to sleep upon the couches which the conductor ingeniously arranged for us, by some skillful adjustment of the car-seats. In the morning, after due ablutions made at the tank of the locomotive, we were hospitably welcomed by the agent, General Dickison, and his son, to a much-needed share of their breakfast in the stationary box-car, which served both as bedroom, parlor, and dining-room.

To our great delight we found that General Dickison knew the tree which I was in search of; and it was arranged that his son should conduct me to the locality, not far distant. So striking an evergreen tree could not fail of notice. The people

of the district knew it by the name of " Stinking Cedar " or
"Savine " — the unsavory adjective referring to a peculiar
unpleasant smell which the wounded bark exhales. The timber is valued for fence-posts and the like, and is said to be
as durable as red cedar. I may add that, in consequence of
the stir we made about it, the people are learning to call it
Torreya. They are proud of having a tree which, as they
have rightly been told, grows nowhere else in the world.

My desire for a sight of it was soon gratified. Making our
way into the woods north of the railroad track, along the
ridges covered with a mixed growth of Pines and deciduous
trees, I soon discerned a thrifty young Torreya, and afterwards several of larger size, some of them with male flowers
just developed.

As we approached the first one, I told my companion that
I expected to find under its shade a peculiar low herb, which
I described, but had never yet seen growing wild. And there,
indeed, it was, — greatly to the wonderment of my companion
— the botanically curious little *Croomia pauciflora*, just as it
was found by Mr. Croom, when he also discovered the tree,
nearly forty years ago, probably at a station several miles
farther south. I was a pupil and assistant of the lamented
Torrey when Mr. Croom brought to him specimens, both of
the tree and of the herb, both new genera. The former, as I
have stated, was named for Dr. Torrey by his correspondent
Arnott. The latter was dedicated to its discoverer by Dr.
Torrey. I well remember Mr. Croom's remark upon the occasion, that if his name was deemed worthy of botanical
honors, it was gratifying to him, and becoming to the circumstances, that it should be borne by the unpretending herb
which delighted to shelter itself under the noble Torreya. It
is not, as Mr. Croom then supposed, exclusively so found ;
for it grows also in the central and upper portions of Alabama
and Georgia, where Torreya is unknown, but where I fancy
it may once have flourished. I cannot here detail the reasons
for this supposition.

There is a second Torreya in Japan, founded on Thunberg's
Taxus nucifera, of which I saw original specimens at the

British Museum, in the winter of 1838–9, and then identified the genus. There is likewise in Japan a second Croomia, very probably in company with the Torreya. A third Torreya inhabits California, but it has no associate Croomia. I have formerly treated of the peculiar distribution of these genera and species between the United States and Japan, have collocated a large number of equally striking similar instances, and have offered certain speculations in explanation of them. The views maintained have been more and more confirmed, and are now adopted by the leading philosophical botanists.

The few hours devoted to this first search for Torreya, pleasant as they were, yet were too scantily rewarded to satiate my interest. I saw no tree with trunk over six inches in diameter, and found no female blossoms. It was necessary to hasten back to the railway car, to await the expected summons to the steamboat. I bore with me, besides my botanical specimens, a stick of Torreya, suitable for a staff, which I propose to make over to the President of the Torrey Botanical Club, for the official baton. Before long the whistle of the steamboat announced its approach to the landing, and offered us a prospect of a much-needed dinner ; the water had fallen sufficiently to allow us to be conveyed to the wharf upon a hand-car, and so we embarked for Apalachicola via Bainbridge. That is, we went up the Flint River about forty miles and thence back in the night, past the place of embarkation.

I will not here give any account of a delightful ten days' episode, beginning with the voyage down the brimming river, bordered with almost unbroken green of every tint, from the dark background of Long-leaved Pines to the tender new verdure of the Liquidambar and other deciduous trees in their freshest development, interspersed with the deep and lustrous hue of *Magnolia grandiflora*, and, when the banks were low, dominated by weird, naked trunks of Southern Cypress (Taxodium), their branches hung with long tufts and streamers of the gray and sombre Southern Moss (Tillandsia) below, while above they were just putting forth their delicate foliage.

Along the lower part of the river, occasional Palmettoes gave a still more tropical aspect. Then followed a week and more at dead and dilapidated, but still charming, Apalachicola, where the post-office opens on Monday evenings, when the steamboat arrives, and closes for a week the next morning, when she departs, — where the climate, thanks to the embracing Gulf, is as delicious in summer as it is bland in winter ; where game, the best of fish, and the most luscious oysters are to be had almost for nothing, and blackberries come early in April when the oranges are gone; and where, far from the crowd and bustle of the world, with Bill Fuller for caterer, and his wife Adeline for cook, the choicest fare is to be enjoyed at the cheapest rate. Then there was the pleasure of renewing our acquaintance with Dr. Chapman, and botanizing with him over some of the ground which he has explored so long and so well, of gathering, under his guidance, the stately *Sarracenia Drummondii* in its native habitat, and, not least, acquiring from him fuller information respecting the localities where Torreya grows.

The return voyage up the river was not less enjoyable than the descent. It was so timed that the bold bluff of Aspalaga, where the tree was first found, was reached after sunrise. But it was sad to see that the Torreya trees, which overhung the river here in former days, had been cut away, perhaps for steamboat fuel. So I did not land ; but leaving the boat a few miles above, at the upper Chattahoochee landing, while it made the run to Bainbridge and back, I had a long day to devote to Torreya. Following Dr. Chapman's directions, I repaired to the wooded bluff to the north of the road, where I soon found an abundance of the trees, of various ages, interspersed among other growth. The largest tree I saw grew near the bottom of a deep ravine ; its trunk just above the base measured almost four feet in circumference, and was proportionally tall. But it was dominated by the noblest *Magnolia grandiflora* I ever set eyes on, with trunk seven and a half feet in girth.

After long search one tree was found with female flowers, or rather with forming fruit, from which a few specimens were gathered. Seedlings and young trees are not uncommon, and some old stumps were sprouting from the base, in the manner of the Californian Redwood. So this species may be expected to endure, unless these bluffs should be wantonly disforested — against which their distance from the river and the steepness of the ground offer some protection. But any species of very restricted range may be said to hold its existence by a precarious tenure. The known range of this species is not more than a dozen miles in length along these bluffs, although Dr. Chapman has heard of its growing further south, where the bluff trends away from the river. At least the Yew-tree grows there, which Mr. Croom found with the Torreya near Aspalaga, and I heard of it (identifying it by the description) as growing five or six miles away.

Returning to the boat at nightfall, I brought with me thirty or forty seedling Torreyas, which, being too far advanced to be safely sent far north this spring, have been successfully consigned to the excellent Mr. Berckman's care, at Augusta, Georgia. I hope that one or more of them may in due time be planted upon the grave of Torrey.

A word or two of Mr. Croom and his sad fate. His name merely is known to botanists as the discoverer of *Torreya taxifolia* and of *Croomia pauciflora*, and as the author of a monograph of Sarracenia, in which the handsomest species, *S. Drummondii*, was originally described and figured. He was the first, after Chapman in 1836, to find this in blossom, Drummond having seen and collected the leaves only, in a winter visit to Apalachicola. Of the botanists who remember and personally knew him, only Dr. Chapman and myself survive. Mr. Croom, originally, I believe, of Newbern, Lenoir County, North Carolina, had a plantation at Quincy, Florida, and another at Mariana, opposite Aspalaga; and it was in passing from one to the other that he discovered the tree of which I have been discoursing, as well as the herbaceous plant which bears his name. He was an accomplished and most amiable young man, full of enterprise and zeal for botany, and much was expected from him. But, just as he was entering upon his chosen field, and had made prepara-

tions for a thorough exploration of Florida, in connection
with his friend Dr. Chapman, he was lost at sea, with his
wife and all his children, in the foundering of the ill-fated
" Home," between New York and Charleston.

I have been told that two seedling Torreyas which Mr.
Croom planted near his house at Quincy, and which had
become stately trees, have recently been demolished by the
present proprietor; also that a tree of Mr. Croom's planting still flourishes in the grounds of the state-house at Tallahassee.

NOTES ON THE HISTORY OF *HELIANTHUS TUBE-ROSUS*, THE SO-CALLED JERUSALEM ARTICHOKE.

UNDER this heading the botanical editor of the Journal [1] proposes to offer a few explanatory remarks, introductory to the subjoined letter which he received from Mr. Trumbull in answer to a recent inquiry.

Linnæus, in the "Species Plantarum," gave to *Helianthus tuberosus* the "habitat in Brasilia." In his earlier "Hortus Cliffortianus" the habitat assigned was "Canada." M. Alphonse De Candolle, in his "Géographie Botanique," ii. 824 (1855), refers to this as "decidedly an error, at least as to Canada properly so called," assigns good reasons for the opinion that it did not come from Brazil, nor from Peru (to which the name under which it appeared in cultivation in the Farnese garden seemed to refer), but in all probability from Mexico or the United States. He adds that Humboldt did not meet with it in any of the Spanish colonies.

About this time I received from my friend and correspondent, the late Dr. Short of Kentucky, some long and narrow tubers of *Helianthus doronicoides*, Lam., with the statement that he and some of his neighbors found them good food for hogs, and, if I rightly remember, had planted them for that purpose. They were planted here in the Botanic Garden; after two or three years it was found that some of the tubers produced were thicker and shorter ; some of these were cooked along with Jerusalem artichokes, and found to resemble them in flavor, although coarser. Consequently, in the second edition of my "Manual of the Botany of the Northern United States" (1856), it is stated that *H. doronicoides* is most probably the original of *H. tuberosus*. This opinion was

---

[1] American Journal of Science and Arts, 3 ser., xiii. 347. (1877.)

strengthened year after year by the behavior of the tubers, and by the close similarity of the herbage and flowers of the two plants, as they grew side by side; indeed, as the two patches were allowed to run together in a waste or neglected place, they have become in a measure confounded. Wishing to obtain an unmixed stock, I applied last autumn to Professor J. M. Coulter of Hanover, Indiana, and received from him a good number of tubers from wild plants of the neighborhood, which will now be grown. Some of these were slender, some thicker and shorter, and a few were to all appearance identical with Jerusalem artichokes. If they were really all from one stock, as there is reason to believe, the question of the origin of *Helianthus tuberosus* is wellnigh settled.

We were now interested to know whether our Indians, at least those of the Mississippi Valley, where *H. doronicoides* belongs, were known to cultivate these tubers or to use them for food. Recently a note in the " American Agriculturist " called attention to a sentence in Dr. Palfrey's " History of New England," i. 27, stating that the Indians of that region raised, among other articles of food, " a species of Sunflower, whose esculent tuberous root resembled the artichoke in taste." The venerable historian found himself at the moment unable to refer me to the sources of this statement; but as it was now certain that some record of the kind existed, I applied to Mr. Trumbull, who obligingly and promptly supplied the information required, and placed it at my disposal in the following letter.

HARTFORD, CONN., March 26, 1877.

MY DEAR PROFESSOR GRAY: I cannot refer you to the authority (*totidem verbis*) for Dr. Palfrey's statement that the Indians of New England cultivated " a species of Sunflower, whose esculent tuberous root resembled the artichoke in taste," but there can be, I think, little doubt of the fact. The historical evidence that " artischoki sub terrâ " were cultivated in Canada and in some parts of New England before the coming of Europeans is tolerably clear. The only question, if there be any, is as to species; and this does not appear to have been raised for more than half a century after the

"Jerusalem artichoke" was known to English and Continental botanists.

I can discover no authority whatever, before 1700, for ascribing to the *Helianthus tuberosus*, either a Brazilian or a Mexican origin, except — and the exception is unimportant — in C. Bauhin's identification (in his Pinax, 277) with "Helianthemum Indicum tuberosum" (*H. tuberosus*, L.), of a plant that he had described in his earlier "Prodromus" (ed. 1671, p. 70) as "*Chrysanthemon latifolium Brasilianum*," from a dried specimen sent to him "eo nomine" from the garden of Contarini.

The first trace I find of this species, in Europe, is in the 2d part (cap. 6) of Fabio Colonna's "Ecphrasis minus cognitarum stirpium," published at Rome in 1616. He described it from a plant growing in the garden of Cardinal Farnese. The Sunflower was already well known to European botanists, and had been described and figured by Dodoens (1563) and Lobel (1576) as *Chrysanthemum Peruvianum* and *Flos solis Peruvianus.* With reference to these descriptions, probably, Colonna gave the new species the name of *Aster Peruanus, tuberosâ radice*, otherwise *Solis flos Farnesianus.* (He gave a more particular description of the plant in his annotations to Recchi's Hernandez, "Plant. Mexic. Hist.," 1651, pp. 878, 881, as *Peruanus Solis flos ex Indiis tuberosus.*)

The author of the "Descriptio variorum plantarum, in Horto Farnesiano," published under the name of Tobias Aldinus (Rome, 1625), gave some account of the roots, which he calls "Tubera Indica," of the *Solis flos tuberosus, seu Flos Farnesianus Fabii Columnæ* (p. 91). It may be observed that several of the rarer plants in the Farnese garden, at this time, were from "Canada" and "Virginia." The Passion Flower (admirably figured by Aldinus) is described under its Virginian name, "Maracot" (the "Maracocks" of John Smith and Strachey); and a *Campanula Americana* is otherwise named "*Campanula Virginiana, seu ex Virginiis insulis.*"

C. Bauhin, in his "Pinax" (first published in 1623), ed.

1671, p. 276, notes that the "*Helianthemum Indicum tube-rosum*" is called "Chrysanthemum è Canada, quibusdam. Canada et Artischoki sub terrâ, aliis. Gigantea, Burgundis." P. Laurenberg, "Apparat. plant." (Rostock, 1632), names the species "Adenes Canadenses or Flos Solis glandulosus." Ant. Vallot, "Hortus Regis, Paris," 1665 (as cited by Bauhin), gives the names "Canada and Artischoki sub terrâ," and "Canadas," and describes also "*Helenium Canadense altis-simum, Vosacan dictum*," which Tournefort distinguishes as "Corona Solis rapunculi radice" (Inst. Herb. 490), and which became *H. strumosus,* L. "Vosacan," by the way, is a French fashion of writing the Algonkin word "wassakone" or "was-sakwân," which means a "bright yellow flower." The modern Chippeways give this name to the flowers of the Pumpkin and Squash.

Under whatever name the Jerusalem Artichoke was described, there seems to have been a general agreement among European botanists that it came from Canada. F. Schuyl, "Catal. Horti Lugd. Bat." (Heidelberg, 1672), varies the specific name to "*Chrysanthemum Canadense Arumosum.*" P. Amman, "Charac. Plant. Nat." (1676), has "*Helenium Canadense.*"

It was introduced to England about 1617. In that year, Mr. John Goodyer, of Maple Durham, Hampshire, "received two small roots thereof, from Mr. Franquevill of London," which were planted, and enabled him, before 1621, to "store Hampshire." He wrote an account of the plant, under date of October 17, 1621, for T. Johnson, — who printed it in his edition of Gerard, 1636 (p. 753). Before this the species had been figured and described by J. Parkinson, in "Para-disus Terrestris" (London, 1629), as "Battatas de Canada," and in his "Theatre of Plants," 1640 (p. 1383), he has the figure — a good one — without the description, under the names "Battatas de Canada, the French Battatas, or Hierusalem Artichoke." Johnson, in Gerard (p. 753), refers to Parkin-son's description, and gives the name as "Flos Solis Pyrami-dalis, Jerusalem Artichoke." It already grew "well and plentifully in many parts of England."

The notices by early voyagers, of ground-nuts, eaten by the Indians, are generally so brief and so vague, that it is not easy to distinguish the three or four species mentioned under that name or its equivalents. The *Solanum tuberosum*, *Apios tuberosa*, *Aralia trifolia*, and a *Cyperus* (*articulatus?*) were all "ground-nuts," or "earth-nuts." We find, however, in a few instances, unmistakable mention of the roots already known in Europe as "Canadian."

Brereton, in his account of Gosnold's voyage to New England in 1602, notes the "great store of ground-nuts" found on all the Elizabeth Islands. They grow "forty together on a string, some of them as big as a hen's egg" (Purchas, iv. 1651). These, doubtless, were the roots of *Apios tuberosa*. But when Champlain, a few years later (1605–6), was in the same region, he observed that the Almouchiquois Indians near Point Mallebarre (Nauset harbor, probably,) had "force des racines qu'els cultivent, lesquelles ont le goût d'artichaut" (Voyages, ed. 1632, p. 84). And it is to these roots, evidently, that Lescarbot alludes, "Histoire de la Nouv. France," 1612 (p. 840): there is, he says, in the country of the Armouchiquois (*i. e.*, New England, west and south of Maine), a certain kind of roots "grosses comme naveaux, très excellentes a manger, ayans un goût retirant aux cardes, mais plus agréable, lesquelles plantées multiplient en telle façon que c'est merveille;" and he thinks these must be the "Afrodilles" described by Pliny.

Sagard-Theodat (Hist. du Canada, 1636, p. 785) mentions the cultivation of the Sunflower by the Hurons — who extracted oil from its seeds, — and names also the "roots that we [the French] call Canadiennes or Pommes de Canada, and that the Hurons call ' Orasqueinta,' which are not very (*assez peu*) common in their country. They eat them raw, as well as cooked, as they eat another sort of root resembling parsnips [*Sium lineare?*], which they call Sondhratates, and which are much better; but they seldom gave us these, and only when they received some present from us or when we visited them in their cabins." He goes on to speak of "patates, fort grosses et très-excellentes," some of which he had

obtained from an English vessel captured by the French; but
none of these were to be found in the Huron country, nor
could the Indians tell him the name of them; and he regretted
that he had not brought some with him, for planting, since
" this root, being cut in pieces and planted, quickly grows
and multiplies, it is said, like the pommes de Canada " (pp.
781, 782). It is plain that the Huron roots first mentioned
were, or that Sagard believed them to be, " Jerusalem Artichokes," — already known as " Canadian."

I find no mention of the artichoke in Virginia or the
southern colonies before it was cultivated by Anglo-Ameri-
cans. The author of " A Perfect Description of Virginia,"
printed in 1649, says that the English planters have (*inter
alia*) " roots of several kindes, Potatoes, Sparagus, Carrets,
. . . and Hartichokes." Beverly (Hist. of Virginia, 1722,
p. 254) mentions " Batatas Canadensis, or Jerusalem Artichoke," as planted by some of the English, for brewing beer.
Yet, the name of one of the esculent roots mentioned by Hariot (Brief and True Report, etc., 1585) ought to belong to
some species of Sun Flower — and if to any, to *H. tuberosus.* Hariot names three tuberous roots found in Virginia:
" Openauk, a kind of roots of round form, some of the bignes
of walnuts, some far greater, which are found in moist and
marish grounds growing many together one by another in
ropes, or as though they were fastened with a string. Being
boiled or sodden, they are very good meate." [C. Bauhin
(Prodromus, 89) identifies these with *Solanum tuberosum
esculentum,* — and has been followed by later writers. The
description seems to me to indicate *Apios tuberosa.*] " Kais-
hucpenauk, a white kind of roots about the bignes of hen egs
and nere of that forme: their taste was not so good to our
seeming as of the other, and therefore their place and manner of growing not so much cared for by vs: the inhabitants
notwithstanding vsed to boile and eat many." These may
be " Virginia potatoes," but their name, if Hariot recorded it
correctly, means " Sun-tubers." The etymology is perfectly
clear. The other roots described by Hariot, " Okeepenauk
are also of round shape, found in dry grounds: some are of

the bignes of a man's head," etc.   These must be the "Tubera terræ maxima," of Clayton, "vulgo Tuckahoo," which Gronovius (Fl. Virgin. 205) refers to *Lycoperdon solidum*, L., and for which Rafinesque (Med. Fl. ii. 270) proposed a new genus Tucahus.   Kalm describes them (Travels, i. 225) as "Truffles."   Fries (El. Fung. ii. 39) assigns them to his *Pachyma cocos*.

Writing in haste and with frequent interruptions, it has been possible to do little more than copy, without condensing or arranging, such notes as I had before me.   They have extended to such a length that I must not add even an apology for the superfluous matter.   Yours truly,

J. H. TRUMBULL.

It would be interesting to know whence came the French name of these Helianthus tubers, " Topinambour," it being the only thing in the case which, as Mr. Trumbull remarks, " looks to a Brazilian origin, as it seems to be derived (and so Littré gives it) from the Topinamboux Indians of Brazil." The English name, " Jerusalem Artichoke," comes, as is well known, from the Italian " Girasola," *i. e.* Sunflower.

As to the annual Sunflower, or *Helianthus annuus*, said by Linnæus to come from Peru and Mexico, I have for some years been convinced that its original is the *H. lenticularis* of Douglas, which again is probably only a larger form of *H. petiolaris* of Nuttall, natives of the western part of the Mississippi Valley and of the plains to and beyond the Rocky Mountains.   It is an interesting confirmation of this opinion that Sagard (as mentioned in the above communication) and Champlain found this Sunflower in cultivation by the Huron Indians, for the sake of the oil of its seeds, which they used for hair-oil.

FOREST GEOGRAPHY AND ARCHÆOLOGY.[1]

IT is the forests of the Northern temperate zone which we are to traverse. After taking some note of them in their present condition and relations, we may inquire into their pedigree; and, from a consideration of what and where the component trees have been in days of old, derive some probable explanation of peculiarities which otherwise seem inexplicable and strange.

In speaking of our forests in their present condition, I mean not exactly as they are to-day, but as they were before civilized man had materially interfered with them. In the district we inhabit such interference is so recent that we have little difficulty in conceiving the conditions which here prevailed, a few generations ago, when the "forest primeval" — described in the first lines of a familiar poem — covered essentially the whole country, from the Gulf of St. Lawrence and Canada to Florida and Texas, from the Atlantic to beyond the Mississippi. This, our Atlantic forest, is one of the largest and almost the richest of the temperate forests of the world. That is, it comprises a greater diversity of species than any other, except one.

In crossing the country from the Atlantic westward, we leave this forest behind us when we pass the western borders of those organized States which lie along the right bank of the Mississippi. We exchange it for prairies and open plains, wooded only along the watercourses, — plains which grow more and more bare and less green as we proceed westward, with only some scattering Cottonwoods (i. e. Poplars) on the immediate banks of the traversing rivers, which are themselves far between.

[1] A lecture delivered before the Harvard University Natural History Society, April 18, 1878. (American Journal of Science and Arts, 3 ser., xvi. 85, 183.)

In the Rocky Mountains we come again to forest, but only in narrow lines or patches ; and if you travel by the Pacific Railroad you hardly come to any : the eastern and the interior-desert plains meet along the comparatively low level of the divide which here is so opportune for the railway; but both north and south of this line the mountains themselves are fairly wooded. Beyond, through all the wide interior basin, and also north and south of it, the numerous mountain chains seem to be as bare as the alkaline plains they traverse, mostly north and south; and the plains bear nothing taller than sage-brush. But those who reach and climb these mountains find that their ravines and higher recesses nourish no small amount of timber, though the trees themselves are mostly small and always low.

When the western rim of this great basin is reached there is an abrupt change of scene. This rim is formed of the Sierra Nevada. Even its eastern slopes are forest-clad in great measure, while the western bear in some respects the noblest and most remarkable forest of the world, — remarkable even for the number of species of evergreen trees occupying a comparatively narrow area, but especially for their wonderful development in size and altitude. Whatever may be claimed for individual Eucalyptus-trees in certain sheltered ravines of the southern part of Australia, it is probable that there is no forest to be compared for grandeur with that which stretches, essentially unbroken, — though often narrowed, and nowhere very wide, — from the southern part of the Sierra Nevada in lat. 36° to Puget Sound beyond lat. 49°, and not a little farther.

Descending into the long valley of California, the forest changes, dwindles, and mainly disappears. In the Pacific coast ranges it resumes its sway, with altered features, some of them not less magnificent and of greater beauty. The Redwoods of the coast, for instance, are little less gigantic than the Big Trees of the Sierra Nevada, and far handsomer, and a thousand times more numerous. And several species, which are merely or mainly shrubs in the drier Sierra, become lordly trees in the moister air of the northerly coast ranges. Through most of California these two Pacific forests are separate; in the northern part of that State they join, and form one rich woodland belt, skirting the Pacific, backed by the Cascade Mountains, and extending through British Columbia into our Alaskan territory.

So we have two forest regions in North America, — an Atlantic and a Pacific. They may take these names, for they are dependent upon the oceans which they respectively border. Also we have an intermediate isolated region or isolated lines of forest, flanked on both sides by bare and arid plains, — plains which on the eastern side may partly be called prairies, on the western, deserts.

This mid-region mountain forest is intersected by a transverse belt of arid and alkaline plateau, or eastward of grassy plain — a hundred miles wide from north to south, — through which passes the Union Pacific Railroad. This divides the Rocky Mountain forest into a southern and a northern portion. The southern is completely isolated. The northern, in a cooler and less arid region, is larger, broader, more diffused. Trending westward, on and beyond the northern boundary of the United States, it approaches, and here and there unites with, the Pacific forest. Eastward, in northern British territory, it makes a narrow junction with northwestward prolongations of the broad Atlantic forest.

So much for these forests as a whole, their position, their limits. Before we glance at their distinguishing features and component trees, I should here answer the question, why they occupy the positions they do; — why so curtailed and separated at the south, so much more diffused at the north, but still so strongly divided into eastern and western. Yet I must not consume time with the rudiments of physical geography and meteorology. It goes without saying that trees are nourished by moisture. They starve with dryness and they starve with cold. A tree is a sensitive thing. With its great spread of foliage, its vast amount of surface which it cannot diminish or change, except by losing that whereby it lives, it is completely and helplessly exposed to every atmospheric change; or at least its resources for adaptation are very limited; and it cannot flee for shelter. But trees are social, and their gregarious habits give a certain mutual support. A tree by itself is doomed, where a forest, once established, is comparatively secure.

Trees vary as widely as do other plants in their constitution; but none can withstand a certain amount of cold and other exposure, nor make head against a certain shortness of summer. Our high northern regions are therefore treeless; and so are the summits of high mountains in lower latitudes. As we ascend them we walk at first under Spruces and Fir-trees or Birches; at 6000 feet on the White Mountains of New Hampshire, at 11,000 or 12,000 feet on the Colorado Rocky Mountains, we walk through or upon them; sometimes upon dwarfed and depressed individuals of the same species that made the canopy below. These depressed trees retain their hold on life only in virtue of being covered all winter by snow. At still higher altitudes the species are wholly different; and for the most part these humble alpine plants of our temperate zone — which we cannot call trees, because they are only a foot or two or a span or two high — are the same as those of the arctic zone, of northern Labrador, and of Greenland. The arctic and the alpine regions are equally unwooded from cold.

As the opposite extreme, under opposite conditions, look to equatorial America, on the Atlantic side, for the wildest and most luxuriant forest-tract in the world, where winter is unknown, and a shower of rain falls almost every afternoon. The size of the Amazon and Orinoco — brimming throughout the year — testifies to the abundance of rain and its equable distribution.

The other side of the Andes, mostly farther south, shows the absolute contrast, in the want of rain, and absence of forest; happily it is a narrow tract. The same is true of great tracts either side of the equatorial regions, the only district where great deserts reach the ocean.

It is also true of great continental interiors out of the equatorial belt, except where cloud-compelling mountain-chains coerce a certain deposition of moisture from air which could give none to the heated plains below. So the broad interior of our country is forestless from dryness in our latitude, as the high northern zone is forestless from cold.

Regions with distributed rain are naturally forest-clad. Regions with scanty rain, and at one season, are forestless or sparsely wooded, except they have some favoring compensations. Rainless regions are desert.

The Atlantic United States in the zone of variable weather and distributed rains, and the Gulf of Mexico as a caldron for brewing rain, and no continental expanse between that great caldron and the Pacific, crossed by a prevalent southwest wind in summer, is greatly favored for summer as well as winter rain.

And so this forest region of ours, with an annual rainfall of fifty inches on the lower Mississippi, fifty-two inches in all the country east of it bordering the Gulf of Mexico, forty-five to forty-one in all the proper Atlantic district from east Florida to Maine and the whole region drained by the Ohio, — diminished only to thirty-four inches on the whole upper Mississippi and Great Lake region, — with this amount of rain, fairly distributed over the year, and the greater part not in the winter, our forest is well accounted for.

The narrow district occupied by the Pacific forest has a much more unequal rainfall, — more unequal in its different parts, most unequal in the different seasons of the year, very different in the same place in different years.

From the Gulf of Mexico to the Gulf of St. Lawrence, the amount of rain decreases moderately and rather regularly from south to north ; but, as less is needed in a cold climate, there is enough to nourish forest throughout. On the Pacific coast, from the Gulf of California to Puget Sound, the southerly third has almost no rain at all ; the middle portion has less than our Atlantic least ; the northern third has about our Atlantic average.

Then, New England has about the same amount of rainfall in winter and in summer ; Florida and Alabama about one half more in the three summer than in the three winter months, — a fairly equable distribution. But on the Pacific coast there is no summer rain at all, except in the northern portion, and there little. And the winter rain, of forty-four inches on the northern border, diminishes to less than one half before reaching the Bay of San Francisco; dwindles to twelve, ten, and eight inches on the southern coast, and to four inches before we reach the United States boundary below San Diego.

Taking the whole year together, and confining ourselves to the coast, the average rainfall for the year, from Puget Sound to the border of California, is from eighty inches at the north to seventy at the south, *i. e.*, seventy on the northern edge of California; thence it diminishes rapidly to thirty-six, twenty (about San Francisco), twelve, and at San Diego to eight inches.

The two rainiest regions of the United States are the Pacific coast north of latitude forty-five, and the northeastern coast and borders of the Gulf of Mexico. But when one is rainy the other is comparatively rainless. For while this Pacific rainy region has only from twelve to two inches of its rain in the summer months, Florida, out of its forty to sixty, has twenty to twenty-six in summer, and only six to ten in the winter months.

Again, the diminution of rainfall, as we proceed inland from the Atlantic and Gulf shores, is gradual ; the expanse that is or was forest-clad is very broad, and we wonder only that it did not extend farther west than it does.

On the other side of the continent, at the north, the district so favored with winter rain is but a narrow strip, between the ocean and the Cascade Mountains. East of the latter, the amount abruptly declines, — for the year, from eighty inches to sixteen; for the winter months, from forty-four and forty to eight and four inches ; for the summer months, from twelve and four to two and one.

So we can understand why the Cascade Mountains abruptly separate dense and tall forest on the west from treelessness on the east. We may conjecture, also, why this north Pacific forest is so magnificent in its development.

Equally, in the rapid decrease of rainfall southward, in its corresponding restriction to one season, in the continuation of the Cascade Mountains as the Sierra Nevada, cutting off access of rain to the interior, in the unbroken stretch of coast ranges near the sea, and the consequent small and precarious rainfall in the great interior valley of California, we see reasons why the Californian forest is mainly attenuated southward into two lines, — into two files of a narrow but lordly procession, advancing southward along the coast ranges, and along the western flank of the Sierra Nevada, leaving the long valley between comparatively bare of trees.

By the limited and precarious rainfall of California we may account for the limitations of its forest. But how shall we account for the fact that this district of comparatively little rain produces the largest trees in the world — not only produces, alone of all the world, those two peculiar Big Trees which excite our special wonder, — their extraordinary growth might be some idiosyncrasy of a race, — but also produces Pines and Fir-trees whose brethren we know, and whose capabilities we can estimate, upon a scale only less gigantic? Evidently there is something here wonderfully favorable to the development of trees, especially of coniferous trees ; and it is not easy to determine what it can be.

Nor, indeed, does the rainfall of the coast of Oregon, great as it is, fully account for the extraordinary development of its forest; for the rain is nearly all in winter, very little in summer. Yet here is more timber to the acre than in any other part of North America, or perhaps in any other part of the world. The trees are never so enormous in girth as some of the Californian, but are of equal height, at least on the average, three hundred feet being common, and they stand almost within arms' length of each other.

The explanation of all this may mainly be found in the great climatic differences between the Pacific and the Atlantic sides of the continent ; and the explanation of these differences is found in the difference in the winds and the great ocean currents.

The winds are from the ocean to the land all the year round, from northwesterly in summer, southwesterly in winter. And the great Pacific Gulf-stream sweeps toward and along the coast, instead of bearing away from it, as on our Atlantic side. The winters are mild and short, and are to a great extent a season of growth, instead of suspension of growth as with us. So there is a far longer season available to tree vegetation than with us, during all of which trees may either grow or accumulate the materials for growth. On our side of the continent and in this latitude, trees use the whole autumn in getting ready for a six-months winter, which is completely lost time.

Finally, as concerns the west coast, the lack of summer rain is made up by the moisture-laden ocean winds, which regularly every summer afternoon wrap the coast-ranges of mountains, which these forests affect, with mist and fog. The Redwood, one of the two California Big Trees, — the handsomest and far the most abundant and useful, — is restricted to these coast ranges, bathed with soft showers fresh from the ocean all winter, and with fogs and moist ocean air all summer. It is nowhere found beyond the reach of these fogs. South of Monterey, where this summer condensation lessens, and winter rains become precarious, the Redwoods disappear, and the general forest becomes restricted to favorable stations on mountain sides and summits. . . . The whole coast is bordered by a line of mountains, which condense the moisture of the sea-breezes upon their cool slopes and summits. These winds, continuing eastward, descend dry into the valleys, and warming as they descend, take up moisture instead of dropping any. These valleys, when broad, are sparsely wooded or woodless, except at the north, where summer rain is not very rare.

Beyond stretches the Sierra Nevada, all rainless in summer, except local hail-storms and snowfalls on its higher crests and peaks. Yet its flanks are forest-clad; and, between the levels of 3000 and 9000 feet, they bear an ample growth of the largest coniferous trees known. In favored spots of this forest, and only there, are found those groves of the giant Sequoia, near kin of the Redwood of the coast ranges, whose trunks are from fifty to ninety feet in circumference, and whose height is from two hundred to three hundred and twenty-five feet. And in reaching these wondrous trees you ride through miles of Sugar Pines, Yellow Pines, Spruces, and Firs, of such magnificence in girth and height, that the Big Trees, when reached — astonishing as they are — seem not out of keeping with their surroundings.

I cannot pretend to account for the extreme magnificence of this Sierra forest. Its rainfall is in winter, and of unknown but large amount. Doubtless most of it is in snow, of which fifty or sixty feet fall in some winters, and — different from the coast and from Oregon, where it falls as rain, and at a temperature which does not suspend vegetable action — here the winter must be complete cessation. But with such great snowfall the supply of moisture to the soil should be abundant and lasting.

Then the Sierra — much loftier than the coast ranges, rising from 7000 or 8000 to 11,000 and 14,000 feet — is refreshed in summer by the winds from the Pacific, from which it takes the last drops of available moisture ; and mountains of such altitude, to which moisture from whatever source or direction must necessarily be attracted, are always expected to support forests, — at least when not cut off from sea-winds by interposed chains of equal altitude. Trees such mountains will have. The only and the real wonder is, that the Sierra Nevada should rear such immense trees !

Moreover, we shall see that this forest is rich and superb only in one line ; that, beyond one favorite tribe, it is meagre enough. Such for situation, and extent, and surrounding conditions, are the two forests — the Atlantic and Pacific — which are to be compared.

In order to come to this comparison, I must refrain from all account of the intervening forest of the Rocky Mountains — only saying that it is comparatively poor in the size of its trees and the number of species ; that few of its species are peculiar, and those mostly in the southern part, and of the Mexican plateau type ; that they are common to the mountain chains which lie between, stretched north and south *en echelon*, all through that arid or desert region of Utah and Nevada, of

which the larger part belongs to the great basin between the
Rocky Mountains and the Sierra Nevada; that most of the
Rocky Mountain trees are identical in species with those of the
Pacific forest, except far north, where a few of our eastern
ones are intermingled. I may add that the Rocky Mountains
proper get from twelve to twenty inches of rain in the year,
mostly in winter snow, some in summer showers.

But the interior mountains get little, and the plains or valleys between them less; the Sierra arresting nearly all the
moisture coming from the Pacific, the Rocky Mountains all
coming from the Atlantic side.

Forests being my subject, I must not tarry on the woodless
plain — on an average 500 miles wide — which lies between
what forest there is in the Rocky Mountains and the western
border of our eastern wooded region. Why this great sloping
plain should be woodless — except where some Cottonwoods
and their like mark the course of the traversing rivers — is,
on the whole, evident enough. Great interior plains in temperate latitudes are always woodless, even when not very arid.
This of ours is not arid to the degree that the corresponding
regions west of the Rocky Mountains are. The moisture from
the Pacific which those could otherwise share is — as we have
seen — arrested on or near the western border by the coast-
ranges and again by the Sierra Nevada ; and so the interior
(except for the mountains) is all but desert.

On the eastern side of the continent, the moisture supplied
by the Atlantic and the Gulf of Mexico meets no such obstruc-
tion. So the diminution of rainfall is gradual instead of
abrupt. But this moisture is spread over a vast surface, and it
is naturally bestowed, first and most on the seaboard district,
and least on the remote interior. From the lower Mississippi
eastward and northward, including the Ohio River basin, and
so to the coast, and up to Nova Scotia, there is an average
of forty-seven inches of rain in the year. This diminishes
rather steadily westward, especially northwestward, and the
western border of the ultra-Mississippian plain gets less than
twenty inches.

Indeed, from the great prevalence of westerly and southerly

winds, what precipitation of moisture there is on our western plains is not from Atlantic sources, nor much from the Gulf. The rain-chart plainly shows that the water raised from the heated Gulf is mainly carried northward and eastward. It is this which has given us the Atlantic forest-region ; and it is the limitation of this which bounds that forest at the west. The line on the rain-chart indicating twenty-four inches of annual rain is not far from the line of the western limit of trees, except far north, beyond the Great Lakes, where, in the coolness of high latitudes, as in the coolness of mountains, a less amount of rainfall suffices for forest-growth.

We see, then, why our great plains grow bare as we proceed from the Mississippi westward; though we wonder why this should take place so soon and so abruptly as it does. But, as already stated, the general course of the wind-bearing rains from the Gulf and beyond is such as to water well the Mississippi Valley and all eastward, but not the district west of it.

It does not altogether follow that, because rain or its equivalent is needed for forest, therefore wherever there is rain enough, forest must needs cover the ground. At least there are some curious exceptions to such a general rule, — exceptions both ways. In the Sierra Nevada we are confronted with a stately forest along with a scanty rainfall, with rain only in the three winter months. All summer long, under those lofty trees, if you stir up the soil you may be choked with dust. On the other hand, the prairies of Iowa and Illinois, which form deep bays or great islands in our own forest region, are spread under skies which drop more rain than probably ever falls on the slopes of the Sierra Nevada, and give it at all seasons. Under the lesser and brief rains we have the loftiest trees we know; under the more copious and well-dispersed rain, we have prairies, without forest at all.

There is little more to say about the first part of this paradox; and I have not much to say about the other. The cause or origin of our prairies — of the unwooded districts this side of the Mississippi and Missouri — has been much discussed, and a whole hour would be needed to give a fair account of the different views taken upon this knotty question. The only settled thing about it is, that the prairies are not directly due to a deficiency of rain. That the rain-charts settle, as Professor Whitney well insists.

The prairies which indent or are inclosed in our Atlantic forest region, and the plains beyond this region, are different things. But as the one borders — and in Iowa and Nebraska passes into — the other, it may be supposed that common causes have influenced both together, perhaps more than Professor Whitney allows.

He thinks that the extreme fineness and depth of the usual prairie soil will account for the absence of trees; and Mr. Lesquereux equally explains it by the nature of the soil, in a different way. These and other excellent observers scout the idea that immemorial burnings, in autumn and spring, have had any effect. Professor Shaler, from his observations in the border land of Kentucky, thinks that they have, — that there are indications there of comparatively recent conversion of Oak-openings into prairie, and now — since the burnings are over — of the reconversion of prairie into woodland.

I am disposed, on general considerations, to think that the line of demarcation between our woods and our plains is not where it was drawn by Nature. Here, when no physical barrier is interposed between the ground that receives rain enough for forest and that which receives too little, there must be a debatable border, where comparatively slight causes will turn the scale either way. Difference in soil and difference in exposure will here tell decisively. And along this border, annual burnings — for the purpose of increasing and improving Buffalo-feed — practised for hundreds of years by our nomade predecessors, may have had a very marked effect. I suspect that the irregular border line may have in this way been rendered more irregular, and have been carried farther eastward wherever nature of soil or circumstances of exposure predisposed to it.

It does not follow that trees would reoccupy the land when the operation that destroyed them, or kept them down, ceased. The established turf or other occupation of the soil, and the sweeping winds, might prevent that. The difficulty of re-foresting bleak New England coasts, which were originally well wooded, is well known. It is equally but probably not more difficult to establish forest on an Iowa prairie, with proper selection of trees.

The difference in the composition of the Atlantic and Pacific forests is not less marked than that of the climate and geographical configuration to which the two are respectively adapted.

With some very notable exceptions, the forests of the whole northern hemisphere in the temperate zone (those that we are concerned with) are mainly made up of the same or similar kinds. Not of the same species; for rarely do identical trees occur in any two or more widely separated regions. But all round the world in our zone, the woods contain Pines and Firs and Larches, Cypresses and Junipers, Oaks and Birches, Willows and Poplars, Maples and Ashes, and the like. Yet with all these family likenesses throughout, each region has some peculiar features, some trees by which the country may at once be distinguished.

Beginning by a comparison of our Pacific with our Atlantic forest, I need not take the time to enumerate the trees of the latter, as we all may be supposed to know them, and many of the genera will have to be mentioned in drawing the contrast to which I invite your attention. In this you will be impressed most of all, I think, with the fact that the greater part of our familiar trees are " conspicuous by their absence " from the Pacific forest.

For example, it has no Magnolias, no Tulip-tree, no Papaw, no Linden or Basswood, and is very poor in Maples ; no Locust-tree — neither Flowering Locust nor Honey Locust — nor any Leguminous tree ; no Cherry large enough for a timber-tree, like our wild Black Cherry ; no Gum-trees (Nyssa nor Liquidambar), nor Sorrel-tree, nor Kalmia ; no Persimsom, or Bumelia ; not a Holly ; only one Ash that may be called a timber-tree ; no Catalpa, or Sassafras ; not a single Elm, nor Hackberry ; not a Mulberry, nor Planer-tree, nor Maclura ; not a Hickory, nor a Beech, nor a true Chestnut, nor a Hornbeam; barely one Birch-tree, and that only far north, where the differences are less striking. But as to Coniferous trees, the only missing type is our Bald Cypress, the so-called Cypress of our southern swamps, and that deficiency is made up by other things. But as to ordinary trees, if you ask what takes the place in Oregon and California of all these missing kinds, which are familiar on our side of the continent, I must answer, nothing, or nearly nothing. There is the Madroña (Arbutus) instead of our Kalmia (both really trees in some places); and there is the California Laurel instead of our southern Red Bay tree. Nor in any of the genera common to the two does the Pacific forest equal the Atlantic in species. It has not half as many Maples, nor Ashes, nor Poplars, nor Walnuts, nor Birches, and those it has are of smaller size and inferior quality: it has not half as many Oaks; and these and the Ashes are of so inferior economical value, that (as we are told) a passable wagon-wheel cannot be made of California wood, nor a really good one in Oregon.

This poverty of the western forest in species and types may be exhibited graphically, in a way which cannot fail to strike the eye more impressively than when we say that, whereas the Atlantic forest is composed of sixty-six genera and one hundred and fifty-five species, the Pacific forest has only thirty-one genera and seventy-eight species.[1] In the appended diagrams, the short side of the rectangle is proportional to the number of genera, the long side to the number of species.

Now the geographical areas of the two forests are not very different. From the Gulf of Mexico to the Gulf of St. Lawrence about twenty degrees of latitude intervene. From the southern end of California to the peninsula of Alaska there

---

[1] We take in only timber trees, or such as attain in the most favorable localities to a size which gives them a clear title to the arboreous rank. The subtropical southern extremity and Keys of Florida are excluded. So also are one or two trees of the Arizonian region which may touch the evanescent southern borders of the Californian forest. In counting the coniferous genera, Pinus, Larix, Picea, Abies, and Tsuga are admitted to this rank, but Cupressus and Chamæcyparis are taken as one genus.

are twenty-eight degrees, and the forest on the coast runs some degrees north of this; the length may therefore make up for the comparative narrowness of the Pacific forest region. How can so meagre a forest make so imposing a show? Surely not by the greater number and size of its individuals, so far as deciduous (or more correctly non-coniferous) trees are concerned; for on the whole they are inferior to their eastern brethren in size if not in number of individuals. The reason is, that a large proportion of the genera and species are coniferous trees; and these, being evergreen (except the Larches), of aspiring port and eminently gregarious habit, usually dominate where they occur. While the east has almost three times as many genera and four times as many species of non-coniferous trees as the west, it has slightly fewer genera and almost one half fewer species of coniferous trees than the west. That is, the Atlantic coniferous forest is represented by eleven genera and twenty-five species; the Pacific by twelve genera and forty-four species. This relative preponderance may also be expressed by the diagrams, in which the smaller inclosed rectangles, drawn on the same scale, represent the coniferous portions of these forests.

Indeed, the Pacific forest is made up of conifers, with non-coniferous trees as occasional undergrowth or as scattered individuals, and conspicuous only in valleys or in the sparse tree-growth of plains, on which the Oaks at most reproduce the features of the " Oak-openings " here and there bordering the Mississippi prairie region. Perhaps the most striking contrast between the west and the east, along the latitude usually traversed, is that between the spiry evergreens which the traveler leaves when he quits California, and the familiar woods of various-hued round-headed trees which give him the feeling of home when he reaches the Mississippi. The Atlantic forest is particularly rich in these, and is not meagre in coniferous trees. All the glory of the Pacific forest is in its coniferous trees; its desperate poverty in other trees appears in the annexed diagram.

These diagrams could be made more instructive, and the relative richness of the forests round the world in our latitude could be most simply exhibited, by the addition of two or three similar ones. Two would serve, one for Europe, the other for northeast Asia. A third would be the Himalay-Altaian region, geographically intermediate between the other two, as the Arizona-Rocky Mountain district is intermediate between our eastern and western. Both are here left out of view, partly for the same, partly for special reasons pertaining

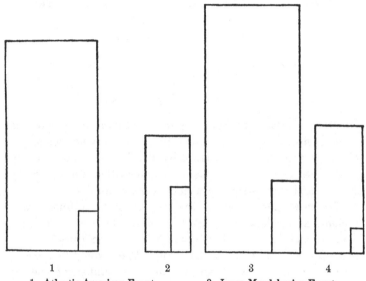

1. Atlantic American Forest.   3. Japan-Mandchurian Forest.
2. Pacific American Forest.   4. European Forest.

to each, which I must not stop to explain. These four marked specimens will simply and clearly exhibit the general facts.

Keeping as nearly as possible to the same scale, we may count the indigenous forest trees of all Europe at thirty-three genera and eighty-five species ; and those of the Japan-Mandchurian region, of very much smaller geographical area, at sixty-six genera and one hundred and sixty-eight species. I here include in it only Japan, eastern Mandchuria, and the adjacent borders of China. The known species of trees must be rather roughly determined ; but the numbers here given are not exaggerated, and are much more likely to be sensibly increased by further knowledge than are those of any of the other regions. Properly to estimate the surpassing richness of this Japan-Mandchurian forest, the comparative smallness of geographical area must come in as an important consideration.

To complete the view, let it be noted that the division of these forests into coniferous and non-coniferous is, for the

European, non-coniferous . . . . . 26 genera, 68 species.
"   coniferous   7   "   17   "

Total,   33 genera, 85 species.

Japan-Mandchurian, non-coniferous . 47 genera, 123 species.
"   "   coniferous . . 19   "   45   "

Total,   66 genera, 168 species.

In other words, a narrow region in eastern Asia contains twice as many genera and about twice as many species of indigenous trees as are possessed by all Europe ; and as to coniferous trees, the former has more genera than the latter has species, and over twice and a half as many species.

The only question about the relation of these four forest-regions, as to their component species, which we can here pause to answer, is to what extent they contain trees of identical species. If we took the shrubs, there would be a small number, if the herbs a very considerable number, of species common to the two New World and to the two Old World areas respectively, at least to their northern portions, even after excluding arctic-alpine plants. The same may be said, in its degree, of the north European flora compared with the Atlantic North American, of the northeast Asiatic compared with the northern part of the Pacific North American, and also in a peculiar way (which I have formerly pointed out and shall have soon to mention) of the northeastern Asiatic flora in its relations to the Atlantic North American. But as to the forest trees there is very little community of species. Yet this is not absolutely wanting. The Red Cedar (*Juniperus Virginiana*) among coniferous trees, and *Populus tremuloides* among the deciduous, extend across the American continent specifically unchanged, though hardly developed as forest trees on the Pacific side. There are probably, but not certainly, one or two instances on the northern verge of these two forests. There are as many in which eastern and western species are suggestively similar. The Hemlock Spruce of the northern Atlantic States and the Yew of Florida are extremely like corresponding trees of the Pacific forest; indeed the Yew-trees of all four regions may come to be regarded as forms of one polymorphous species. The White Birch of Europe and that of Canada and New England are in similar case ; and so is the common Chestnut (in America confined to the Atlantic States), which on the other side of the world is also represented in Japan. A link in the other direction is seen in one Spruce-tree (called in Oregon Menzies Spruce) which inhabits northeast Asia, while a peculiar form of it represents the species in the Rocky Mountains.

But now other and more theoretical questions come to be asked, such as these : —

Why should our Pacific forest-region, which is rich and in some respects unique in coniferous, be so poor in deciduous trees ?

Then the two Big Trees, Sequoias, as isolated in character as in location, — being found only in California, and having no near relatives anywhere, — how came California to have them ?

Such relatives as the Sequoias have are also local, peculiar, and chiefly of one species to each genus. Only one of them is American, and that solely eastern, the Taxodium of our Atlantic States and the plateau of Mexico. The others are Japanese and Chinese.

Why should trees of six related genera, which will all thrive in Europe, be restricted naturally, one to the eastern side of the American continent, one genus to the western side and very locally, the rest to a small portion of the eastern border of Asia ?

Why should coniferous trees most affect and preserve the greatest number of types in these parts of the world ?

And why should the northeast Asian region have, in a

comparatively small area, not only most coniferous trees, but a notably larger number of trees altogether than any other part of the northern temperate zone? Why should its only and near rival be in the antipodes, namely, here in Atlantic North America? In other words, why should the Pacific and the European forests be so poor in comparison, and why the Pacific poorest of all in deciduous, yet rich in coniferous trees?

The first step toward an explanation of the superior richness in trees of these antipodal regions, is to note some striking similarities of the two, and especially the number of peculiar types which they divide between them. The ultimate conclusion may at length be ventured, that this richness is normal, and that what we really have to explain is the absence of so many forms from Europe on the one hand, from Oregon and California on the other. Let me recall to mind the list of kinds (*i. e.* genera) of trees which enrich our Atlantic forest but are wanting to that of the Pacific. Now almost all these recur, in more or less similar but not identical species, in Japan, north China, etc. Some of them are likewise European, but more are not so. Extending the comparison to shrubs and herbs, it more and more appears, that the forms and types which we count as peculiar to our Atlantic region, when we compare them, as we first naturally do, with Europe and with our West, have their close counterparts in Japan and north China; some in identical species (especially among the herbs), often in strikingly similar ones, not rarely as sole species of peculiar genera or in related generic types. I was a very young botanist when I began to notice this; and I have from time to time made lists of such instances. Evidences of this remarkable relationship have multiplied year after year, until what was long a wonder has come to be so common that I should now not be greatly surprised if a Sarracenia or a Dionæa, or their like, should turn up in eastern Asia. Very few of such isolated types remain without counterparts. It is as if Nature, when she had enough species of a genus to go round, dealt them fairly, one at least to each quarter of our zone; but when she had only

two of some peculiar kind, gave one to us and the other to
Japan, Mandchuria, or the Himalayas; when she had only one,
divided these between the two partners on the opposite sides
of the table. The result, as to the trees, is seen in these four
diagrams. As to number of species generally, it cannot be
said that Europe and Pacific North America are at all in
arrears. But as to trees, either the contrasted regions have
been exceptionally favored, or these have been hardly dealt
with. There is, as I have intimated, some reason to adopt
the latter alternative.

We may take it for granted that the indigenous plants of
any country, particularly the trees, have been selected by
climate. Whatever other influences or circumstances have
been brought to bear upon them, or the trees have brought to
bear on each other, no tree could hold its place as a member
of any forest or flora which is not adapted to endure even
the extremes of the climate of the region or station. But
the character of the climate will not explain the remarkable
paucity of the trees which compose the indigenous European
forest. That is proved by experiment, sufficiently prolonged
in certain cases to justify the inference. Probably there is no
tree of the northern temperate zone which will not flourish in
some part of Europe. Great Britain alone can grow double
or treble the number of trees that the Atlantic States can.
In all the latter we can grow hardly one tree of the Pacific
coast. England supports all of them, and all our Atlantic
trees also, and likewise the Japanese and north Siberian species, which do thrive here remarkably in some part of the
Atlantic coast, especially the cooler temperate ones. The
poverty of the European sylva is attributable to the absence
of our Atlantic American types, to its having no Magnolia,
Liriodendron, Asimina, Negundo, no Æsculus, none of that
rich assemblage of Leguminous trees represented by Locusts,
Honey-Locusts, Gymnocladus, and Cladrastis (even its Cercis, which is hardly European, is like the Californian one
mainly a shrub); no Nyssa, nor Liquidambar; no *Ericaceæ*
rising to a tree; no Bumelia, Catalpa, Sassafras, Osage
Orange, Hickory, or Walnut; and as to Conifers, no Hem-

lock Spruce, Arbor-Vitæ, Taxodium, or Torreya. As compared with northeastern Asia, Europe wants most of these same types, also the Ailantus, Gingko, and a goodly number of coniferous genera. I cannot point to any types tending to make up the deficiency, that is, to any not either in east North America or in northeast Asia, or in both. Cedrus, the true Cedar, which comes near to it, is only north African and Asian. I need not say that Europe has no Sequoia, and shares no special type with California.

Now the capital fact is, that many and perhaps almost all of these genera of trees were well represented in Europe throughout the later Tertiary times. It had not only the same generic types, but in some cases even the same species, or what must pass as such, in the lack of recognizable distinctions between fossil remains and living analogues. Probably the European Miocene forest was about as rich and various as is ours of the present day, and very like it. The Glacial period came and passed, and these types have not survived there, nor returned. Hence the comparative poverty of the existing European sylva, or, at least, the probable explanation of the absence of those kinds of trees which make the characteristic difference.

Why did these trees perish out of Europe but survive in America and Asia? Before we inquire how Europe lost them, it may be well to ask, how it got them. How came these American trees to be in Europe? And among the rest, how came Europe to have Sequoias, now represented only by our two Big Trees of California? It actually possessed two species and more; one so closely answering to the Redwood of the coast ranges, and another so very like the *Sequoia gigantea* of the Sierra Nevada, that, if such fossil twigs with leaves and cones had been exhumed in California instead of in Europe, it would confidently be affirmed that we had resurrected the veritable ancestors of our two giant trees. Indeed, so it may probably be. " Cœlum non animam mutant," etc., may be applicable even to such wide wanderings and such vast intervals of time. If the specific essence has not changed, and even if it has suffered some change, genealogical connection is to be inferred in all such cases.

That is, in these days it is taken for granted that individuals of the same species, or with a certain likeness throughout, had a single birthplace, and are descended from the same stock, no matter how widely separated they may have been either in space or time, or both. The contrary supposition may be made, and was seriously entertained by some not very long ago. It is even supposable that plants and animals originated where they now are, or where their remains are found. But this is not science ; in other words it is not conformable to what we now know, and is an assertion that scientific explanation is not to be sought.

Furthermore, when species of the same genus are not found almost everywhere, they are usually grouped in one region, as are the Hickories in the Atlantic States, the Asters and Goldenrods in North America and prevailingly on the Atlantic side, the Heaths in western Europe and Africa. From this we are led to the inference that all species closely related to each other have had a common birthplace and origin. So that, when we find individuals of a species or of a group widely out of the range of their fellows, we wonder how they got there. When we find the same species all round the hemisphere, we ask how this dispersion came to pass.

Now, a very considerable number of species of herbs and shrubs, and a few trees, of the temperate zone are found all round the northern hemisphere ; many others are found part way round, — some in Europe and eastern Asia ; some in Europe and our Atlantic States ; many, as I have said, in the Atlantic States and eastern Asia ; fewer (which is curious) common to the Pacific States and eastern Asia, nearer though these countries be.

We may set it down as useless to try to account for this distribution by causes now in operation and opportunities now afforded, *i. e.,* for distribution across oceans by winds and currents, and birds. These means play their part in dispersion from place to place, by step after step, but not from continent to continent, except for few things and in a subordinate way.

Fortunately we are not obliged to have recourse to overstrained suppositions of what might possibly have occurred now and then, in the lapse of time, by the chance conveyance of seeds across oceans, or even from one mountain to another. The plants of the top of the White Mountains and of Labrador are mainly the same; but we need not suppose that it is so because birds have carried seeds from the one to the other.

I take it that the true explanation of the whole problem comes from a just general view, and not through piecemeal suppositions of chances. And I am clear that it is to be found by looking to the north, to the state of things at the arctic zone, — first, as it now is, and then as it has been.

North of our forest-regions comes the zone unwooded from cold, the zone of arctic vegetation. In this, as a rule, the species are the same round the world ; as exceptions, some are restricted to a part of the circle.

The polar projection of the earth down to the northern tropic, as here exhibited, shows to the eye — as our maps do not — how all the lands come together into one region, and how natural it may be for the same species, under homogeneous conditions, to spread over it. When we know, moreover, that sea and land have varied greatly since these species existed, we may well believe that any ocean-gaps, now in the way of equable distribution, may have been bridged over. There is now only one considerable gap.

What would happen if a cold period was to come on from the north, and was to carry very slowly the present arctic climate, or something like it, down far into the temperate zone ? Why, just what has happened in the Glacial period, when the refrigeration somehow pushed all these plants before it down to southern Europe, to middle Asia, to the middle and southern part of the United States ; and, at length receding, left some parts of them stranded on the Pyrenees, the Alps, the Apennines, the Caucasus, on our White and Rocky Mountains, or wherever they could escape the increasing warmth as well by ascending mountains as by receding northward at lower levels. Those that kept together at a low level, and made good their retreat, form the main body of present arctic vegetation. Those that took to the mountains had their line of retreat cut off, and hold their positions on the mountain-tops under cover of the frigid climate due to elevation. The conditions of these on different continents or different mountains are similar, but not wholly alike. Some species proved better adapted to one, some to another, part of the world ; where less adapted, or less adaptable, they have perished ; where better adapted, they continue, — with or without some change ; and hence the diversification of alpine plants, as well as the general likeness through all the northern hemisphere.

All this exactly applies to the temperate-zone vegetation, and to the trees that we are concerned with. The clew was seized when the fossil botany of the high arctic regions came to light ; when it was demonstrated that in the times next preceding the Glacial period — in the latest Tertiary — from Spitzbergen and Iceland to Greenland and Kamtschatka, a climate like that we now enjoy prevailed, and forests like those of New England and Virginia, and of California, clothed the land. We infer the climate from the trees ; and the trees give sure indications of the climate.

I had divined and published the explanation long before I knew of the fossil plants. These, since made known, render the inference sure, and give us a clear idea of just what the climate was. At the time we speak of, Greenland, Spitzbergen, and our arctic sea-shore had the climate of Pennsylvania and Virginia now. It would take too much time to enumerate the sorts of trees that have been identified by their leaves and fruits in the arctic later Tertiary deposits.

I can only say, at large, that the same species have been found all round the world ; that the richest and most extensive finds are in Greenland ; that they comprise most of the sorts which I have spoken of as American trees which once lived in Europe, — Magnolias, Sassafras, Hickories, Gum-trees, our identical Southern Cypress (for all we can see of difference), and especially Sequoias, not only the two which obviously answer to the two Big Trees now peculiar to California, but several others ; that they equally comprise trees now peculiar to Japan and China, three kinds of Gingko-trees, for instance, one of them not evidently distinguishable

from the Japanese species which alone survives ; that we have
evidence, not merely of Pines and Maples, Poplars, Birches,
Lindens, and whatever else characterize the temperate-zone
forests of our era, but also of particular species of these, so
like those of our own time and country, that we may fairly
reckon them as the ancestors of several of ours. Long gene-
alogies always deal more or less in conjecture ; but we appear
to be within the limits of scientific inference when we announce that our existing temperate trees came from the north,
and within the bounds of nigh probability when we claim not
a few of them as the originals of present species. Remains
of the same plants have been found fossil in our temperate
region, as well as in Europe.

Here, then, we have reached a fair answer to the question
how the same or similar species of our trees came to be so dispersed over such widely separated continents. The lands all
diverge from a polar centre, and their proximate portions —
however different from their present configuration and extent,
and however changed at different times — were once the home
of those trees, where they flourished in a temperate climate.
The cold period which followed, and which doubtless came on
by very slow degrees during ages of time, must have long
before its culmination brought down to our latitude, with
the similar climate, the forest they possess now, or rather the
ancestors of it. During this long (and we may believe first)
occupancy of Europe and the United States, were deposited in
pools and shallow waters the cast leaves, fruits, and occasionally
the branches, which are imbedded in what are called Miocene
Tertiary or later deposits, most abundant in Europe, from
which the American character of the vegetation of the period
is inferred. Geologists give the same name to these beds, in
Greenland and southern Europe, because they contain the
remains of identical or very similar species of plants; and
they used to regard them as of the same age on account of
this identity. But in fact this identity is good evidence that
they cannot be synchronous. The beds in the lower latitudes
must be later, and were forming when Greenland probably
had very nearly the climate which it has now.

Wherefore the high, and not the low, latitudes must be assumed as the birthplace of our present flora; [1] and the present arctic vegetation is best regarded as a derivative of the temperate. This flora, which when circumpolar was as nearly homogeneous round the high latitudes as the arctic vegetation is now, when slowly translated into lower latitudes, would preserve its homogeneousness enough to account for the actual distribution of the same and similar species round the world, and for the original endowment of Europe with what we now call American types. It would also vary or be selected from by the increasing differentiation of climate in the divergent continents, and on their different sides, in a way which might well account for the present diversification. From an early period, the system of the winds, the great ocean currents (however they may have oscillated north and south), and the general proportions and features of the continents in our latitude (at least of the American continent) were much the same as now, so that species of plants, ever so little adapted or predisposed to cold winters and hot summers, would abide and be developed on the eastern side of continents, therefore in the Atlantic States and in Japan and Mandchuria; those with preference for milder winters would incline to the western sides; those disposed to tolerate dryness would tend to interiors, or to regions lacking summer rain. So that, if the same thousand species were thrust promiscuously into these several districts, and carried slowly onward in the way supposed, they would inevitably be sifted in such a manner that the survival of the fittest for each district might explain the present diversity.

Besides, there are re-siftings to take into the account. The Glacial period or refrigeration from the north, which at its inception forced the temperate flora into our latitude, at its culmination must have carried much or most of it quite beyond. To what extent displaced, and how far superseded by

[1] This takes for granted, after Nordenskjöld, that there was no preceding Glacial period, as neither palæontology nor the study of arctic sedimentary strata afford any evidence of it. Or if there was any, it was too remote in time to concern the present question.

the vegetation which in our day borders the ice, or by ice itself, it is difficult to form more than general conjectures — so
different and conflicting are the views of geologists upon the
Glacial period. But upon any, or almost any, of these views,
it is safe to conclude that temperate vegetation, such as preceded the refrigeration and has now again succeeded it, was
either thrust out of northern Europe and the northern Atlantic States, or was reduced to precarious existence and diminished forms. It also appears that, on our own continent
at least, a milder climate than the present, and a considerable
submergence of land, transiently supervened at the north, to
which the vegetation must have sensibly responded by a northward movement, from which it afterward receded.

All these vicissitudes must have left their impress upon the
actual vegetation, and particularly upon the trees. They furnish probable reason for the loss of American types sustained
by Europe.

I conceive that three things have conspired to this loss.
First, Europe, hardly extending south of latitude 40°, is all
within the limits generally assigned to severe glacial action.
Second, its mountains trend east and west, from the Pyrenees
to the Carpathians and the Caucasus beyond, near its southern
border; and they had glaciers of their own, which must have
begun their operations, and poured down the northward
flanks, while the plains were still covered with forest on the
retreat from the great ice-wave coming from the north. Attacked both on front and rear, much of the forest must have
perished then and there. Third, across the line of retreat of
those which may have flanked the mountain-ranges, or were
stationed south of them, stretched the Mediterranean, an impassable barrier. Some hardy trees may have eked out their
existence on the northern shore of the Mediterranean and the
Atlantic coast. But we doubt not, Taxodium and Sequoias,
Magnolias and Liquidambars, and even Hickories and the
like, were among the missing. Escape by the east, and rehabilitation from that quarter until a very late period, was
apparently prevented by the prolongation of the Mediterranean to the Caspian, and thence to the Siberian ocean. If we

accept the supposition of Nordenskjöld, that anterior to the
Glacial period, Europe was " bounded on the south by an
ocean extending from the Atlantic over the present deserts of
Sahara and Central Asia to the Pacific," all chance of these
American types having escaped from or reëntered Europe
from the south and east, is excluded. Europe may thus be
conceived to have been for a time somewhat in the condition
in which Greenland is now, and indeed to have been connected with Greenland in this or in earlier times. Such a
junction, cutting off access of the Gulf Stream to the Polar sea,
would, as some think, other things remaining as they are, almost of itself give glaciation to Europe. Greenland may be
referred to, by way of comparison, as a country which having
undergone extreme glaciation, bears the marks of it in the extreme poverty of its flora, and in the absence of the plants
to which its southern portion, extending six degrees below the
arctic circle, might be entitled. It ought to have trees, and
might support them. But since destruction by glaciation, no
way has been open for their return. Europe fared much better, but suffered in its degree in a similar way.

Turning for a moment to the American continent for a
contrast, we find the land unbroken and open down to the
tropic, and the mountains running north and south. The
trees, when touched on the north by the on-coming refrigeration, had only to move their southern border southward, along
an open way, as far as the exigency required; and there was
no impediment to their due return. Then the more southern
latitude of the United States gave great advantage over Europe. On the Atlantic border, proper glaciation was felt
only in the northern part, down to about latitude 40°. In
the interior of the country, owing doubtless to greater dryness and summer heat, the limit receded greatly northward
in the Mississippi Valley, and gave only local glaciers to the
Rocky Mountains ; and no volcanic outbreaks or violent
changes of any kind have here occurred since the types of our
present vegetation came to the land. So our lines have been
cast in pleasant places, and the goodly heritage of forest trees
is one of the consequences.

The still greater richness of northeast Asia in arboreal vegetation may find explanation in the prevalence of particularly favorable conditions, both ante-glacial and recent. The trees of the Miocene circumpolar forest appear to have found there a secure home; and the Japanese islands, to which most of these trees belong, must be remarkably adapted to them. The situation of these islands — analogous to that of Great Britain, but with the advantage of lower latitude and greater sunshine, — their ample extent north and south, their diversified configuration, their proximity to the great Pacific gulf-stream, by which a vast body of warm water sweeps along their accentuated shores, and the comparatively equable diffusion of rain throughout the year, all probably conspire to the preservation and development of an originally ample inheritance.

The case of the Pacific forest is remarkable and paradoxical. It is, as we know, the sole refuge of the most characteristic and widespread type of Miocene *Coniferæ*, the Sequoias; it is rich in coniferous types beyond any country except Japan; in its gold-bearing gravels are indications that it possessed, seemingly down to the very beginning of the Glacial period, Magnolias and Beeches, a true Chestnut, Liquidambar, Elms, and other trees now wholly wanting to that side of the continent, though common both to Japan and to Atlantic North America.[1] Any attempted explanation of this extreme paucity of the usually major constituents of forests, along with a great development of the minor, or coniferous, element, would take us quite too far, and would bring us to mere conjectures.

Much may be attributed to late glaciation;[2] something to the tremendous outpours of lava which, immediately before

[1] See, especially, " Report on the Fossil Plants of the auriferous gravel deposits of the Sierra Nevada," by L. Lesquereux ; " Mem. Mus. Comp. Zoölogy," vi. No. 2. — Determinations of fossil leaves, etc., such as these, may be relied on to this extent by the general botanist, however wary of specific and many generic identifications. These must be mainly left to the expert in fossil botany.

[2] Sir Joseph Hooker, in an important lecture delivered to the Royal Institution of Great Britain, April 12, insists much on this.

the period of refrigeration, deeply covered a very large part of the forest-area ; much to the narrowness of the forest-belt, to the want of summer rain, and to the most unequal and precarious distribution of that of winter.

Upon all these topics questions present themselves which we are not prepared to discuss. I have done all that I could hope to do in one lecture if I have distinctly shown that the races of trees, like the races of men, have come down to us through a prehistoric (or pre-natural historic) period ; and that the explanation of the present condition is to be sought in the past, and traced in vestiges and remains and survivals ; that for the vegetable kingdom also there is a veritable Archæology.

THE PERTINACITY AND PREDOMINANCE OF WEEDS.[1]

A WEED is defined by the dictionaries to be "Any useless or troublesome plant." "Every plant which grows in a field other than that of which the seed has been (intentionally) sown by the husbandman is a weed," says the "Penny Cyclopædia," as cited in Worcester's Dictionary. The "Treasury of Botany" defines it as "Any plant which obtrusively occupies cultivated or dressed ground, to the exclusion or injury of some particular crop intended to be grown. Thus, even the most useful plants may become weeds if they appear out of their proper place. The term is sometimes applied to any insignificant-looking or unprofitable plants which grow profusely in a state of nature; also to any noxious or useless plant." We may for present purposes consider weeds to be plants which tend to take prevalent possession of soil used for man's purposes, irrespective of his will; and, in accordance with usage, we may restrict the term to herbs. This excludes predominant indigenous plants occupying ground in a state of nature. Such become weeds when they conspicuously intrude into cultivated fields, meadows, pastures, or the ground around dwellings. Many are unattractive, but not a few are ornamental; many are injurious, but some are truly useful. White Clover is an instance of the latter. Bur Clover (*Medicago denticulata*) is in California very valuable as food for cattle and sheep, and very injurious by the damage which the burs cause to wool. In the United States, and perhaps in most parts of the world, a large majority of the weeds are introduced plants, brought into the country directly or indirectly by man. Some such as Dandelion, Yarrow, and probably the common Plantain and the common Purslane, are importations as weeds, although the species naturally occupy some part of the country.

[1] American Journal of Science and Arts, 3 ser., xviii. 161. (1879.)

Why weeds are so pertinaceous and aggressive, is too large and loose a question : for any herb whatever when successfully aggressive becomes a weed ; and the reasons of predominance may be almost as diverse as the weeds themselves. But we may inquire, whether weeds have any common characteristic which may give them advantage, and why the greater part of the weeds of the United States, and probably of similar temperate countries, should be foreigners.

As to the second question, this is strikingly the case throughout the Atlantic side of temperate North America, in which the weeds have mainly come from Europe ; but it is not so, or hardly so, west of the Mississippi in the region of prairies and plains. So that the answer we are accustomed to give must be to a great extent the true one, namely, that, as the district here in which weeds from the Old World prevail was naturally forest-clad, there were few of its native herbs which, if they could bear the exposure at all, were capable of competition on cleared land with emigrants from the Old World. It may be said that these same European weeds, here prepotent, had survived and adapted themselves to the change from forest to cleared land in Europe, and therefore our forest-bred herbs might have done the same thing here. But in the first place the change must have been far more sudden here than in Europe ; and in the next place, we suppose that most of the herbs in question never were indigenous to the originally forest-covered regions of the Old World ; but rather, as western and northern Europe became agricultural and pastoral, these plants came with the husbandmen and the flocks, or followed them, from the woodless or sparsely wooded regions farther east where they originated. This, however, will not hold for some of them, such as Dandelion, Yarrow, and Ox-eye Daisy. It may be said that our weeds might have come to a considerable extent from the bordering more open districts on the west and south. But there was little opportunity until recently, as the settlement of the country began on the eastern border ; yet a certain number of our weeds appear to have been thus derived : for instance, *Mollugo verticillata, Erigeron Canadense,*

Xanthium, *Ambrosia artemisiæfolia, Verbena hastata, V. urticifolia*, etc., *Veronica peregrina, Solanum Carolinense*, various species of Amarantus and Euphorbia, *Panicum capillare*, etc. Of late, and in consequence of increased communication with the Mississippi region and beyond — especially by railroads — other plants are coming into the eastern States as weeds, step by step, by somewhat rapid strides; such as ˙*Dysodia chrysanthemoides, Matricaria discoidea*, and *Artemisia biennis.* Fifty years ago *Rudbeckia hirta*, which flourished from the Alleghanies westward, was unknown farther east. Now since twenty years, it is an abundant and conspicuous weed in grass-fields throughout the eastern States, having been accidentally desimated with Red Clover seed from the western States.

There are also native American weeds, doubtless indigenous to the region, such as *Asclepias Cornuti, Antennaria margaritacea*, and *A. plantaginifolia*, and in enriched soils *Phytolacca decandra*, which have apparently become strongly aggressive under changed conditions. These are some of the instances which may show that predominance is not in consequence of change of country and introduction to new soil.

In many cases it is easy to explain why a plant, once introduced, should take a strong and persistent hold and spread rapidly. In others we discern nothing in the plant itself which should give it advantage. *Lespedeza striata* is a small and insignificant annual, with no obvious provision for dissemination. It is a native of China and Japan. In some unexplained way it reached Alabama and Georgia, and was first noticed about thirty-five years ago; it has spread rapidly since, especially over old fields and along roadsides, and it is now very abundant up to Virginia and Tennessee, throughout the middle and upper districts, reaching even to the summits of the mountains of moderate elevation. In the absence of better food it is greedily eaten by cattle and sheep. The voiding by them of undigested seeds must be the means of dissemination; but one cannot well understand why it should spread so widely and rapidly, and take such complete possession of the ground. It is one of the few weeds which are accounted a blessing.

Professor Claypole, of Antioch College, Ohio, has recently contributed to the "Third Report of the Montreal Horticultural Society" (1877–8) an interesting essay, "On the Migration of Plants from Europe to America, with an Attempt to explain Certain Phenomena connected therewith." The phenomena which he would explain are the abundant migration of numerous weeds from Europe to the shore of North America, while others fail to come, and the general failure of North American weeds to invade Europe. We have offered a fairly good explanation of the first. And Professor Claypole goes far toward explaining the second when he notes that seed is (or formerly was) mainly brought from the Old World to the New, and the same may be said of cattle and other emigration ; that the cooler and shorter summer of the north of Europe renders the ripening of some seed precarious, etc. He does not mention the fact that American plants by chance reaching Europe have to compete with a vegetable world in comparatively stable equilibrium of its species, while European weeds coming — or which formerly came — to the United States found the course of nature disturbed by man and new-made fields for which they could compete with advantage. But this ingenious hypothesis is that weeds have a peculiarly "plastic nature, one capable of being moulded by and to the new surroundings," by which the plant "ere long adapts itself, if the change is not too great or sudden, to its new situation, takes out a new lease of life, and continues in the strictest sense a weed ; that the plants of the European flora possess more of this plasticity, are less unyielding in their constitution, can adapt themselves more readily to new surroundings," and that it is "the lack of this plasticity in the American flora which incapacitates it from securing a foothold and obtaining a living in the different conditions of the New World ; " that although "in the Miocene era the European and American floras were very much alike," yet "since that era the European flora has been vastly altered, while the American flora still retains a Miocene aspect, and is therefore the elder of the two ; that this long persistence of type in the American flora may have induced, by habit, a rigidity or

indisposition to change ; " that " the European is thus better
able to adapt itself to the strange climate and conditions —
that is to emigrate — than the American : and thus, being
more plastic or adaptable, it succeeds in the New World,
while the less adaptable American flora fails in the Old
World."

So far as we know, the greater plasticity of European
as compared with American plants is purely hypothetical.
"More plastic" would mean of greater variability, which, if
true, might be determined by observation. Because Europe
once had more species or types in common with North America than it now has, it does not seem to follow that the former
has " a younger plant-life," or that its existing plants are more
recent than those of the American flora. And as already intimated, so refined an hypothesis is hardly necessary for the
probable explanation of the predominance of Old World weeds
in the Atlantic United States.

Mr. Henslow, in his remarkable memoir, " On the Self-
Fertilization of Plants," derives from different but equally
theoretical premises an opposite conclusion, — namely, that
weeds or intrusive and dominant plants in general, and of
great emigrating capabilities, have " a longer ancestral life-
history than their less aggressive relatives." He also maintains that weeds, and plants best fitted for domination in the
manner of weeds, possess a common characteristic to which
this dominance may be attributed, namely, that they are in
general self-fertilized plants. A rapid generalizer might find
confirmation of this in the converse, which is obviously true,
that plants with blossoms very specially adapted for cross-fertilization by particular insects, and therefore dependent on
such special aid, are comparatively local and unaggressive ;
yet some of these are widely distributed. It will also be
understood that self-fertilization may give advantage to an
intruding plant at the outset by enabling an exceptionally
well-fitted individual to initiate a favored race. And self-
fertilization, with its sureness, would always be most advantageous unless cross-fertilization brings some compensatory
advantage greater on the whole than that of immediate sureness to fertilize.

But the test of the theory is, whether weeds and emigrating
herbs in general are more self-fertilizing or less subject to
cross-fertilization than the majority of related plants, and
whether many or any of them are actually self-fertilized
through a succession of generations. It seemed to us that, in
a limited way, the weeds which Europe has given to North
America might answer this question. To keep within bounds
and to have a case with all the data unquestionable, we will
collate the weeds of European parentage which evince a dominating character in the United States east of the Mississippi,
referring for the purpose to the "Manual of Botany of the
Northern United States" and Chapman's "Flora of the
Southern States." The latter, however, adds not a single
weed from Europe of any predominance. We include only
those which have taken a strong hold and become prominent
either by their general diffusion over the area or by taking
marked possession of certain districts. For examples of the
latter take *Echium vulgare* in Virginia, *Ranunculus bulbosus*
and *Leontodon autumnale* in eastern New England, and *Genista tinctoria*, which covers certain tracts in the eastern part
of Massachusetts, although nearly unknown elsewhere. We
must include several species which as weeds came from Europe, although they are probably, some of them undoubtedly,
indigenous to some part of the United States.

The following are the herbaceous plants naturalized from
Europe and of an aggressive character in the Atlantic United
States. Herbs of recent introduction, and those of however
ancient naturalization which have not either spread widely or
increased greatly over a considerable district, are omitted.

The eighteen species in italic type, nearly half of them
Grasses, are probably indigenous to some portions of North
America. In some cases the introduced and the indigenous
plants have come into contact.

Ranunculus bulbosus.	Raphanus Raphanis-	Silene inflata.
Ranunculus acris.	trum.	Lychnis Githago.
Nasturtium officinale.	Capsella Bursa-pastoris.	Stellaria media.
Sisymbium officinale.	Reseda Luteola.	*Portulaca oleracea.*
Brassica Sinipistrum.	Saponaria officinalis.	Malva rotundifolia.

Genista tinctoria.
Trifolium arvense.
Trifolium agrarium.
Trifolium repens.
Daucus Carota.
Pastinaca sativa.
Conium maculatum.
Tussilago Farfara.
Inula Helenium
Gnaphalium uliginosum.
Anthemis Cotula.
Achillœa Millefolium.
Tanacetum vulgare.
Leucanthemum vulgare.
Cirsium arvense.
Cirsium lanceolatum.
Lappa officinalis.
Cichorium Intybus.
Leontodon autumnale.
Taraxacum Dens-leonis.
Plantago major.
Plantago lanceolata.
Anagallis arvensis.
Verbascum Thapsus.
Verbascum Blattaria.
Linaria vulgaris.

Mentha viridis.
Mentha piperita.
Calamintha Nepeta.
Calamintha Clinopodium.
Nepeta Cataria.
Nepeta Glechoma.
Marrubium vulgare.
Galeopsis Tetrahit.
Leonurus Cardiaca.
Lamium amplexicaule.
Echium vulgare.
Symphytum officinale.
Echinospermum Lappula.
Cynoglossum officinale.
Solanum nigrum.
Chenopodium album.
Chenopodium hybridum.
Chenopodium Botrys.
Polygonum aviculare.
Polygonum Convolvulus.
Rumex crispus.
Rumex sanguineus.
Rumex Acetosella.

Allium vineale.
Alopecurus pratensis.
Phleum pratense.
Agrostis vulgaris.
Agrostis alba.
Dactylis glomerata.
Poa annua.
Poa compressa.
Poa pratensis.
Poa trivialis.
Eragrostis poæoides.
Festuca ovina.
Festuca pratensis.
Bromus secalinus.
Lolium perenne.
Triticum repens.
Triticum caninum.
Anthoxanthum odoratum.
Panicum glabrum.
Panicum sanguinale.
Panicum Crus-galli.
Setaria glauca.
Setaria virdis.

The plants of this list, regarded as weeds, are of very various character ; and several of them, such as White Clover and most of the Grasses, where most dominant, do not fall under the ordinary definition of weeds at all, but under that of plants useful to the farmer. Some, like Purslane, are only garden weeds ; some belong to pastures and meadows ; others affect roadsides. The fewness of European corn-weeds is remarkable. Ches and Corn-cockle (*Lychnis Githago*) are the only ones on the list. Corn Poppy, Bluebottle and Knapweed (*Centaurea Cyanus* and *C. nigra*) and Larkspur are conspicuously wanting; but the last two are not wholly unknown in some parts of the country.

But the only question before us is, whether these plants introduced from Europe are or are not self-fertilized, or more habitually so than others, so that this may be accounted an element of their predominance. Apparently this question must be answered in the negative. The question is not whether they are self-fertilizable. The great majority of plants are so, even of those specially adapted for intercrossing. The plants of this list appear to belong to the *juste milieu*. Only one (*Rumex Acetosella*) is completely dioecious; a few are incompletely dioecious or polygamous; the two species of Plantago are dichogamous to the extent of necessary dioicism or monoicism; a large number of the corolline species are either proterandrous or proterogynous, including two or three anemophilous species; and all the Grasses (which form the last quarter of the list) are anemophilous and more or less dichogamous, and therefore not rarely cross-fertilized. Of those which are not anemophilous we notice none which are not habitually visited by insects (except perhaps *Gnaphalium uliginosum*), and which therefore are almost as likely to be cross-fertilized as close-fertilized; while in not a few (such as the *Compositæ* generally and most of the other *Gamopetalæ*) the arrangements which favor intercrossing are explicit. There is no cleistogamous and therefore necessarily self-fertilized plant in the list, except *Lamiuum amplexicaule*, which also cross-fertilizes freely.

In California the prevalent weeds are largely different from those of the Atlantic States, and, as would be expected, are mostly of indigenous species or immigrants from South America; yet the common weeds of the Old World, especially of southern Europe, are coming in. The well-established and aggressive ones, such as *Brassica nigra*, *Silene Gallica*, *Erodium cicutarium*, *Malva borealis*, *Medicago denticulata*, *Marrubium vulgare*, and *Avena sterilis*, were perhaps introduced by way of western South America. They are mostly plants capable of self-fertilization, but also with adaptations (of dichogamy and otherwise) which must secure occasional crossing.

We cannot avoid the conclusion that self-fertilization is neither the cause nor a perceptible cause of the prepotency of the European plants which are weeds in North America.

A cursory examination brings us to a similar conclusion as respects the indigenous weeds of the Atlantic States, those

herbs which under new conditions, have propagated most
abundantly and rapidly, and competed most successfully in
the strife for the possession of fields that have taken the
place of forest. The most aggressive of these in the Northern States are *Epilobium spicatum* in the newest clearings,
which is dichogamous (proterandrous) to a degree which
practically forbids self-fertilization; and in older fields, *Asclepias Cornuti*, which is specially adapted for cross-fertilization by flying insects; *Antennaria plantaginifolia* and *A.
margaritacea*, which are dioecious; and next to these perhaps the two wild Strawberries, then *Erigeron annuum* and
E. strigosum, with certain Asters and Goldenrods, all insect-
visited and dichogamous, and *Verbena hastata, V. urticifolia*,
etc., the frequent natural hybridization of which testifies to
habitual intercrossing.

Those who suppose that only conspicuous or odorous flowers
are visited by flying insects should see how bees throng the
small, greenish, and to us odorless blossoms of Ampelopsis or
Virginia Creeper and of its Japanese relative.

THE FLORA OF NORTH AMERICA.[1]

IN the remarks which I have to offer to this Section, you will understand the word " Flora " to be written with a capital initial. I am to speak of the attempts made in my own day, and still making, to provide our botanists with a compendious systematic account of the phænogamous vegetation of the whole country which the American Association calls its own.

I shall make no effort to avoid the personal turn which my narrative is likely to take. In fact, it will be seen that I have partly a personal object in drawing up this statement.

Only two Floras of North America have ever been published as completed works, that of Michaux and that of Pursh. A third was begun (by Dr. Torrey, assisted by a young man who is no longer young), by the publication in the summer of 1838 of a first fasciculus; the first volume of 700 pages was issued two years afterward; and 500 pages of the second volume appeared in 1841 and in the early part of 1843. The time for continuing it in the original form has long ago passed by. Its completion in the form in which I have undertaken it anew is precarious. Precarious in the original sense of the word, for it is certainly to be prayed for: precarious, too, in the current sense of the word as being uncertain ; yet not so, according to an accepted definition, namely : " uncertain, because depending upon the will of another ; " for it is not our will but our power that is in question ; and it is only by the combined powers and efforts of all of us interested in Botany that the desired end can possibly be attained.

---

[1] A paper read to the Botanists at the meeting of the American Association for the Advancement of Science, at Montreal, August 25, 1882. American Journal of Science and Arts, 3 ser., xxiv. 321.

It were well to consider for a moment how and why it is that a task which has twice been — it would seem — easily accomplished has now become so difficult.

The earliest North American Flora, that of the elder Michaux, appeared in the year 1803. It was based entirely upon Michaux's own collections and observations, does not contain any plants which he had not himself gathered or seen, is not, therefore, an exhaustive summary of the botany of the country as then known, and so was the more readily prepared. Michaux came to this country in 1785, returned to France in 1796, left it again in Baudin's expedition to Australia in 1800, and died of fever in Madagascar in 1802. The Flora purports to be edited by his son, F. A. Michaux, who signed the classical Latin preface. The finish of the specific characters, and especially the capital detailed characters of the new genera, reveal the hand of a master ; and tradition has it that these were drawn up by Louis Claude Richard, who was probably the ablest botanist of his time. This tradition is confirmed by the fact that Richard's herbarium (bequeathed to his son, and now belonging to Count Franqueville) contains an almost complete set of the plants described, and I found that the specimens of Michaux supplied to Willdenow's herbarium at Berlin were ticketed and sent by Richard. Not only the younger Richard but Kunth also habitually cited the new genera of the work as of Richard, and some others have followed this example. Singularly enough, however, there is no reference whatever to Richard in any part of the Flora, nor in the elaborate preface. The most venerable botanist now living told me that there was a tradition at Paris that Richard performed a similar work for Persoon's " Synopsis Plantarum," and that he declined all mention of his name in the Synopsis and in the Flora, because the two works — contrary to the French school — were arranged upon the Linnæan Artificial System. He had his way, and the tradition may be preserved in history ; but his name cannot be cited for the genera Elytraria, Micranthemum, Elodea, Stipulicida, Dichromena, Oryzopsis, Erianthus, and the like. For, by the record these are of Michaux, " Flora Boreali-Americana," and not of Richard.

Michaux's explorations extended from Hudson's Bay, which he reached by way of the Saguenay, to Florida, as far, at least as St. Augustine and Pensacola; he was the first botanical explorer of the higher Alleghany Mountains, and, crossing these mountains in Tennessee, he reached the Mississippi in Illinois, and was as far south as Natchez. His original itinerary, which I once consulted, is preserved by the American Philosophical Society, at Philadelphia, to which it was presented by his son. It ought to be printed. That little journal shows that it was not Michaux's fault that the first Flora of North America was restricted to the district east of the Mississippi River. He had a scheme for crossing the continent to the Pacific. He warmly solicited the government at Washington to undertake such an exploration, and offered to accompany it as naturalist. This may have been the germ or the fertilizing idea of the expedition of Lewis and Clark, which was sent out a few years afterward by Jefferson, to whom, if I rightly remember, Michaux addressed his enterprising proposal.

Leaving out the Cryptogams of lower rank than the Ferns, we find that the Flora of Michaux, published at the beginning of this century, say eighty years ago, contains 1530 species, in 528 genera. No very formidable number; as to species (speaking without a count) little over half as many as are described in my " Manual of the Botany of the Northern States," which covers less than half of Michaux's area.

Eleven years afterward, namely, in the year 1814 (the preface is dated December, 1813), appeared the second Flora of North America, namely, the "Flora Americæ Septentrionalis," by Frederick Pursh. This was not confined to the author's own collections, but aimed at completeness, or to give "a systematic arrangement and description of the plants of North America, containing, besides what have been described by preceding authors, many new and rare species, collected during twelve years' travels and residence in that country."

It appears that Pursh was born at Tobolsk, in Siberia, of what parentage we do not know. He himself tells us, in his preface, that he was educated in Dresden, and that he came to this country — to Baltimore and Philadelphia — at the close of the last century, when he must have been only twenty-five years old. He was able to make the acquaintance not only of Muhlenberg, who survived until 1815, and of William Bartram, who died in 1823, but also of the veteran, Humphrey Marshall, who died in 1805. His early and principal patron was Dr. Benjamin Smith Barton, who supplied the means for most of the travels which he was able to undertake, and who, as Pursh states, "for some time previous had been collecting materials for an American Flora." Pursh's personal explorations were not extensive. From 1802 till 1805 he was in charge of the gardens of William Hamilton, near Philadelphia. In the spring of the latter year, as he says, he "set out for the mountains and western territories of the southern States, beginning at Maryland and extending to the Carolinas (in which tract the interesting high mountains of Virginia and Carolina took my particular attention), returning late in the autumn through the lower countries along the sea-coast to Philadelphia." But, in tracing his steps by his collections [1] and by other indications, it appears that he did not reach the western borders of Virginia nor cross its southern boundary into the mountains of North Carolina. The Peaks of Otter and Salt-pond Mountain (now Mountain Lake) were the highest elevations which he attained. Pursh's preface continues: "The following season, 1806, I went in like manner over the northern States, beginning with the mountains of Pennsylvania and extending to those of New Hampshire (in which tract I traversed the extensive and highly interesting country of the Lesser and Great Lakes), and returning as before by the sea-coast." The diary of this expedition, found among Dr. Barton's papers and collection in possession of the American Philosophical Society, has recently been printed by the late Mr. Thomas Potts James. It shows that the journey was not as extended or as thorough as would be supposed; that it was from Philadelphia directly north to the Pokono Mountains, thence to Onondaga, and to Oswego, — the only point on the Great Lakes reached, — thence back to Utica,

---

[1] In herb. Barton and herb. Lambert.

down the Mohawk Valley to Saratoga, and north to the upper part of Lake Champlain and to the lesser Green Mountains in the vicinity of Rutland, but not beyond. Discouraged by the lateness of the season, and disheartened — as he had all along been — by the failure and insufficiency of remittances from his patron, Pursh turned back from Rutland on the 22d of September, reached New York on the 1st of October, and Philadelphia on the 5th. The next year (1807) Pursh took charge of the Botanic Garden which Dr. Hosack had formed at New York and afterward sold to the State, which soon made it over to Columbia Collège.[1] In 1810, he made a voyage to the West Indies for the recovery of his health. Returning in the autumn of 1811, he landed at Wiscasset, in Maine, "had an opportunity of visiting Professor Peck of Cambridge College, near Boston," and of seeing the alpine plants which Peck had collected on the White Mountains.[2] At the end of the latter year or early in 1812 he went to England with his collections and notes; and at the close of 1813, under the auspices of Lambert, he produced his Flora, consulting, the while, the herbaria of Clayton, Pallas, Plukenet, Catesby, Morison, Sherard, Walter, and that of Banks. Evidently such consultations and the whole study must have

---

[1] Expecting, no doubt, that it would be kept up. But the Elgin Botanic Garden was soon discontinued. It occupied the block of ground now covered by the buildings of the College, and the surrounding tract — now so valuable — from which the college derives an ample revenue. *Noblesse oblige*, and it may be expected that the College, so enriched, will, before long, provide itself with a botanical professorship, and see to the careful preservation and maintenance of the precious Torrey Herbarium, which it possesses along with other subsidiary herbaria.

[2] It is at Wiscasset, therefore, that Pursh's "*Plantago cucullata*, Lam. . . . in wet rocky situations, Canada and Province of Maine," is to be sought. Mr. Pringle has recently found the related *P. Cornuti* (which may be the plant meant), in Lower Canada, not far from the other side of Maine.

It must have been in Professor Peck's herbarium (no longer extant), that Pursh saw what he took to be *Alchemilla alpina*, which he marks "*v. s.*" and refers to from memory only, probably mistakenly. For it has not since been detected either in Vermont or New Hampshire, or anywhere in North America ; and Pursh's Journal makes it certain that he did not reach any alpine region in the Green Mountains.

been rapid. The despatch is wonderful. One can hardly understand the ground of the statement made by Lambert to my former colleague, Dr. Torrey, that he was obliged to shut Pursh up in his house in order to keep him at his work.

I know not how Pursh was occupied for the next four years, nor when he came to Canada. But he died here at Montreal, in 1820, at the early age of forty-six. More is probably known of him here. If I rightly remember, his grave has been identified, and a stone placed upon it inscribed to his memory.[1] A tradition has come down to us — and it is partly confirmed by a statement which Lambert used to make, in reference to the vast quantity of beer he had to furnish during the preparation of the Flora — that, in his latter days, our predecessor was given to drink, and that his days were thereby shortened.

In Pursh's Flora we begin to have plants from the Great Plains, the Rocky Mountains, and the Pacific coast, although the collections were very scanty. The most important one which fell into Pursh's hands was that of about 150 specimens, gathered by Lewis and Clark on their homeward journey from the mouth of the Columbia River. A larger collection, more leisurely made on the outward journey, was lost. Menzies in Vancouver's voyage had botanized on the Pacific coast, both in California and much farther north. Some of his plants were seen by Pursh in the Banksian herbarium, and taken up. I may here say that in the winter of 1838–39 I had the pleasure of making the acquaintance of the venerable Menzies, then about ninety-five years old.

In the Supplement, Pursh was able to include a considerable

---

[1] In the Canadian Naturalist, Principal Dawson gives a brief account of the transference of the remains of Pursh from a grave-yard below Montreal, in which they were interred, to the beautiful Mount Royal Cemetery, where they rest in a lot purchased for the purpose and under a neat and durable granite monument, provided by the naturalists of Montreal and their friends. A small company of botanists, led by Dr. Dawson, visited the spot shortly after the reading of this paper. We learned that Pursh had botanized largely in Canada, in view of a Canadian Flora, and that his collections were consumed by a fire at Quebec shortly before his death, to his extreme discouragement.

number of species, collected by Bradbury on the upper Missouri, in what was then called Upper Louisiana, — much to the discontent of Nuttall, who was in that region at the same time, and who, indeed, partly and imperfectly anticipated Pursh in certain cases, through the publication by the Frasers of a catalogue of some of the plants collected by Nuttall.

To come now to the extent of Pursh's Flora, published nearly sixty-nine years ago. It contains 740 genera of Phænogamous and Filicoid plants, and 3076 species, — just about double the number of species contained in Michaux's Flora of eleven years before.

I must omit all mention of more restricted works, even such as Nuttall's " Genera of North American Plants," which came only four years after Pursh's Flora; also the " Flora Boreali-Americana" of Sir William Hooker, which began in 1829, but was restricted to British America. I cannot say how early it was that my revered master, Dr. Torrey, conceived the idea of the Flora which he at length undertook. But he once told me that he had invited Nuttall to join him in the production of such a work, and that Nuttall declined. This must have been as early as the year 1832, that is, half a century ago. My correspondence with Dr. Torrey began in the summer of 1830, when I was a young medical student, and three or four years afterward I joined him at New York and became, for a short time, his assistant, for all the rest of his life, his botanical colleague. He was very much occupied with his duties as professor, chiefly of chemistry; he had not yet abandoned the idea of completing his " Flora of the Northern and Middle States," the first volume of which was finished in 1824, while yet free from all professional cares. Although working in the direction of the larger undertaking, the " Flora of North America " did not assume definite shape before the year 1835. I believe that some of the first actually-prepared manuscript for it was written by myself in that or the following year. I was then and for a long time expecting to accompany the South Pacific Exploring Expedition, as originally organized under the command of Commodore Ap Catesby Jones, but which was subject to long delay and many vicissitudes ; during which, having plentiful leisure, I tried my 'prentice hand upon some of the earlier natural orders. Before the expedition, as modified, was ready to sail, under the command of Captain Wilkes, I had accepted Dr. Torrey's proposal that I should be his associate in the work upon which I had made a small beginning as a volunteer. Two parts, or half, of the first volume (360 pages), of this Flora, were printed and issued in July and October, 1838.

It was thought at first, in all simplicity, that the whole task could be done at something like this rate. But, apart from other considerations, it soon became clear that there had been no proper identification of the foundation-species of the earlier botanists, from Linnæus downward ; and that our Flora could not go on satisfactorily without this. Dr. Torrey had, indeed, some years before, made a hasty visit to Hooker at Glasgow, to London, and to Paris ; but the taking of a few notes upon some particular plants in the herbaria of Hooker, Lambert, and Michaux, and the acquisition, from Hooker, of a good set of the Arctic plants of the British explorers, was about all that had been done. I proposed to attempt something more ; so, taking advantage of a favorable opportunity, I sailed for Liverpool in November, 1838, and devoted a good part of the ensuing year to the examination of the principal herbaria, which I need not here specify, in Scotland (where the important one of Sir William Hooker still remained), England, France, Switzerland, and Germany, namely those which contained the specimens upon which most of the then-published North American species had been directly or indirectly founded, especially those of Linnæus and Gronovius, of Walter, of Aiton's " Hortus Kewensis," Michaux, Wildenow, Pursh, and the later ones of De Candolle and Hooker.

After my return the work made good progress ; the remaining half of the first volume was brought out in the spring of the year 1840, and by the spring of 1843 the five hundred pages of the second volume, mostly occupied by the vast order *Compositæ*, had been issued. But meanwhile I had in my turn to assume professorial duties and incident engagements, — with the result that, although the study of North Ameri-

can plants was at no time pretermitted, either by Dr. Torrey
while he lived, or by myself, we were unable to continue the
publication during my associate's lifetime ; and it was only
recently, in the spring of 1878, that I succeeded in bringing
out, in a changed form, another instalment of the work, completing the *Gamopetalæ.*

In the interval I had made two year-long visits to Europe
for botanical investigation, the first partly relating to the botany of the South Pacific, the second wholly in view of the
North American flora. And since this last publication still
another visit — the fourth and we may suppose the last — of
the same character and the same duration, has been successfully accomplished.

The serious question, in which we are all concerned, arises,
whether this work can be carried through to completion, and
the older parts (wholly out of print and out of date), reëlaborated, — I will not say by my hands, but in my time, or soon
enough to render the whole a reasonably full and homogeneous representation of the North American flora, as known in
this latter part of the nineteenth century. And it brings us
to consider why the undertaking to which so much time has
been devoted should be so slow of accomplishment.

If this slowness is a constant wonder and disappointment to
most people interested in the matter, I can only add that it is
hardly less so to myself. It is a constant surprise — if one
may so say — that the work does not get on faster.

Of course the undertaking has become more and more formidable with the enlargement of geographical boundaries and
of the number of species discovered. As to the increase in
the number of species to be treated, we have by no means yet
reached the end. The area, that of our continent down to the
Mexican line, we trust is definitely fixed, at least for our day.
And since we cannot be rid of the peninsula and keys of Florida, which entails upon us a considerable number of tropical
species, mostly belonging to the West Indies — the southern
boundary is now as natural a one as we can have.

The area which Pursh's Flora covered was, we may say, the
United States east of the Mississippi, with Canada to Labra-

dor, to which was added a couple of hundred of species known to him outside these limits northwestward.

Torrey and Gray's Flora took the initiative in annexing Texas, ten years before its political incorporation into the Union; although the only plants we then possessed from it were certain portions of Drummond's collections. California was also annexed at the same time, on account of Douglas's collections, and those of Nuttall, who had just returned from his visit to the western coast, which he reached by a tedious journey across the continent over ground in good part new to the botanist. Douglas had already made remarkably full collections along a more northern line. The British arctic explorers, both by sea and land, had well developed the botany of the boreal regions, and Sir William Hooker was bringing out the results in his Flora of British America. Of course our knowledge of the whole interior and western region was small indeed, compared with the present; and the botany of a vast region from the western part of Texas to the Californian coast was absolutely unknown, and so remained until after the publication of the Flora was suspended.

As to the number of species which Torrey and Gray had to deal with, I can only say that a rapid count gives us for the first volume about 2200 *Polypetalæ;* that there are one hundred and nine species in the small orders which in the second volume precede the *Compositæ;* and that there are of the *Compositæ* 1054. So one may fairly conclude that if the work had been pushed on to completion, say in the year 1850, the 3076 species of Pursh's Flora in the year 1814 might have been just about doubled. Probably more rather than less; for if we reckon from the number of the *Compositæ,* and on the estimate that they constitute one-eighth of the phænogamous plants of North America, instead of 6150, there would have been 8430 species known in the year specified.

It most concerns us to know the number of species which, after the lapse of thirty years more — years in which exploration has been active, and has left no considerable part of our great area wholly unvisited — the now revived Flora has to deal with. We can make an estimate which cannot be far wrong. In the year 1878, my colleague, Mr. Watson, finished and published his " Bibliographical Index to the Polypetalæ of North America," covering, that is, the same ground as the first volume of Torrey and Gray's Flora, completed in 1840. In it the 2200 species of the latter date are increased to 3038. The " *Gamopetalæ* after *Compositæ* " in the " Synoptical Flora," brought out in the same year, contains 1656 species. The two together must make up half of our phænogamous botany, that is, adding the increase of the last four years, about 5000 species. And so Mr. Watson adopts the estimate of 10,000 species of our known Phænogams and Ferns. My impression is that the species of *Compositæ* have increased at a rate which, unless they exceed the eight part of our Phænogams, will warrant a still higher estimate. The number of introduced species of various orders, which will have to be enumerated and most of them described, is, unhappily, fast increasing ; [1] and new indigenous species are almost daily coming to us from some part or other of our wide territory. So that the 10,000 species of this estimate may before long rise to eleven or twelve thousand. Only the experienced botanist can form a just idea of what is involved in the accurate discrimination and proper coördination of 10,000 to 12,000 species, and in the putting of the results into the language and form which may make our knowledge available to learners or to succeeding botanists.

Moreover, there is of late an *embarras des richesses* which is becoming serious as respects labor and time. The continued and ever increasing influx of material to Cambridge, beneficial as it ever is, is accountable for this retardation of progress in a greater degree than almost any one would suppose. The herbarium, upon whose materials this work is mainly done, and which has been, like the Temple, full forty and six years in building, has received the contributions of two generations of botanists, and the Torrey herbarium goes back one generation farther. Still the number of American

[1] I say "unhappily," for they adulterate the natural character of our flora, and raise difficult questions as to how much of introduction and settlement should give to these denizens the rights of adopted citizens.

specimens annually coming to it is greater than in most former years. Apart from the mere selection and care of these,
consider how in other ways it affects the rate of progress of
the Flora. The incoming of additional specimens may at a
glance settle doubt as to the validity of a species; but new
specimens are as apt to raise questions as to settle them ; more
commonly they raise the question as to the limitation and
right definition of the species concerned, not rarely, also, that
of their validity. When one has only single specimens of related species, the case may seem clear and the definition easy.
The acquisition of a few more, from a different region or
grown under different conditions, almost always calls for some
reconsideration, not rarely for reconstruction. People generally suppose that species, and even genera, are like coin from
the mint, or bank notes from the printing press, each with its
fixed marks and signature, which he that runs may read, or
the practised eye infallibly determine. But in fact species
are judgments — judgments of variable value, and often very
fallible judgments, as we botanists well know. And genera
are more obviously judgments, and more and more liable to
be affected by new discoveries. Judgments formed to-day —
perhaps with full confidence, perhaps with misgiving — may
to-morrow, with the discovery of new materials or the detection of some before unobserved point of structure, have to be
weighed and decided anew. You see how all this bears upon
the question of time and labor in the preparation of the Flora
of a great country. If even in Old Europe the work has to
be done over and over, how much more so in America, where
new plants are almost daily coming to hand. It is true that
these fall into their ranks, or are adjustable into their proper
or probable places, but not without painstaking and tedious
examination.

Of our Flora, it may indeed be said, that " If it were done,
when 't is done, then 't were well it were done quickly."
But I may have made it clear that, in the actual state of the
case, it is likely to be done slowly. At least you will understand why thus far it has been done slowly. As to the future,
if it depended wholly upon me, the completion would obvi-

ously be hopeless. I need not say that our dependence, for the actual elaboration, must largely be upon associates, upon the few who have the training and the vast patience, and the access to herbaria and libraries, requisite for this kind of work, but above all upon my associate in the herbarium at Cambridge, to whom, being present with us, I will not further allude.

Of course we rely, very much indeed, upon the continued coöperation of all the cultivators of botany in the country; and it is gratifying to know that their number is increasing, new ones not less zealous than the old, and better equipped, are taking the place of those that have passed away, and some of them extending their exploration over the remotest parts of the land, and into districts where there is most to be discovered. All can help on the work, and all are doing so, by communication of specimens and of observations. Those within the range of the published Manuals and Floras get on — or should get on — with only occasional help from us. They should send us notes and specimens to any amount; but they should not ask us to stop to examine and name their plants, except in special cases, which we are always ready enough to take up. Those who collect in regions as yet destitute of such advantages may claim more aid, and we take great pains to render it: partly on our own account, that we may assort their contributions into their proper places, partly for the encouragement of such correspondents, who otherwise would not know what they have obtained, and who naturally like to know when they have made interesting discoveries.

But the scattered and piecemeal study of plants is neither very satisfactory nor safe. And it involves great loss of time, besides interrupting that continuity and concentration of attention which the proper study of any group of plants demands. As respects the orders of plants which are yet to be elaborated for the Flora, and as to plants which require critical study or minute examination, necessarily consuming much time, it is better to defer their complete determination until the groups to which they severally belong are regularly taken in hand.

The coöperation of all our botanical associates is solicited in this regard, as a matter of common interest and advantage. For we are all equally concerned in forwarding the progress of the Flora of North America; and we may confidently expect from our botanical associates their sympathy, their forbearance, and their continued aid.

GENDER OF NAMES OF VARIETIES.[1]

AMONG other subordinate questions in Natural-history nomenclature, it has been asked whether names of varieties, like those of species, should conform in gender to the genus, or whether they may not as well conform to the word *varietas*, and so always be feminine.

Linnæus introduced the current practice of numbering varieties by the letters of the Greek alphabet a, β, γ, etc. But to some varieties, evidently to the more important, he gave names. These names, when adjectives, were always (so far as we know) made to agree in gender with the generic name, *e. g.*: *Viburnun Opulus, β roseum. Asparagus officinalis, a maritimus, β altilis. Mesembryanthemum ringers, a canium, β felinum.*

In our days named varieties play a more and more important part; and all botanists, as a rule, appear to have followed the Linnean model, with now and then a divergence which is readily explained, and which may be said to be accidental, such as *Ripogonum album*, var. *leptostachya*, Benth.

This is as one writes "form a *albiflora*" or "var. *albiflora*," a white-flowered form or variety. But that this is not the pattern nor the true construction of varietal names appears at once on reference to ordinary cases. Thus, for example, in "*Nasturium amphibium, a indivisum*, DC. Syst.," it is not an individual variety of the species that is meant, but a name which stands in the same grammatical relation to Nasturtium that amphibium does, and to write *N. amphibium, a indivisa*, is obviously wrong. We should say that it makes no difference whether the word variety, or its abbreviation var. is expressed or understood. When the conditions of the case seem to call for it, we should write *N. amphibium*, var. *a indivisum*, just as, if it were ever needful, we might write

[1] American Journal of Science and Arts, 3 ser., xxvii. 396. (1884.)

" *Nasturtium,* spec. *amphibium,*" and just as L. C. Richard
(a good model), in Michaux's Flora writes, *Viburnum den-
tatum,* var. α *glabellum,* β *semi-tomentosum. Rhus Toxicodendron,* var. α *vulgare,* β *quercifolium.*
The editor of the " Gardener's Chronicle " (March 22, p.
373), having put this kind of question to M. Alphonse De
Candolle (whom we should consider the highest living authority upon nomenclatural matters), understands him to
reply that " the insertion of the abbreviation *var.* for *varietas,*
which is feminine, demands a feminine termination; but if
the word *var.* be omitted, then the rule would be for the variety to follow the specific name; " — meaning probably the
generic name, for in one of the examples given, *Thymus
Serpyllum,* β *montanus,* it does not follow the specific.
From this point of view, namely : that where the nature of
the group (in this case variety) is expressed, the adjective
name should be feminine, but where only understood, it
might be masculine or neuter — we must commend the editor's closing remark : —
" Perhaps the simplest and most easily recollected rule
would be to make the varietal name feminine in all cases,
whether the *var.* of *varietas,* were expressed, or understood.
This at least would be intelligible, and would conduce to uni-
formity of practice."
It would also be logical, and the logic also would require
all specific names to be feminine; for the word understood,
species, is feminine.
Now we do not suppose that M. De Candolle would tolerate
a double set of genders for the names of varieties. His doctrine is that the " var. " should be discarded and the Greek
letters only employed, not only for numbering the varieties,
but for designating the fact that the name they are prefixed
to is a variety.
It is not difficult to perceive why it has come to pass that
" English writers generally use the abbreviation var.," and
that some continental botanical writers follow the practice.
One reason is, that it enables us to cite an author's variety by
its name without having to concern ourselves with its Greek
number, whether it is β or γ or δ, which otherwise we should

have to attend to. Another is, that our sense of good form revolts at beginning sentences and paragraps without capitals. In our books, varieties usually stand in independent paragraphs. Even in Latin we do not like to begin a paragraph —

" *a indivisum* foliis omnibus integerrimis serratisve, non aut vix basi auriculatis."

In English we can still less abide it. So we prefix " Var.," and either number our varieties with Greek letters or, preferentially, leave them out.

But, we did not suppose that by the employment of the word " var. " we had interfered with the relation of the name of the variety to that of its genus. Var. *indivisum*, in this case, we should construe the phrase: " Varietas cujus nomen est *indivisum*. ' Var. *indivisum* ' stands on the same ground as ' species amphibium.' " The latter rank we rarely need to express, because we always prefix the generic name or its initial. The former may often come in a shape which renders the designating prefix " var. " necessary, or at least most convenient.

We may indeed, quite correctly write, var. *albiflora*, a white flowered variety, var. *longifolia*, a long-leaved variety; but that is not according to the Linnean pattern nor to the regular practice, nor to the strict analogy of the varietal name with the specific.

Moreover, if the gender of the word which designates the grade of the name is to govern the gender of the name, at least when expressed, as by *var.*, then all subspecies must be made feminine. Now this term subspecies is coming largely into use. And it has to be expressed in every case, in this wise: *Ranunculus aquatilis, L.* Subsp. *heterophyllus.* Subsp. *hederaceus*, etc.

If the proposition which we deprecate is adopted, these names would have to be written *heterophylla* and *hederacea* by an author who ranked them as subspecies but *heterophyllus* and *hederaceus* by one who took them as varieties and simply numbered them by Greek letters. Obviously the propositions in the " Gardener's Chronicle " has not been thoroughly worked out.

CHARACTERISTICS OF THE NORTH AMERICAN FLORA.[1]

WHEN the British Association, with much painstaking, honors and gratifies the cultivators of science on this side of the ocean by meeting on American soil, it is but seemly that a corresponding member for the third of a century should endeavor to manifest his interest in the occasion and to render some service, if he can, to his fellow-naturalists in Section D. I would attempt to do so by pointing out, in a general way, some of the characteristic features of the vegetation of the country which they have come to visit, — a country of " magnificent distances," but of which some vistas may be had by those who can use the facilities which are offered for enjoying them. Even to those who cannot command the time for distant excursions, and to some who may know little or nothing of botany, the sketch which I offer may not be altogether uninteresting. But I naturally address myself to the botanists of the Association, to those who, having crossed the wide Atlantic, are now invited to proceed westward over an almost equal breadth of land ; some, indeed, have already journeyed to the Pacific coast, and have returned ; and not a few, it is hoped, may accept the invitation to Philadelphia, where a warm welcome awaits them — warmth of hospitality, rather than of summer temperature, let us hope ; but Philadelphia is proverbial for both. There opportunities may be afforded for a passing acquaintance with the botany of the Atlantic border of the United States, in company with the botanists of the American Association, who are expected to muster in full force.

What may be asked of me, then, is to portray certain out-

[1] An Address to the Botanists of the British Association for the Advancement of Science at Montreal ; read to the Biological Section, August 29, 1884. (American Journal of Science and Arts, 3 ser., xxviii. 323.)

lines of the vegetation of the United States and the Canadian
Dominion, as contrasted with that of Europe; perhaps also to
touch upon the causes or anterior conditions to which much
of the actual differences between the two floras may be as-
cribed. For, indeed, however interesting or curious the facts
of the case may be in themselves, they become far more instructive when we attain to some clear conception of the dependent relation of the present vegetation to a preceding state
of things, out of which it has come.

As to the Atlantic border on which we stand, probably the
first impression made upon the botanist or other observer coming from Great Britain to New England or Canadian shores,
will be the similarity of what he here finds with what he left
behind. Among the trees the White Birch and the Chestnut
will be identified, if not as exactly the same, yet with only
slight differences — differences which may be said to be no
more essential or profound than those in accent and intona-
tion between the British speech and that of the " Americans."
The differences between the Beeches and Larches of the two
countries are a little more accentuated; and still more those
of the Hornbeams, Elms, and the nearest resembling Oaks.
And so of several other trees. Only as you proceed westward
and southward will the differences overpower the similarities,
which still are met with.

In the fields and along open roadsides the likeness seems
to be greater. But much of this likeness is the unconscious
work of man, rather than of Nature, the reason of which is
not far to seek. This was a region of forest, upon which the
aborigines, although they here and there opened patches of
land for cultivation, had made no permanent encroachment.
Not very much of the herbaceous or other low undergrowth
of this forest could bear exposure to the fervid summer's sun;
and the change was too abrupt for adaptive modification. The
plains and prairies of the great Mississippi Valley were then
too remote for their vegetation to compete for the vacancy
which was made here when forest was changed to grain-fields
and then to meadow and pasture. And so the vacancy came
to be filled in a notable measure by agrestial plants from Eu-

rope, the seeds of which came in seed-grain, in the coats and fleece and in the imported fodder of cattle and sheep, and in the various but not always apparent ways in which agricultural and commercial people unwittingly convey the plants and animals of one country to another. So, while an agricultural people displaced the aborigines which the forests sheltered and nourished, the herbs, purposely or accidentally brought with them, took possession of the clearings, and prevailed more or less over the native and rightful heirs to the soil, — not enough to supplant them, indeed, but enough to impart a certain adventitious Old World aspect to the fields and other open grounds, as well as to the precincts of habitations. In spring-time you would have seen the fields of this district yellow with European Buttercups and Dandelions, then whitened with the Ox-eye Daisy, and at midsummer brightened by the cerulean blue of Chicory. I can hardly name any native herbs which in the fields and at the season can vie with these intruders in floral show. The common Barberry of the Old World is an early denizen of New England. The tall Mullein, of a wholly alien race, shoots up in every pasture and new clearing, accompanied by the common Thistle, while another imported Thistle, called in the United States "the Canada Thistle," has become a veritable nuisance, at which much legislation has been leveled in vain.

According to tradition the wayside Plantain was called by the American Indian "White-Man's foot," from its springing up wherever that foot had been planted. But there is some reason for suspecting that the Indian's ancestors brought it to this continent. Moreover there is another reason for surmising that this long-accepted tradition is fictitious. For there was already in the country a native Plantain, so like *Plantago major* that the botanists have only of late distinguished it. (I acknowledge my share in the oversight.) Possibly, although the botanists were at fault, the aborigines may have known the difference. The cows are said to know it. For a brother botanist of long experience tells me that, where the two grow together, cows freely feed upon the undoubtedly native species, and leave the naturalized one untouched.

It has been maintained that the ruderal and agrestial Old World plants and weeds of cultivation displace the indigenous ones of newly settled countries in virtue of a strength which they have developed through survival in the struggle of ages, under the severe competition incident to their former migrations. And it does seem that most of the pertinacious weeds of the Old World which have been given to us may not be indigenous even to Europe, at least to western Europe, but belong to campestrine or unwooded regions farther east; and that, following the movements of pastoral and agricultural people, they may have played somewhat the same part in the once forest-clad western Europe that they have been playing here. But it is unnecessary to build much upon the possibly fallacious idea of increased strength gained by competition. Opportunity may count for more than exceptional vigor; and the cases in which foreign plants have shown such superiority are mainly those in which a forest-destroying people have brought upon newly-bared soil the seeds of an open-ground vegetation.

The one marked exception that I know of, the case of recent and abundant influx of this class of Old World plants into a naturally treeless region, supports the same conclusion. Our associate, Mr. John Ball, has recently called attention to it. The pampas of southeastern South America beyond the Rio Colorado, lying between the same parallels of latitude in the south as Montreal and Philadelphia in the north, and with climate and probably soils fit to sustain a varied vegetation, and even a fair proportion of forest, are not only treeless, but excessively poor in their herbaceous flora. The district has had no trees since its comparatively recent elevation from the sea. As Mr. Darwin long ago intimated: " Trees are absent not because they cannot grow and thrive, but because the only country from which they could have been derived — tropical and sub-tropical South America — could not supply species to suit the soil and climate." And as to the herbaceous and frutescent species, to continue the extract from Mr. Ball's instructive paper recently published in the Linnæan Society's Journal, " in a district raised from the sea during the latest

geological period, and bounded on the west by a great mountain range mainly clothed with an alpine flora requiring the
protection of snow in winter, and on the north by a warm-
temperate region whose flora is mainly of modified sub-tropical
origin — the only plants that could occupy the newly-formed
region were the comparatively few which, though developed
under very different conditions, were sufficiently tolerant of
change to adapt themselves to the new environment. The
flora is poor, not because the land cannot support a richer one,
but because the only regions from which a large population
could be derived are inhabited by races unfit for emigration."

Singularly enough, this deficiency of herbaceous plants is
being supplied from Europe, and the incomers are spreading
with great rapidity; for lack of other forest material even
Apple-trees are running wild and forming extensive groves.
Men and cattle are, as usual, the agents of dissemination.
But colonizing plants are filling, in this instance, a vacancy
which was left by nature, while ours was made by man. We
may agree with Mr. Ball in the opinion that the rapidity with
which the intrusive plants have spread in this part of South
America " is to be accounted for, less by any special fitness
of the immigrant species, than by the fact that the ground is
to a great extent unoccupied."

The principle applies here also ; and in general, that it is
opportunity rather than specially acquired vigor that has
given Old World weeds an advantage may be inferred from
the behavior of our weeds indigenous to the country, the
plants of the unwooded districts — prairies or savannas west
and south, — which, now that the way is open, are coming in
one by one into these eastern parts, extending their area continually, and holding their ground quite as pertinaciously as
the immigrant denizens. Almost every year gives new examples of the immigration of campestrine western plants into
the eastern States. They are well up to the spirit of the
age ; they travel by railway. The seeds are transported,
some in the coats of cattle and sheep on the way to market,
others in the food which supports them on the journey, and
many in a way which you might not suspect, until you

consider that these great roads run east and west, that the
prevalent winds are from the west, that a freight-train left
unguarded was not long ago blown on for more than one
hundred miles before it could be stopped, not altogether on
down grades, and that the bared and mostly unkempt borders
of these railways form capital seed-beds and nursery-ground
for such plants.

Returning now from this side issue, let me advert to another
and, I judge, a very pleasant experience which the botanist
and the cultivator may have on first visiting the American
shores. At almost every step he comes upon old acquaintances,
upon shrubs and trees and flowering herbs, mostly peculiar to
this country, but with which he is familiar in the grounds and
gardens of his home. Great Britain is especially hospitable
to American trees and shrubs. There those both of the eastern and western sides of our continent flourish side by side.
Here they almost wholly refuse such association. But the
most familiar and longest-established representatives of our
flora (certain western annuals excepted) were drawn from the
Atlantic coast. Among them are the Virginia Creeper or
Ampelopsis, almost as commonly grown in Europe as here,
and which, I think, displays its autumnal crimson as brightly
there as along the borders of its native woods where you will
everywhere meet with it; the Red and Sugar Maples, which
give the notable autumnal glow to our northern woods, but
rarely make much show in Europe, perhaps for lack of sharp
contrast between summer and autumn; the ornamental Eri-
caceous shrubs, Kalmias, Azaleas, Rhododendrons, and the
like, specially called American plants in England, although
all the Rhododendrons of the finer sort are half Asiatic, the
hardy American species having been crossed and recrossed
with more elegant but tender Indian species.

As to flowering herbs, somewhat of the delight with which
an American first gathers wild Primroses and Cowslips and
Foxgloves and Daisies in Europe, may be enjoyed by the
European botanist when he comes upon our Trilliums and
Sanguinaria, Cypripediums and Dodecatheon, our species of
Phlox, Coreopsis, etc., so familiar in his gardens; or when,

crossing the continent, he comes upon large tracts of ground yellow with Eschscholtzia or blue with Nemophilas. But with a sentimental difference; in that Primroses, Daisies, and Heaths, like nightingales and larks, are inwrought into our common literature and poetry, whereas our native flowers and birds, if not altogether unsung, have attained at the most to only local celebrity.

Turning now from similarities, and from that which interchange has made familiar, to that which is different or peculiar, I suppose that an observant botanist upon a survey of the Atlantic border of North America (which naturally first and mainly attracts our attention) would be impressed by the comparative wealth of this flora in trees and shrubs. Not so much so in the Canadian Dominion, at least in its eastern part; but even here the difference will be striking enough on comparing Canada with Great Britain.

The *Coniferæ*, native to the British Islands, are one Pine, one Juniper, and a Yew; those of Canada proper are four or five Pines, four Firs, a Larch, an Arbor-Vitæ, three Junipers, and a Yew, fourteen or fifteen to three. Of Amentaceous trees and shrubs, Great Britain counts one Oak (in two marked forms), a Beech, a Hazel, a Hornbeam, two Birches, an Alder, a Myrica, eighteen Willows, and two Poplars, — twenty-eight species in nine genera, and under four natural orders. In Canada there are at least eight Oaks, a Chestnut, a Beech, two Hazels, two Hornbeams of distinct genera, six Birches, two Alders, about fourteen Willows and five Poplars, also a Plane tree, two Walnuts and four Hickories; say forty-eight species, in thirteen genera, and belonging to seven natural orders. The comparison may not be altogether fair; for the British flora is exceptionally poor, even for islands so situated. But if we extend it to Scandinavia, so as to have a continental and an equivalent area, the native *Coniferæ* would be augmented only by one Fir, the *Amentaceæ* by several more Willows, a Poplar, and one or two more Birches; no additional orders nor genera.

If we take in the Atlantic United States east of the Mississippi, and compare this area with Europe, we should find the species and the types increasing as we proceed southward, but about the same numerical proportion would hold.

But, more interesting than this numerical preponderance — which is practically confined to the trees and shrubs — will be the extra-European types, which, intermixed with familiar Old-World forms, give peculiar features to the North American flora, — features discernible in Canada, but more and more prominent as we proceed southward. Still confining our survey to the Atlantic district, that is, without crossing the Mississippi, the following are among the notable points:

1. Leguminous Trees of peculiar types. Europe abounds in leguminous shrubs or under-shrubs, mostly of the Genisteous tribe, which is wanting in all North America, but has no Leguminous tree of more pretence than the Cercis and Laburnum. Our Atlantic forest is distinguished by a Cercis of its own, three species of Locust, two of them fine trees, and two Honey Locusts, the beautiful Cladrastis, and the stately Gymnocladus. Only the Cercis has any European relationship. For relatives of the others we must look to the Chino-Japanese region.

2. The great development of the *Ericaceæ* (taking the order in its widest sense), along with the absence of the Ericeous tribe, that is, of the Heaths themselves. We possess on this side of the Mississippi thirty genera, and not far from ninety species. All Europe has only seventeen genera and barely fifty species. We have most of the actual European species, excepting their Rhododendrons and their Heaths, — and even the latter are represented by some scattered patches of Calluna, of which it may be still doubtful whether they are chance introductions or sparse and scanty survivals; and besides we have a wealth of peculiar genera and species. Among them the most notable in an ornamental point of view are the Rhododendrons, Azaleas, Kalmias, Andromedas, and Clethras; in botanical interest, the endemic *Monotropeæ*, of which there is only one species in Europe, but seven genera in North America, all but one absolutely peculiar; and in edible as well as botanical interest, the unexampled development and diversification of the genus Vaccinium (along with

the allied American type, Gaylussacia) will attract attention. It is interesting to note the rapid falling away of *Ericaceœ* westward in the valley of the Mississippi as the forest thins out.

3. The wealth of this flora in *Compositœ* is a most obvious feature; one especially prominent at this season of the year, when the open grounds are becoming golden with Solidago, and the earlier of the autumnal Asters are beginning to blossom. The *Compositœ* form the largest order of Phænogamous plants in all temperate floras of the northern hemisphere, are well up to the average in Europe, but are nowhere so numerous as in North America, where they form an eighth part of the whole. But the contrast between the *Compositœ* of Europe and Atlantic North America is striking. Europe runs to Thistles, to *Inuloideœ*, to *Anthemideœ*, and to *Cichoriaceœ*. It has very few Asters, and only two Solidagoes, no Sunflowers, and hardly anything of that tribe. Our Atlantic flora surpasses all the world in Asters and Solidagoes, as also in Sunflowers and their various allies, is rich in *Eupatoriaceœ*, of which Europe has extremely few, and is well supplied with *Vernoniaceœ* and *Helenioideœ*, of which she has none; but is scanty in all the groups that predominate in Europe. I may remark that if our larger and most troublesome genera, such as Solidago and Aster, were treated in our systematic works even in the way that Nyman has treated Hieracium in Europe, the species of these two genera (now numbering seventy-eight and one hundred and twenty-four respectively) would be at least doubled.

4. Perhaps the most interesting contrast between the flora of Europe and that of the eastern border of North America is in the number of generic and even ordinal types here met with which are wholly absent from Europe. Possibly we may distinguish these into two sets of differing history. One will represent a tropical element, more or less transformed, which has probably acquired or been able to hold its position so far north in virtue of our high summer temperature. (In this whole survey the peninsula of Florida is left out of view, regarding its botany as essentially Bahaman and Cuban, with a

certain admixture of northern elements.) To the first type I
refer such trees and shrubs as Asimina, sole representative of
the *Anonaceæ* out of the tropics, and reaching even to lat. 42°;
Chrysobalanus, representing a tropical suborder; Pinckneya,
representing as far north as Georgia the Cinchoneous tribe;
the Baccharis of our coast, reaching even to New England;
Cyrilla and Cliftonia, the former actually West Indian;
Bumelia, representing the tropical order *Sapotaceæ;* Big-
nonia and Tecoma of the *Bignoniaceæ;* Forestiera in *Ole-
aceæ;* Persea of the *Laurineæ;* and finally the *Cactaceæ.*
Among the herbaceous plants of this set I will allude only to
some of peculiar orders. Among them I reckon Sarracenia,
of which the only extra-North American representative is
tropical-American, the *Melastomaceæ*, represented by Rhexia;
Passiflora (our species being herbaceous), a few representatives of *Loasaceæ* and *Turneraceæ*, also of *Hydrophyllaceæ;*
our two genera of *Burmanniaceæ;* three genera of *Hæmo-
doraceæ;* Tillandsia in *Bromeliaceæ;* two genera of *Ponte-
deriaceæ;* two of *Commelynaceæ;* the outlying Mayaca and
Xyris, and three genera of *Eriocaulonaceæ.* I do not forget
that one of our species of Eriocaulon occurs on the west coast
of Ireland and in Skye, wonderfully out of place, though on
this side of the Atlantic it reaches Newfoundland. It may be
a survival in the Old World; but it is more probably of
chance introduction.

The other set of extra-European types, characteristic of the
Atlantic North American flora, is very notable. According to
a view which I have much and for a long while insisted on, it
may be said to represent a certain portion of the once rather
uniform flora of the arctic and less boreal zone, from the late
Tertiary down to the incoming of the Glacial period, and
which, brought down to our lower latitudes by the gradual
refrigeration, has been preserved here in eastern North
America and in the corresponding parts of Asia, but was lost
to Europe. I need not recapitulate the evidence upon which
this now generally accepted doctrine was founded; and to
enumerate the plants which testify in its favor would amount
to an enumeration of the greater part of the genera or sub-

ordinate groups of plants which distinguish our Atlantic flora
from that of Europe. The evidence, in brief, is that the
plants in question, or their moderately differentiated representatives, still coexist in the flora of eastern North America
and that of the Chino-Japanese region, the climates and conditions of which are very similar ; and that the fossilized representatives of many of them have been brought to light in
the late tertiary deposits of the arctic zone wherever explored.
In mentioning some of the plants of this category I include
the Magnolias, although there are no nearly identical species,
but there is a seemingly identical Liriodendron in China, and
the Schizandras and Illiciums are divided between the two
floras ; and I put into the list Menispermum, of which the
only other species is eastern Siberian, and is hardly distinguishable from ours. When you call to mind the series of
wholly extra-European types which are identically or approximately represented in the eastern North American and in the
eastern Asiatic temperate floras, such as Trautvetteria and
Hydrastis in *Ranunculaceæ ;* Caulophyllum, Diphylleia,
Jeffersonia and Podophyllum in *Berberideæ ;* Brasenia and
Nelumbium in *Nymphœaceæ ;* Stylophorum in *Papaveraceæ ;*
Stuartia and Gordonia in *Ternstrœmiaceæ ;* the equivalent
species of Xanthoxylum, the equivalent and identical species
of Vitis, and of the poisonous species of Rhus (one, if not
both, of which you may meet with in every botanical excursion, and which it will be safer not to handle) ; the Horse-
chestnuts, here called Buckeyes ; the Negundo, a peculiar off-
shoot of the Maple tribe; when you consider that almost
every one of the peculiar Leguminous tree mentioned as
characteristic of our flora is represented by a species in China
or Mandchuria or Japan, and so of some herbaceous *Leguminosæ ;* when you remember that the peculiar small order of
which Calycanthus is the principal type has its other representative in the same region ; that the species of Philadelphus,
of Hydrangea, of Itea, Astilbe, Hamamelis, Diervilla, Trios-
teum, Mitchella which carpets the ground under evergreen
woods, Chiogenes, creeping over the shaded bogs ; Epigæa,
choicest woodland flower of early spring; Elliottia ; Shortia

(the curious history of which I need not rehearse) ; Styrax of
cognate species ; Nyssa, the Asiatic representatives of which
affect a warmer region ; Gelsemium, which under the name
of Jessamine is the vernal pride of the southern Atlantic
States; Pyrularia and Buckleya, peculiar Santalaceous shrubs ;
Sassafras and Benzoins of the Laurel family ; Planera and
Maclura ; Pachysandra of the Box tribe ; the great development of the *Juglandaceæ* (of which the sole representative
in Europe probably was brought by man into southeastern
Europe in pre-historic times) ; our Hemlock Spruces, Arbor-
Vitæ, Chamæcyparis, Taxodium, and Torreya, with their east
Asian counterparts, the *Roxburghiaceæ*, represented by
Croomia, — and I might much further extend and particularize
the enumeration, — you will have enough to make it clear that
the peculiarities of the one flora are the peculiarities of the
other, and that the two are in striking contrast with the flora
of Europe. This contrast is susceptible of explanation. I have ventured to regard the two antipodal floras thus compared as the
favored heirs of the ante-glacial high-northern flora, or rather
as the heirs who have retained most of their inheritance.
For, inasmuch as the present arctic flora is essentially the
same round the world, and the Tertiary fossil plants entombed
in the strata beneath are also largely identical in all the longitudes, we may well infer that the ancestors of the present
northern temperate plants were as widely distributed throughout their northern home. In their enforced migration southward, geographical configuration and climatic differences
would begin to operate. Perhaps the way into Europe was
less open than into the lower latitudes of America and eastern
Asia, although there is reason to think that Greenland was
joined to Scandinavia. However that be, we know that Europe
was fairly well furnished with many of the vegetable types
that are now absent, possibly with most of them. Those that
have been recognized are mainly trees and shrubs, which
somehow take most readily to fossilization, but the herbaceous
vegetation probably accompanied the arboreal. At any rate,
Europe then possessed Torreyas, and Gingkos, Taxodium and

Glyptostrobus, Libocedrus, Pines of our five-leaved type, as well as the analogues of other American forms, several species of Juglans answering to the American forms, and the now peculiarly American genus Carya, Oaks of the American types, Myricas of the two American types, one or two Planer-trees, species of Populus answering to our Cottonwoods and our Balsam-poplar, a Sassafras and the analogues of our Persea and Benzoin, a Catalpa, Magnolias, and a Liriodendron, Maples answering to ours, and also a Negundo, and such peculiarly American Leguminous genera as the Locust, Honey Locust, and Gymnocladus. To understand how Europe came to lose these elements of her flora, and Atlantic North America to retain them, we must recall the poverty of Europe in native forest trees, to which I have already alluded. A few years ago, in an article on this subject, I drew up a sketch of the relative richness of Europe, Atlantic North America, Pacific North America, and the eastern side of temperate Asia in genera and species of forest trees.[1] In that sketch, as I am now convinced, the European forest elements were somewhat underrated. I allowed only thirty-three genera and eighty-five species, while to our Atlantic American forest were assigned sixty-six genera and one hundred and fifty-five species. I find from Nyman's Conspectus that there are trees on the southern and eastern borders of Europe which I had omitted; that there are good species which I had reckoned as synonyms, and some that may rise to arboreal height which I had counted as shrubs. But on the other hand and for the present purpose it may be rejoined that the list contained several trees, of as many genera, which were probably carried from Asia into Europe by the hand of man. On Nyman's authority I may put into this category *Cercis Siliquastrum*, *Ceratonia Siliqua*, *Diospyros Lotus*, *Styrax officinalis*, the Olive, and even the Walnut, the Chestnut, and the Cypress. However this may be, it seems clear that the native forest flora of Europe is exceptionally poor, and that it has lost many species and types which once belonged to it. We must suppose that the herbaceous flora has suffered in the same

[1] American Journal of Science and Arts, 3 ser., xvi. 85.

way. I have endeavored to show how this has naturally come about. I cannot state it more concisely than in the terms which I used six years ago.

" I conceive that three things have conspired to this loss of American, or as we might say, of normal types sustained by Europe. First, Europe, extending but little south of lat. 40°, is all within the limits of severe glacial action. Second, its mountains trend east and west, from the Pyrenees to the Carpathians and the Caucasus beyond : they had glaciers of their own, which must have begun their work and poured down the northward flanks while the plains were still covered with forest on the retreat from the great ice forces coming from the north. Attacked both on front and rear, much of the forest must have perished then and there.

"Third, across the line of retreat of whatever trees may have flanked the mountain ranges, or were stationed south of them, stretched the Mediterranean, an impassable barrier. . . . Escape by the east, and rehabilitation from that quarter until a very late period, was apparently prevented by the prolongation of the Mediterranean to the Caspian, and probably thence to the Siberian Ocean. If we accept the supposition of Nordenskjöld that, anterior to the Glacial period, Europe was ' bounded on the south by an ocean extending from the Atlantic over the present deserts of Sahara and Central Asia to the Pacific,' all chance of these American types having escaped from and reëntered Europe from the south and east seems excluded. Europe may thus be conceived to have been for a time somewhat in the condition in which Greenland is now. . . . Greenland may be referred to as a country which, having undergone extreme glaciation, bears the marks of it in the extreme poverty of its flora, and in the absence of the plants to which its southern portion, extending six degrees below the arctic circle, might be entitled. It ought to have trees and it might support them. But since their destruction by glaciation no way has been open for their return. Europe fared much better, but has suffered in its degree in a similar way."

Turning to this country for a contrast, we find the continent on the eastern side unbroken and open from the arctic circle to the tropic, and the mountains running north and south. The vegetation when pressed on the north by on-coming refrigeration had only to move its southern border southward to enjoy its normal climate over a favorable region of great extent; and, upon the recession of glaciation to the present limit, or in the oscillations which intervened, there was no physical impediment to the adjustment. Then, too, the more southern latitude of this country gave great advantage over Europe. The line of terminal moraines, which marks the limit of glaciation, rarely passes the parallel of 40° or 39°. Nor have any violent changes occurred here, as they have on the Pacific side of the continent, within the period under question. So, while Europe was suffering hardship, the lines of our Atlantic American flora were cast in pleasant places, and the goodly heritage remains essentially unimpaired.

The transverse direction and the massiveness of the mountains of Europe, while they have in part determined the comparative poverty of its forest-vegetation, have preserved there a rich and widely distributed alpine flora. That of Atlantic North America is insignificant. It consists of a few arctic plants, left scattered upon narrow and scattered mountaintops, or in cool ravines of moderate elevation; the maximum altitude is only about 6000 feet in lat. 44°, on the White Mountains of New Hampshire, where no winter snow outlasts mid-summer. The best alpine stations are within easy reach of Montreal. But as almost every species is common to Europe, and the mountains are not magnificent, they offer no great attraction to a European botanist.

Farther south, the Appalachian Mountains are higher, between lat. 36° and 34° rising considerably above 6000 feet; they have botanical attractions of their own, but they have no alpine plants. A few sub-alpine species linger on the cool shores of Lake Superior, at a comparatively low level. Perhaps as many are found nearly at the level of the sea on Anticosti, in the Gulf of St. Lawrence, abnormally cooled by the Labrador current.

The chain of the great fresh-water lakes, which are discharged by the brimming St. Lawrence, seems to have little effect upon our botany, beyond the bringing down of a few northwestern species. But you may note with interest that they harbor sundry maritime species, mementoes of the former saltness of these interior seas. *Cakile Americana,* much like the European Sea Rocket, *Hudsonia tomentosa* (a peculiar Cistaceous genus imitating a Heath), *Lathyrus maritimus,* and *Ammophila arenaria,* are the principal. Salicornia, Glaux, *Scirpus maritimus, Ranunculus Cymbalaria,* and some others, may be associated with them. But these are widely diffused over the saline soil which characterizes the plains beyond our wooded region.

I have thought that some general considerations like these might have more interest for the biological section at large than any particular indications of our most interesting plants, and of how and where the botanist might find them. Those who in these busy days can find time to herborize will be in the excellent hands of the Canadian botanists. At Philadelphia their brethren of the United States will be assembled to meet their visitors, and the Philadelphians will escort them to their classic ground, the Pine Barrens of New Jersey. To have an idea of this peculiar phytogeographical district, you may suppose a long wedge of the Carolina coast to be thrust up northward quite to New York harbor, bringing into a comparatively cool climate many of the interesting low-country plants of the south, which, at this season, you would not care to seek in their sultry proper homes. Years ago, when Pursh and Leconte and Torrey used to visit it, and in my own younger days, it was wholly primitive and upsoiled. Now, when the shore is lined with huge summer hotels, the Pitch Plnes carried off for firewood, the bogs converted into Cranberry-grounds, and much of the light sandy or gravelly soil planted with wine-yards or converted into Melon and Sweetpotato patches, I fear it may have lost some of its botanical attractions. But large tracts are still nearly in a state of nature. *Drosera filiformis,* so unlike any European species, and the beautiful Sabbatias, the yellow Fringed Orchis,

Lachnanthes and Lophiola, the larger Xyrises and Eriocaulons, the curious grass Amphicarpum with cleistogamous flowers at the root, the showy species of Chrysopsis, and many others, must still abound. And every botanist will wish to collect *Schizœa pusilla*, rarest, most local, and among the smallest of Ferns.

If only the season would allow it, there is a more southern station of special interest, — Wilmington, on the coast of North Carolina. Carnivorous plants have, of late years, excited the greatest interest, both popular and scientific ; and here, of all places, carnivorous plants seem to have their most varied development. For this is the only and the very local home of Dionæa; here grow almost all the North American species of Drosera ; here or near by are most of the species of Sarracenia, of the bladder-bearing Utricularias, — one of which the president of our Section has detected in fish-catching, — and also the largest species of Pinguicula.

But at this season a more enjoyable excursion may be made to the southern portion of the Alleghany or Appalachian Mountains, which separate the waters of the Atlantic side from those of the Mississippi. These mountains are now easily reached from Philadelphia. In Pennsylvania, where they consist of parallel ridges without peaks or crests, and are of no great height, they are less interesting botanically than in Virginia ; but it is in North Carolina and the adjacent borders of Tennessee that they rise to their highest altitude and take on more picturesque forms. On their sides the Atlantic forest, especially its deciduous-leaved portion, is still to be seen to greatest advantage, nearly in pristine condition, and composed of a greater variety of genera and species than in any other temperate region, excepting Japan. And in their shade are the greatest variety and abundance of shrubs, and a good share of the most peculiar herbaceous genera. This is the special home of our Rhododendrons, Azaleas, and Kalmias ; at least here they flourish in greatest number and in most luxuriant growth. *Rhododendron maximum* (which is found in a scattered way even as far north as the vicinity of Montreal) and *Kalmia latifolia* (both called

Laurels) even become forest trees in some places ; more commonly they are shrubs, forming dense thickets on steep mountain-sides, through which the traveler can make his way only by following old bear-paths, or by keeping strictly on the dividing crests of the leading ridges.

Only on the summits do we find *Rhododendron Catawbiense*, parent of so many handsome forms in English grounds, and on the higher wooded slopes the yellow and the flame-colored *Azalea calendulacea ;* on the lower, the pink *A. nudiflora* and more showy *A. arborescens*, along with the common and widespread *A. viscosa.* The latter part of June is the proper time to explore this region, and, if only one portion can be visited, Roan Mountain should be preferred.

On these mountain-tops we meet with a curious anomaly in geographical distribution. With rarest exceptions, plants which are common to this country and to Europe extend well northward. But on these summits from southern Virginia to Carolina, yet nowhere else, we find — undoubtedly indigenous and undoubtedly identical with the European species — the Lily-of-the-Valley !

I have given so much of my time to the botany of the Atlantic border that I can barely touch upon that of the western regions.

Between the wooded country of the Atlantic side of the continent and that of the Pacific side lies a vast extent of plains which are essentially woodless, except where they are traversed by mountain-chains. The prairies of the Atlantic States bordering the Mississippi and of the Winnipeg country shade off into the drier and gradually more saline plains, which, with an even and gradual rise, attain an elevation of 5000 feet or more where they abut against the Rocky Mountains. Until these are reached (over a space from the Alleghanies westward of about twenty degrees of longitude) the plains are unbroken. To a moderate distance beyond the Mississippi the country must have been in the main naturally wooded. There is rainfall enough for forest on these actual prairies. Trees grow fairly well when planted ; they are coming up spontaneously under present opportunities ; and there is reason for thinking that all the prairies east of the Mississippi, and of the Missouri up to Minnesota, have been either greatly extended or were even made treeless under Indian occupation and annual burnings. These prairies are flowery with a good number of characteristic plants, many of them evidently derived from the plains farther west. At this season, the predominant vegetation is of *Compositæ*, especially of Asters and Solidagoes, and of Sunflowers, Silphiums, and other Helianthoid *Compositæ*.

The drier and barer plains beyond, clothed with the short Buffalo-grasses, probably never bore trees in their present state, except as now some Cottonwoods (*i. e.* Poplars) on the margins of the long rivers which traverse them in their course from the Rocky Mountains to the Mississippi. Westward, the plains grow more and more saline ; and Wormwoods and *Chenopodiaceæ* of various sorts form the dominant vegetation, some of them *sui generis* or at least peculiar to the country, others identical or congeneric with those of the steppes of central Asia. Along with this common campestrine vegetation there is a large infusion of peculiar American types, which I suppose came from the southward, and to which I will again refer.

Then come the Rocky Mountains, traversing the whole continent from north to south ; their flanks wooded, but not richly so, — chiefly with Pines and Firs of very few species, and with a single ubiquitous Poplar, their higher crests bearing a well-developed alpine flora. This is the arctic flora prolonged southward upon the mountains of sufficient elevation, with a certain admixture in the lower latitudes of types pertaining to the lower vicinity.

There are almost 200 alpine Phænogamous species now known on the Rocky Mountains ; fully three-quarters of which are arctic, including Alaskan and Greenlandian ; and about half of them are known in Europe. Several others are north Asian but not European. Even in that northern portion of the Rocky Mountains which the Association is invited to visit, several alpine species novel to European botany may be met with; and farther south the peculiar forms increase. On the other hand, it is interesting to note how many Old-World species extend their range southward even to lat. 36° or 35°.

I have not seen the Rocky Mountains in the Dominion; but I apprehend that the aspect and character of the forest is Canadian, is mainly coniferous, and composed of very few species. Oaks and other cupuliferous trees, which give character to the Atlantic forest, are entirely wanting, until the southern confines of the region are reached in Colorado and New Mexico, and there they are few and small. In these southern parts there is a lesser amount of forest, but a much greater diversity of genera and species; of which the most notable are the Pines of the Mexican-plateau type.

The Rocky Mountains and the coast ranges on the Pacific side so nearly approach in British America that their forests merge, and the eastern types are gradually replaced by the more peculiar western. But in the United States a broad, arid and treeless, and even truly desert region is interposed. This has its greatest breadth and is best known where it is traversed by the Central Pacific Railroad. It is an immense plain between the Rocky Mountains and the Sierra Nevada, largely a basin with no outlet to the sea, covered with Sagebrush (*i. e.* peculiar species of Artemisia) and other subsaline vegetation, all of grayish hue; traversed, mostly north and south, by chains of mountains, which seem to be more bare than the plains, but which hold in their recesses a considerable amount of forest and of other vegetation, mostly of Rocky Mountains types.

Desolate and desert as this region appears, it is far from uninteresting to the botanist; but I must not stop to show how. Yet even the ardent botanist feels a sense of relief and exultation when, as he reaches the Sierra Nevada, he passes abruptly into perhaps the noblest coniferous forest in the world, — a forest which stretches along this range and its northern continuation, and along the less elevated ranges which border the Pacific coast, from the southern part of California to Alaska.

So much has been said about this forest, about the two

gigantic trees which have made it famous, and its Pines and
Firs which are hardly less wonderful, and which in Oregon
and British Columbia, descending into the plains, yield far
more timber to the acre than can be found anywhere else, — I
have myself discoursed upon the subject so largely on former
occasions, that I may cut short all discourse upon the Pacific-
coast flora and the questions it brings up.

I note only these points. Although this flora is richer than
that of the Atlantic in *Coniferæ* (having almost twice as many
species), richer indeed than any other except that of eastern
Asia, it is very meagre in deciduous trees. It has a fair number of Oaks, indeed, and it has a Flowering Dogwood, even
more showy than that which brightens our eastern woodlands
in spring. But, altogether it possesses only one-quarter of the
number of species of deciduous trees that the Atlantic forest
has; it is even much poorer than Europe in this respect. It
is destitute not only of the characteristic trees of the Atlantic
side, such as Liriodendron, Magnolia, Asimina, Nyssa, Catalpa,
Sassafras, Carya, and the arboreous *Leguminosæ* (Cercis excepted), but it also wants most of the genera which are common throughout all the other northern-temperate floras, having
no Lindens, Elms, Mulberries, Celtis, Beech, Chestnut, Hornbeam, and few and small Ashes and Maples. The shrubbery
and herbaceous vegetation, although rich and varied, is largely
peculiar, especially at the south. At the north we find a fair
number of species identical with the eastern; but it is interesting to remark that this region, interposed between the northeast Asiatic and the north-east American and with coast approximate to the former, has few of those peculiar genera
which, as I have insisted, witness to a most remarkable connection between two floras so widely sundered geographically.
Some of these types, indeed, occur in the intermediate region,
rendering the general absence the more noteworthy. And
certain peculiar types are represented in single identical
species on the coasts of Oregon and Japan, etc., (such as
Lysichiton, Fatsia, Glehnia); yet there is less community
between these floras than might be expected from their
geographical proximity at the north. Of course the high-
northern flora is not here in view.

Now if, as I have maintained, the eastern side of North
America and the eastern side of northern Asia are the favored
heirs of the old boreal flora, and if I have plausibly explained
how Europe lost so much of its portion of a common inheritance, it only remains to consider how the western side of
North America lost so much more. For that the missing types
once existed there, as well as in Europe, has already been indicated in the few fossil explorations that have been made.
They have brought to light Magnolias, Elms, Beeches, Chestnuts, a Liquidambar, etc. And living witnesses remain in the
two Sequoias of California, whose ancestors, along with Taxodium, which is similarly preserved on the Atlantic side, appear to have formed no small part of the Miocene flora of the
arctic regions.

Several causes may have conspired in the destruction; —
climatic differences between the two sides of the continent,
such as must early have been established (and we know that
a difference no greater than the present would be effective);
geographical configuration, probably confining the migration
to and fro to a long and narrow tract, little wider, perhaps,
than that to which it is now restricted; the tremendous out-
pouring of lava and volcanic ashes just anterior to the Glacial
period, by which a large part of the region was thickly covered; and, at length, competition from the Mexican-plateau
vegetation, — a vegetation beyond the reach of general glacial
movement from the north, and climatically well adapted to
the southwestern portion of the United States.

It is now becoming obvious that the Mexican-plateau vegetation is the proximate source of most of the peculiar elements of the Californian flora, as also of the southern Rocky
Mountain-region and of the Great Basin between; and that
these plants from the south have competed with those from
the north on the eastward plains and prairies. It is from this
source that are derived not only our *Cacteæ* but our *Mimoseæ*,
our Daleas and Petalostemons, our numerous and varied *Ona-
graceæ*, our *Loasaceæ*, a large part of our *Compositæ*, especially the *Eupatoriaceæ*, *Helianthoideæ*, *Helenioideæ*, and
Mutisiaceæ, which are so characteristic of the country, the

Asclepiadeœ, the very numerous *Polemoniaceœ, Hydrophyllaceœ, Eriogoneœ,* and the like.

I had formerly recognized this element in our North American flora; but I have only recently come to apprehend its full significance. With increasing knowledge we may in a good measure discriminate between the descendants of the ancient northern flora and those which come from the highlands of the southwest.

BIOGRAPHICAL SKETCHES.

BROWN AND HUMBOLDT.[1]

BEYOND the immediate pale of science, and the circle of its most devoted cultivators, the association of the names of Humboldt and Brown may seem new and strange; — the one a name familiar to the whole civilized world; the other, hardly known to a large portion of his educated countrymen. Yet these names stand together, in the highest place, upon the rolls of almost every Academy of Science in the world; and the common judgment of those competent to pronounce it will undoubtedly be, that although these vacant places upon these honorable rolls may be occupied, they will not be filled, in this, perhaps not in several generations.

Upon the death of Robert Brown, which occurred on the 10th of June last, in his eighty-fifth year, it was remarked that, next to Humboldt, his name adorned the list of a greater number of scientific societies than that of any other naturalist or philosopher. It was Humboldt himself who, many years ago, saluted Brown with the appellation of "Botanicarum facile Princeps"; and the universal consent of botanists recognized and confirmed the title. However the meed of merit in science should be divided between the most profound, and the most active and prolific minds, — between those who divine and those who elaborate, — it will probably be conceded by all that no one since Linnæus has brought such rare sagacity to bear upon the structure, and especially upon the ordinal characters and natural affinities of plants, as did Robert

[1] Proceedings of the American Academy of Arts and Science, iv. 229. (1859.)

Brown. True, he was fortunate in his time and in his opportunities. Men of great genius, happily, often are, or appear to be, through their power of turning opportunities to good account. The whole herbaria of Sir Joseph Banks and the great collections which he himself made around the coast of Australia, in Flinder's expedition, and which he was able to investigate upon the spot in the four years devoted to this exploration, opportunely placed in Brown's able hands as it were the vegetation of a new world, as rich as it was peculiar, — just at the time, too, when the immortal work of Jussieu had begun to be appreciated, and the European and other ordinary forms of vegetation had begun to be understood in their natural relations. The new, various, and singular types which render the botany of New Holland so unlike all other, Mr. Brown had to compare among themselves, — to unravel their intricacies with scarcely a clue to guide him, except that which his own genius enabled him to construct in the process of the research, — and to bring them harmoniously into the general system of botanical natural alliance as then understood, and as he was himself enabled to ascertain and display it. It was the wonderful sagacity and insight which he evinced in these investigations, which, soon after his return from Australia, revealed the master mind in botanical science, and erelong gave him the position of almost unchallenged eminence, which he retained, as if without effort, for more than half a century.

The common observer must wonder at this general recognition, during an era of great names and unequaled activity, of a claim so rarely, and as it were so reluctantly, asserted. For brief and comparatively few — alas! how much fewer than they should have been! — are Mr. Brown's publications. Much the largest of them is the " Prodromus of the Flora of New Holland," issued fifty years ago, which begins upon the one hundred and forty-fifth page, and which stopped short at the end of the first volume. The others are special papers, mostly of small bulk, devoted to the consideration of a particular plant, or a particular group or small collection of plants. But their simple titles seldom foreshow the full import of their contents. Brown delighted to rise from a special case to high and wide generalizations; and was apt to draw most important and always irresistible conclusions from small, selected data, or particular points of structure, which to ordinary apprehension would appear wholly inadequate to the purpose. He had unequaled skill in finding decisive instances. So all his discoveries, so simply and quietly announced, and all his notes and observations sedulously reduced to the briefest expression, are fertile far beyond the reader's expectation. Cautious to excess, never suggesting a theory until he had thoroughly weighed all the available objections to it, and never propounding a view which he did not know how to prove, perhaps no naturalist ever taught so much in writing so little, or made so few statements that had to be recalled, or even recast; and of no one can there be a stronger regret that he did not publish more.

With this character of mind, and while carefully sounding his way along the deep places of a science the philosophy and ground of which were forming, day by day, under his own and a few contemporary hands, Brown could not have been a voluminous writer. He could never have undertaken a "Systema Regni Vegetabilis," content to do his best at the moment, and take upon trust what he had not the means or the time to verify, — like his contemporary, De Candolle, who may worthily be compared with Brown for genius, and contrasted with him for the enthusiastic devotion which constantly impelled him to publication, and to lifelong, unselected herculean labor, over all the field, for the general good.

Nor could Brown ever be brought to undertake a "Genera Plantarum," like that of Jussieu; although his favorable and leisurely position, his vast knowledge, his keen discrimination, and his most compact mode of expression, especially indicated him for the task. Evidently, his influence upon the progress of botany might have been greater, or at least more immediate and more conspicuous. Yet, rightly to estimate that influence now, we have only to compare the "Genera Plantarum" of Endlicher with that of Jussieu, — separated as they are by the half-century which coincided with Brown's career, — and mark how largely the points of difference between the two, so far as they represent inquiry, and genuine advancement in the knowledge of floral structure, actually originated with him. Still, after making due allowance for a mind as scrupulous and cautious as it was clear and profound, also for an unusually retiring disposition, which even in authorship seems to have rendered him as sedulous to avoid publicity as most writers are to gain it, it must be acknowledged that his retentiveness was excessive; and that his guarded published statements sometimes appear as if intended — like the anagrams of the older mathematicians and philosophers — rather to record his knowledge than to reveal it. But this was probably only in appearance, and rather to be attributed to his sensitive regard for entire accuracy, and his extreme dislike of all parade of knowledge, — to the same peculiarity which everywhere led him to condense announcements of great consequence into short paragraphs or foot-notes, and to insert the most important facts in parentheses, which he who runs over the page may read, indeed, but which only the most learned and the most reflecting will be apt to comprehend. In candor it must be said that his long career has left some room for the complaint that he did not feel bound to exert fully and continuously all his matchless gifts in behalf of the science of which he was the most authoritative expositor.

But if thus in some sense unjust to himself and to his high calling, Brown could never be charged with the slightest injustice to any fellow-laborer. He was scrupulously careful, even solicitous, of the rights and claims of others; and in tracing the history of any discovery in which he had himself borne a part, he was sure to award to each one concerned his full due. If not always communicative, he was kind and considerate to all. To adopt the words of one of his intimate associates, "those who knew him as a man will bear unanimous testimony to the unvarying simplicity, truthfulness, and benevolence of his character," as well as to "the singular uprightness of his judgment."

The remaining and the most illustrious name of all — and one in its wide renown strongly in contrast with the last — has only just now been inscribed on our obituary list.

The telegraph of last week brought to us the painful intelligence that the patriarch of science, the universal Humboldt, died at Berlin on the 6th of May. Born in 1769, a year more prolific in great men than any equal period of all preceding time,[1] Humboldt had, before the end of the eighteenth century, exhibited qualities of the very highest order, and obtained a place of acknowledged celebrity in Europe. This, however, was the mere prelude to his career, for with the close of that century he commenced, with Bonpland, his wonderful exploration of Spanish America, which continued during five years. This journey must be considered in all future time as, substantially, the scientific discovery of Spanish America ; and whether we measure its results by the amount of knowledge through the wide fields of astronomy, geography, geology, mineralogy, meteorology, zoölogy, botany, and political economy, or the personal qualities by which this knowledge was collected and reduced to its place in the records of science, we cannot hesitate to rank the expedition amongst the most important and successful ever executed by man.

On his return to Europe, in 1805, Humboldt was employed several years in reducing his immense collection of materials to form for publication. From that time to his death, a period of almost half a century, he resided (except for a short time, in which he made his journey to northern Asia) in Europe, mostly in France and Germany. The last twelve or fifteen years of this great man were principally employed in the production of his " Cosmos," — the crowning labor of his long life, the harvest of his mature wisdom, — a work that could not have been produced by any other man, simply because no other man possessed the treasures, or the key to the treasures, of the various knowledge contained in it.

From his return to Europe to his death, he possessed, indisputably, the first place among philosophers, for the vast ex-

[1] Napoleon, Wellington, Mehemet Ali, Soult, Lannes, Ney, Castlereagh, Chateaubriand, Cuvier, and Humboldt. The name of Metternich is sometimes added to this list, probably incorrectly. That of Canning does not belong here, nor that of Mackintosh, nor of Sir Walter Scott.

tent of his acquirements. Without doubt, at all times during the present century there have been men much greater than Humboldt in each special department of science, but no one to compare with him in the number of subjects in which he had but few superiors, — no one who could, like him, bring all the sciences into one field of view, and compare them as one whole, through their relations and dependences. It was probably this extent of knowledge that led him to generalization rather than particular discovery; to trace connections and relations, rather than to search for new and minute facts or particular laws; to produce the " Cosmos," rather than discover the atomic theory or the cellular formation of organic structures. Many other men have been masters of several specialties; Humboldt alone brought the whole range of the physical and natural sciences into one specialty.

We cannot close this brief notice of the character and career of our illustrious associate without one moment's allusion to his amiable moral nature, his love of justice, and his superiority to all merely personal ends. So strong was his desire to give the influence of his high scientific position to the cause of civilization and the progress of knowledge, by assisting all applicants for his opinion and advice upon scientific subjects, that he permitted a correspondence to be extorted from him which in his last days became a load too great to be borne, and compelled a cry for relief that had hardly subsided when the news of his death reached us.

Such is the faint outline of a man whose name is indelibly written with those who have been most eminent in this wonderful age of scientific activity. The Academy claims the privilege, in common with the learned societies with which he was associated throughout the civilized world, to express its sorrow for his death, and to offer its tribute of honor to his memory.

AUGUSTIN-PYRAMUS DE CANDOLLE.

DE CANDOLLE was born at Geneva on the fourth day of February, 1778 ; he commenced his distinguished career as a botanist in Paris in the later days of the French republic; he continued it at Montpellier until 1816, when he returned to his native Geneva, where he died in September, 1841, — on the fifth day of that month, according to the opening paragraph of his son's preface to this volume,[1] — on the twenty-fifth according to the note by the same excellent authority at the close of the Memoir, p. 489. We cannot account for the discrepancy ; but the former is without doubt the true date.

The twenty-one years which have elapsed since his death have thinned the ranks of those who knew De Candolle, either personally or by correspondence. The " Théorie Elémentaire," the " Organographie," and the " Physiologie Végétale " have played their part, and have long ago passed out of general use. Yet, thanks to their influence, but more especially to the " Prodromus," the name of De Candolle is still perhaps the most prominent one with the cultivators of the science in general the world over, — is associated, not indeed with the profoundest depths, but with a larger amount of botany than any other name except that of Linnæus. These are the personal memoirs of an industrious, highly useful, prosperous, and honored life. Begun at middle age, perhaps mainly for the writer's own satisfaction, or that of his family, and continued at considerable intervals down to his last year, and evidently with a growing expectation of future publication, — they have appeared none too soon to secure the most interested but rapidly narrowing circle of readers. The outer circle, however, is as

[1] *Mémoires et Souvenirs* de Augustin-Pyramus De Candolle. *Ecrit par lui-même.* Geneva and Paris, 1862. (American Journal of Science and Arts, 2 ser., xxxv. 1. 1863.)

wide as ever, embracing all the lovers of botany in our day, to none of whom can the name of De Candolle be indifferent. The memoirs portray not so much the botanist as the man. Indeed, the perusal was rather disappointing to us in the former regard. We expected to get fresh glimpses of his mind at work upon the problems of the time, and to watch the rise and development of the ideas which brought him fame. That could be had, however, only from letters, diaries, or other contemporary records: these are only reminiscences. On this account, too, and perhaps because the record was made with only a dim and distant view to publication, the narrative somehow has not all the vivacity and sprightliness, nor the ready flow of language, nor the affluence of anecdote, which those who personally knew the writer would have expected. There are, however, many favorable specimens of De Candolle's powers of delineation, and some amusing anecdotes or interesting recollections of distinguished savans and others.

The family of De Candolle (to retain the style of orthography which is kept up at Geneva, in which the De is written as a substantial part of the name) is an old and noble one in the Provence ; and a branch of it, reaching Naples in the thirteenth century in the suite of the Anjou princes, flourished there, under a name gradually changed from Candola to Caldora, down to the middle of the sixteenth century. Augustin-Pyramus De Candolle derived one of his baptismal names from his ancestor, Pyramus de Candolle, who, becoming protestant, fled from Provence to Geneva in the year 1591, following an uncle who had already been established there for thirty or forty years. Augustin was the name of his father, in his earlier days a Genevan banker, a member of the state council, military syndic, and, about the time of the outbreak of the French Revolution, Premier Syndic of the little republic. Displaced by an earlier *coup d'état* just as he was about to enter upon the duties of this office, he had retired into the country just in time to escape the worst perils of the woful imitation at Geneva of the reign of terror, in July, 1794, although he was condemned to death for contumacy, and his property in the city for a time sequestrated. The rest of his life was peaceful and long : he attained the age of eighty-four years, and died in 1820.

Augustin-Pyramus, the writer of this autobiography, appears to have been remarkable in his boyhood rather for quickness of learning than for scholarship. His early tastes were for belles-lettres and poetry. Specimens of his poetical productions, both of his youth and of maturer years, are appended to the volume. Of their merit we cannot pretend to judge. At the age of sixteen he happened to attend a few lectures of a short course on botany, given by Vaucher, — who, living to a venerable age, survived his distinguished pupil. Here he learned the names of the parts of the flower, but nothing whatever of classification, having gone into the country for the summer before that portion of the course was reached. But his curiosity was awakened ; and in his leisure hours he began to collect, observe, and even to describe the plants he met with in his rambles, at first without any botanical book whatever to guide him, and without any idea beyond that of amusement or relaxation. The next winter, returning to Geneva and to his college studies, he came to know Saussure, then in his last years, and half paralytic. The veteran physicist, while he endeavored to attract the young man to scientific pursuits, discouraged his predilection for botany. That he regarded as quite unworthy of serious attention. Another summer passed upon the side of the Jura, however, and the perusal of Duhamel's " Physique des Arbres," of the " Researches upon Leaves " of Pastor Bonnet (a friend of his father), also of Hale's " Vegetable Statics," which he painfully translated from the English, and finally the acquisition of the " Linné de l'Europe " of Gilibert — in which the Linnæan artificial classification even then annoyed him by its incongruity with the natural relationships which he already recognized: these had by this time fixed his fate before he was at all aware of it; and perhaps had even determined in some sort his characteristics as a botanist.

An unexpected opportunity to pass the ensuing winter in Paris opened the way. This occurred through an invitation from Dolomieu, who, while young De Candolle was herborizing in the Jura, had been mineralogizing in the Alps, attended by two of De Candolle's school-mates, Picot and Pictet. In the autumn of 1796 the three young men proceeded to Paris, under the auspices of Dolomieu, who secured for De Candolle a lodging immediately over his own apartments, and presented him to Desfontaines and Deleuze at the Jardin des Plantes. No botanical lectures were given at that season of the year; but De Candolle attended the principal scientific courses then in progress; among them those of Fourcroy and Vauquelin upon Chemistry, of Portal and Cuvier upon anatomy, and of Hauy upon mineralogy. It was at this early period that his acquaintance and life-long intimacy with the excellent Delessert family commenced. By a rather ingenious device he contrived to make the acquaintance of Lamarck, but he gained little thereby in the way of botany, Lamarck being just then wholly occupied with the discussion of chemical theories. When De Candolle returned to Geneva in the spring of 1797, Lamarck sent by his hands a volume to Senebier, and so he came to know his amiable countryman, who, in ascertaining the capital fact that plants decompose carbonic acid, may be said to have laid the foundation of modern vegetable physiology. The first genus which De Candolle established (in 1799) was Senebiera.

From his narrative it would appear that, during this summer of 1797, the ambitious young botanist of two years' standing, and only eighteen years old, had not only conceived the idea of writing an elementary work, but actually traced the plan and written some chapters of it! He even states that from this period date the first observations and the conceptions — confused indeed, but correct — of the part which the abortion and the union of organs play in floral structure, — namely, the ideas which principally distinguish the "Théorie Elémentaire," published fifteen years later. How far these ideas were developed, however, we have no means of ascertaining. One would like to see an extract from this early manuscript, in confirmation.

The following winter he began to study law at Geneva. But with the little state now annexed to the great French

Republic, the prospects were not encouraging. A career must be sought elsewhere. De Candolle determined to study medicine, at the same time prosecuting his botanical studies, so as to have a double chance, by falling back upon the former in case the latter should fail to support him. In this view, he returned to Paris in the spring of 1798, just in time to see his patron Dolomieu set out for Egypt, as one of the savans of that famous expedition, and to decline a pressing invitation to accompany him. Taking a lodging in the Rue Copeau, to be near the Jardin des Plantes, he attended the hospitals and medical lectures, which he disliked, but recompensed himself at the Garden of Plants with the courses of Lacépède, Lamarck, Cuvier, and Hauy, omitting the botanical lectures as not to his mind, but sedulously examining the plants of the Garden. He renewed his acquaintance with Lamarck, at whose request he wrote a few articles (under the letter P) for the " Dictionaire Encyclopedique." Lamarck himself by this time had quite abandoned botany.

It was to Desfontaines that De Candolle was indebted for an immediate opportunity of beginning his botanical career. It came about thus : L'Heritier, who appears to have been wealthy, had engaged Redouté, the celebrated flower-painter, to prepare drawings of all the fleshy plants in cultivation, it being impossible to preserve them well in the herbarium. The artist undertaking to publish these drawings, applied to Desfontaines for a botanist to furnish the descriptive letterpress. The kind Desfontaines recommended De Candolle, and moreover offered to direct him in the work. He freely opened to the young botanist his herbarium and library, and allowed him to study by his side ; indeed Desfontaines was his botanical master and fatherly friend. The botanical library of L'Heritier, then much the largest at Paris, was naturally at his service, until the death by assassination, soon afterwards, of its singular owner. De Candolle, thus connecting his name and studies with the work of the unrivaled flower-painter, acquired thereby, as he remarks, more reputation than he deserved, and more instruction than he expected.

In the course of this same summer, of 1798, an invitation

from Alexander Brongniart, the mineralogist (whom De Candolle had slightly known, through Dolomieu, on his first visit to Paris), connected him with a small party of naturalists who made an excursion to Fontainebleau. Besides Dejean, the entomologist, then very young, Cuvier and Dumeril were of the party. In the autumn of the same year he visited Normandy, with less celebrated companions, and formed his first acquaintance with marine vegetation. The next year he made a visit to Holland, to consult the gardens and conservatories of that country, the richest in the " plantes grasses," which then occupied his attention. One result of this journey was that he induced his friend Benjamin Delessert to purchase Burmann's herbarium, and thus to lay the foundation of the important collections and library at the Hotel Delessert which have been so useful to naturalists, and so liberally devoted to their service. During the winter of the following year De Candolle elaborated the " Astragalogia," his first independent work of any considerable consequence, and which was published two years later ; in this he found opportunity to dedicate to his friend Delessert the Leguminous genus Lessertia.

About this time, namely, at the beginning of the century, he became acquainted with Mirbel, who had come up to Paris from the south of France, where he had been a pupil of Ramond. Instead of translating De Candolle's remarks, we may as well give them in the original.

" Il [Mirbel] savait alors peu de botanique, mais il annonçait de l'esprit et des talents. Je me liai avec lui. Il venait souvent déjeuner chez moi. Nous causions botanique ; j'avais deux ou trois ans d'avance sur lui, et j'étais naturellement communicatif ; je lui fis parts de plusieurs idées, nouvelles pour lui, et dont quelques-unes l'étaient pour le science. Elles parurent l'interresser, car j'en retrouvai une grande partie dans les éléments de physiologie qu'il publia peu d'années après ; telles sont la distinction des feuilles séminales et primordiales, l'importance de l'étude des nervures principales des feuilles, etc. Appelé à rendre un compte succinct de cette ouvrage dans le ' Bulletin philomathique,' je me divertis à ne citer que les idées que j'avais suggérées à l'auteur : je n'en revindiquai aucune,

et ne sais pas même s'il s'est aperçu de cette petite malice. Je dois
dire que je ne prétendis point, même alors, que se fût un plagiat
volontaire, mais il arrive souvent dans les sciences qu'on s'appropie,
sans s'en douter, ce qu'on a entendu dire.

"Cette circonstance éveilla ma propre attention sur la justice ri-
goureuse que j'ai désiré rendre a tous : la force de ma mémoire, et
surtout le soin que j'ai eu très-jeune de noter les faits et les idées
nouvelles que j'entendais dans la conversation, m'ont mis à même de
pouvoir, bien des années après une conversation, citer exactement
celui de qui j'avais appris un fait ou une opinion quelconque. Cette
habitude de justice m'a fait beaucoup d'amis, et j'ai eu souvent des
remerciements de gens cités par moi, qui eux-mêmes avaient oublié
ce qu'ils m'avaient dit." (pp. 91, 92.)

To De Candolle's credit it must be said, not only that his
career was remarkably free from controversies about priority
and reclamations, but that his example and precepts, his scrupulous care to render due credit to every contributor, his respect for unpublished names communicated to his own or
recorded in other herbaria, and the like, have been most
influential in establishing both the law and the ethics which
prevail in systematic botany (more fully, or from an earlier
period than in the other departments of natural history),
and which have secured such general coöperation and harmo-
nious relations among its votaries.

In these early days De Candolle was a good deal occupied
with vegetable physiology; the results are contained in his
papers "on the pores in the bark of leaves," *i. e.* stomata; on
the vegetation of the Mistletoe ; and on his experiments relative to the influence of light on certain points, mainly those
which exhibit strikingly the change in the position of their
leaves at night, which has been called the sleep of plants. The
account of these experiments, in which he caused certain plants
to acknowledge an artificial night and day, when read before
the Institute, gave him considerable eclat, — and probably
also the compliment of being named one of the three candi-
dates to fill the vacancy in the Academy of Sciences left by
the death of L'Heritier. A mere compliment, for the contest,
of course, was between Labillardière and Beauvois. In the

canvass De Candolle called upon Adanson, then very aged, and in his dotage more eccentric than ever.

If not chosen into the Institute, which indeed he could not pretend to expect, De Candolle was in that year made a member of that active association, — " la pépinière de l'Academie des Sciences," — the Société Philomathique, and was soon placed on the committee in charge of its Bulletin. This brought him into intimate connection with such colleagues as Brongniart (Alex.), Duméril, Cuvier, Biot, Lacroix, and Sylvestre.

" We met, at each other's lodgings, on Saturday evenings, after the session of the society, to read and to discuss the *morceaux* intended for the Bulletin, and when our labor was finished we took tea together and chatted familiarly. As one by one we exchanged the celibate for the married state, our wives were introduced ; — then we no longer read our extracts, and at length we gave over making the Bulletin, but we kept up our Saturday evening reunions. It was in consequence of this that Cuvier continued long afterwards his Saturday evening receptions ; but I return to the year 1800."

By De Candolle's account he was by about ten years the youngest member of this reunion. Yet he has the name of Biot and Duméril on his list, both of whom survived him for twenty years ; and Biot was really not quite four years his senior, and Duméril only five.

As a member of this select circle of intimate friends and zealous savans, all then pressing on to the very highest distinction, we may well believe that the ambitious young botanist enjoyed, and improved to the full, such golden opportunities, that he learned something of every branch of natural history, and also — what was no less useful at Paris — " à connaître les hommes et les mobiles cachés de bien des choses."

De Candolle sketches the following portraits of three of his associates, Duméril, Cuvier, and Lacroix. And first of

" The excellent Duméril. He was the ideal of the frank character which we attribute to the Picards. He was a sincere and devoted friend, always ready to second and render any service to me and mine. No cloud ever threw a shadow over our alliance, which became closer yet when, at a later period, the friendly connection of my wife with the widowed Madame Say determined the latter to marry Duméril. He was chief demonstrator in the anatomical department at the School of Medicine, but he became professor and member of the Academy of Sciences. Duméril was remarkable rather for the clearness of his ideas, and the variety and accuracy of his knowledge in natural history, than for theoretical principles. He was a practical man, whose elementary works had considerable success, but who, after having had a glimpse of some of the laws of organic symmetry, such as the analogy of the skull to vertebræ, seemed to have collapsed before their immensity. His principal services to science were in the way of teaching, and in the encouragement which he so well knew how to give to the young. The heart in this kind of influence is more essential than the head, and although Duméril's judgment was clear and quick, he was much more remarkable for his moral qualities.

"Cuvier, who was from the beginning the intimate friend of Duméril, was entirely different; and it would be difficult to find two people who were less analogous. Born at Montbéliard and brought up at Stuttgart, Cuvier had something of the gravity and even of the obstinacy of the German. Placed for some time in an inferior position, he was forced from his youth to make up for it by the dignity of his manner; but the world of savans, at least, will never forget his sojourn in Normandy, where he made those beautiful investigations on the molluscs which were the beginning of his fame. Called afterwards to the Jardin des Plantes as assistant to the aged Mertrud, he owed this position to the friendship of Geoffroy; but he soon surpassed his patron. In consequence of this position he was a member of the Institute from its foundation, and quickly acquired the reputation which results from great talent united to a skillful ambition. At the time when the office of secretary was annual he foresaw it would become perpetual, and arranged in such a manner as to fill one secretaryship almost continually, either himself or by others; so that he found himself in position to have it without contest when it became permanent and well paid. These first steps being taken, all places fell to him as of themselves, and we saw him successively Professor of the Ecoles Centrales, of the College de France, at the Jardin des Plantes, Inspector, then Councillor, then Chancellor of the University, Councillor of State, Baron, Peer of France, etc., etc. His talent, his aptitude for knowing and doing everything, made him skillful in every function; he brought to it method, order, facility for administration, a knowledge of details and of the whole, a sincere love of justice, and a disinterestedness which caused him to be noticed and admired.

"Cuvier might justly be compared to Haller, whom he resembled as much as the difference of nation and time would allow. Both astonished by their extraordinary capacity for learning, knowing equally well natural and historical science, greedy of positive facts on all subjects, endowed with wonderful memory and a remarkable spirit of order, capable of great labor, and yet gifted with much facility. But at the side of these admirable qualities it might be observed that neither had an inventive genius; they observed facts well, but never thought to unite them by a theory that would divine or discover others. Their characters corresponded even outside of science: both loved power, and sacrificed precious time to the desire of political advancement; both loved reading to a passion, even at the hours destined ordinarily for meals and domestic intercourse; both were cold and haughty in conversation with those who inspired them with no interest, piquant and profound to those whom they thought worthy of it; finally both had a certain contempt for that class of ideas called liberal, and held to the aristocratic party. The great size of their heads gave them a certain physical resemblance. In one word, it would be difficult to find two celebrated men more exactly alike, and the lovers of metempsychosis might say, if the epochs would permit, that the soul of Haller had passed without change into the body of Cuvier.

"To me personally, Cuvier was wellnigh perfection. . . . Notwithstanding the great difference in our respective views of life and of politics, and even of science in some theoretical matters, our intimacy was never clouded, nor was it disturbed by his quarrel with Geoffroy, although he knew that my opinions inclined toward those of the latter.

"The geometrician Lacroix was a genuine specimen of the philosopher of the eighteenth century, a republican of the school of Condorcet, an enemy to the great and their hangers-on, uniting the gaiety of a child with the moroseness of a disappointed old man, — the ease, grace, and kindness of a warm-hearted gentleman with the gruffness of a grumbler. He was a thoroughly excellent man, but a stranger to the life of the world around him. The character of the misanthrope in Molière, which I supposed purely imaginary, I found completely realized when I knew Lacroix."

An episode of fifteen days, during which De Candolle, to
his great surprise, had political functions to perform, — being
appointed one of the three notables of the department of the
Léman, in a representation of all the departments of the
French Republic, which the First Consul called together, —
gives us the first glimpse of Bonaparte in this narrative ; and
De Candolle's account of the interviews with him, and with
his minister of police, Fouché, is well worth preserving. With
this transient exception, we have only the most incidental
allusions to public affairs during the eventful years of the
Consulate, the Empire, and the Restoration.

We pass by, also, the interesting account which De Candolle gives of the doings of Delessert and himself, in the
establishment and administration of the Philanthropic Society,
which grew out of the introduction by them of Count Rumford's economical soups, distributed to the poor. These honorable undertakings brought the two friends into relations
with Rumford himself when he came to reside at Paris. Indeed Delessert, as we have had occasion to learn, became one
of Count Rumford's executors. The admiration with which
Rumford's writings and economical inventions had inspired
the two young philanthropists was much diminished upon
personal acquaintance.

"It was after his plans," writes De Candolle, "that we had
constructed our furnaces, after his receipts that we made our
soups, upon his advice that we were induced to substitute
such assistance for gifts of money."

So when Rumford was expected at Paris, they congratulated
themselves upon such an acquisition, went to meet him on his
arrival, and brought him to dine with them.

"We found him a dry, methodical man, who spoke of benevolence
as a discipline, and of the poor as we should not have dared to speak
of vagabonds. It is necessary, said he, to punish those who give
alms ; the poor must be forced to work, etc., etc. Great was our
astonishment at hearing such maxims : however, we did our utmost
to profit by his advice in practical matters. I had a good deal of
intercourse with him, one among others odd enough. Mademoiselle
Rath, a Genevese painter, and like ourselves enthusiastic about

Rumford, wished to paint his portrait to be engraved. M. Jay, her relation and my friend, then director of the ' Décade Philosophique,' wished to put it into his journal, and asked me for a notice of M. Rumford to accompany it. Knowing little of his former life, I asked M. Rumford himself for a few notes ; he promised them, and appointed an interview at his house to give them to me. I went : what was my astonishment when he presented an article entirely complete and quite eulogistic. That was not all ; he required me to copy it on the spot, not wishing to leave the manuscript in his writing in my hands. I thought the proceeding rather indelicate, and the distrust not very polite. I deferred however to the wishes of a man for whom I had always had until then the highest respect ; I obeyed : I transmitted to the ' Décade' the written article, with small additions, and I have never mentioned until after the death of Rumford, not even until now, the secret of its origin, thinking that this trait would not raise him in estimation.

" M. Rumford settled in Paris, where he afterwards married Madame Lavoisier, the widow of the celebrated chemist. I saw something of both, and I never knew an odder union. M. Rumford was cold, imperturbable, obstinate, egotistical, prodigiously occupied with the material part of life, and in inventions in the smallest matters. He was engrossed with chimneys, lamps, coffee-pots, and windows made after a peculiar fashion ; and he contradicted his wife twenty times a day about the management of her housekeeping. Madame Lavoisier-Rumford . . . was a woman of very decided character. A widow for twelve or fifteen years, she had been in the habit of having her own way, and did not like to be contradicted. Her mind was broad, her will strong, her character masculine. She was capable of lasting friendship, and I could always congratulate myself on her kindness to me. Her second marriage was soon disturbed by grotesque scenes. Separation was better for both than union. He got a pension, which he needed, but which death prevented his long enjoying. She obtained liberty and the title of Countess : both were satisfied. He could now arrange the house at Auteuil as he liked : she continued to receive a select circle at hers."

Of this racy and unflattering sketch we have only to remark that, however it may have been as to the pension, Rumford's pecuniary means, as shown by his endowments and legacies in this country, were more considerable than De Candolle supposed.

Apropos to reminiscences of distinguished savans, we look forward a year or two in the narrative, and select the following. And first, of a person who was well known to a past generation, and to some who still survive, at Philadelphia.

"Joseph Correa de Serra was then about fifty-five or sixty years old. He was of an ancient family in Portugal, which had produced several literary men. After studying at the University of Coimbra he was transferred to Rome, where he pursued theological studies for a dozen years at the College of the Sapienza, but which he left with a knowledge of many things beside theology. Returning to Portugal, he was made governor to the hereditary Prince, Secretary to the Academy of Sciences, etc., and became a very influential person, both on account of his talents and on account of the position of his pupil, who it was supposed would become king on attaining his majority, as his mother was only regent. Correa was made Minister; and his first act was to overthrow the Inquisition. But the Prince died just as he was coming of age, and Correa was left exposed to the hatred and jealousy of the priests. After a while he obtained permission to go to England, where he lived in the society of the savans of which Sir Joseph Banks' house was the centre. Afterwards he moved to Paris, where he also lived among savans and men of letters, and where he showed the most noble character when the seizure of Portugal by Bonaparte deprived him of all his resources. He possessed the singular faculty of knowing everything apparently without labor. It is only the people of the south who can thus combine great facility with profound idleness. The latter prevented his publishing anything beyond small dissertations, quite below his talents; but in conversation all his various knowledge and his ingenious views were charmingly exhibited. In these days Humboldt and Cuvier often came to my lodgings, where they occasionally met Correa. Although their celebrity was far above his, and justly so, on account of their published works, yet Correa always got the advantage over them; and it was by no means the least of the enjoyments of our sociable little dinners to see the sort of deference, and even fear, which Cuvier and Humboldt exhibited in the announcement of their opinions before Correa, who, with the grace and sly maliciousness of a cat, would at once expose their weak sides. Like them, he was familiar with all the historical and natural sciences, and he used his vast stores of knowledge with a severe logic and rare sagacity. He spent many hours in my herbarium; where the subtle perspicacity which he brought to bear at a glance upon plants, often wholly new to him, taught me much of the art of observing, and especially of combining observations in botany. To such talents he joined a lofty soul and a heart devoted to friendship. It was a great grief to me when, at over sixty years of age, he quitted Europe to rejoin in Brazil the king who had persecuted him ; but he forgot all his wrongs when his sovereign became unfortunate. Correa died when ambassador to the United States."

The following, of a somewhat later period, is abridged from De Candolle's account of the Société d'Arcueil : —

" Its founder was the excellent and illustrious Berthollet, who, then living in his country residence at Arcueil, . . . invited thither, once a month, a few young savans, by way of encouraging their efforts. His colleagues MM. de la Place and Chaptal, also senators and members of the Institute, were, so to say, vice-presidents of this little reunion. Humboldt also had a place, and the parterre was composed of Biot, Thénard, Gay-Lussac, Descotils, Malus, Amédée Berthollet, and myself. Later, Berard and François de la Roche were admitted. [And finally Arago, Poisson, and Dulong, adds the editor, who notes that the last volume of the " Mémoires d'Arcueil " was published in 1817.] The association was devoted to the physical and chemical sciences. I was admitted in view of the applications of vegetable physiology to chemistry ; and I contributed some articles upon this subject to the 'Mémoires d'Arcueil,' namely, my Note on the cause of the direction of stems towards the light, my Memoir on the influence of absolute height upon vegetation, and upon the geographical or topographical distribution of plants, and, later, one upon double flowers, especially of the *Ranunculaceæ.* The first of these writings was a simple and clear solution [although an incorrect one, as it proves. — Eds.] of a problem which was deemed insoluble ; the second reduced to just proportions the exaggerations of Humboldt upon the influence of elevation ; the third was an essay connected with the observations of the degenerescence of organs, to which my ' Théorie Elémentaire ' was devoted. . . .

" We commonly made our rendezvous at Thénard's, and went together to Arcueil, as happy with this run into the country as schoolboys out for a holiday. We walked about in this pleasant villa, and relished the society of our leaders. Nothing can fully describe the good-nature and simplicity of M. Berthollet and even of Madame. They were with us as parents with their children, and we made ourselves at home in the house with perfect *abandon.* M. Berthollet was quite fat and very full-blooded. He feared heat so much that he wore clothes only out of respect to society, and at night he slept entirely uncovered upon his bed. ' What,' said we, ' even in winter?' ' Oh,' he answered, ' when it is very cold I spread my pocket-handkerchief over my feet.' This man, so high in social rank and scientific celebrity, bore contradiction unusually well, and loved above all things truth. When the first works of Berzelius upon definite proportions became known at Paris, I was very much taken with them, and although they were in direct opposition to the principles of statical chemistry he sustained, I did not fear to tell M. Berthollet the high opinion I had of them. Far from taking offense at this preference, he encouraged me to study the writings of Berzelius.

" M. de la Place was of quite a different character. He had the dryness of a geometrician and the haughtiness of a parvenu. Over and above these defects of manner, he was a man of honor and worth. . . . He often seconded me, although in truth he thought very little of natural history. In our meetings he often had little quarrels with M. Berthollet, and would think to silence him by saying, ' But you see, M. Berthollet, what I say to you is mathematics.' ' Eh, par Dieu, what I say to you is physics,' answered the other, ' and that is quite as good.' . . . Humboldt also came from time to time; but he added much of life and interest when he appeared. He affected to pass himself as the creator of the science of botanical geography, to which he has only added certain facts, and the exaggeration of a true theory so as to render it almost false. He never quite pardoned me for having, in the preface to my memoir on the geography of the plants of France, cited those who before him had occupied themselves with geographical botany, although in this exposition I had, in truth, much amplified his share.

" Among the other members of the society of whom I have not yet spoken, I would chiefly mention Thénard, who was then commencing a career which has since become very brilliant. His activity, his ardor, and his uprightness pleased me very much. . . . I could draw, in an anecdote, the contrast between the characters of Thénard and Descotils. . . . It was then very difficult to correspond with England, on account of the continental blockade. I happened to be the first to receive, by a letter from Dr. Marcet, the news of Davy's great discovery in decomposing the fixed alkalies. By a happy chance, it reached me on the morning of the day of our meeting. I hastened to our usual rendezvous, and could not wait for the session to impart so important a discovery. I read my letter to the members present. Thénard was enthusiastic ; he ran about the room like a madman, crying out, ' It is beautiful ! it is admirable ! ' Then turning to me, and laying hold of his arm : ' Look here,' said he, ' I would give this arm to have made this discovery.' Descotils, tranquilly buried in an arm-chair, said also, but in quite another tone : ' It is very fine ; but I would not give the end of my little finger to have made it.' "

We pass over all De Candolle's account of his life and domestic affairs during his residence at Paris, his particular investigations, his excursions, in Switzerland and elsewhere, — even the memorable one in the Jura with Biot and Bonpland. in which he led the party into a position of imminent danger, causing Bonpland to bemoan his hard fate in having to perish on such a mole-hill as the Jura, after having safely climbed Chimborazo (p. 154) ; — his engagement and marriage (the latter in April, 1802), with Mademoiselle Torras, of a Genevan family resident in Paris ; of the foundation of his herbarium by the fortunate acquisition of that of L'Heritier ; — of the first course of lectures which he gave, at the Collége de France, as a substitute for Cuvier, during the temporary absence of the latter, giving a course of vegetable physiology in place of one on general natural history ; — how he prepared to take the degree of M. D. in order to qualify himself as a candidate for the chair of medical natural history at the School of Medicine, then vacant ; but how Richard, who disliked him because he was a pupil of Desfontaines, as De Candolle says, instigated Jussieu to offer himself for this chair, upon which, of course, De Candolle withdrew, but nevertheless wrote and sustained as a thesis for the doctorate, his Essay on the Medical Properties of Plants, compared with their exterior forms and their natural classification. He bore his examination creditably, received his diploma, and, the same evening, a private mock inauguration, which, considering the parties engaged in it, must have been irresistibly comical.

" Duméril invited to his house my family, my comrades of the ' Bulletin Philomathique,' and even some of the Professors of the

Ecole de Medicine. This grave assembly amused themselves in giving me the reception, in full dress, from the 'Malade imaginaire.' It was a curious sight to see Cuvier, Lacroix, Biot, and other learned Academicians rehearsing the scene from Molière in the costumes of the Comédie Française. They had smothered me in an immense sugar-loaf paper cap ornamented all over with little lamps all alight. In the motion of bowing I constantly expected to be set on fire. But the acolyte who conducted me would then press a sponge well filled with water borne on the top of the cap, and the water ran down, not upon the lamps, but upon my head, — the audience laughing uproariously at my surprise."

Let us pass on to more serious matters, and rapidly sketch the outlines of the scientific career now fairly and promisingly opening. For the event which fixed De Candolle in his true field of labor was his arrangement (in 1802) with Lamarck — who had long since abandoned botany — to prepare a new edition of the "Flore Française." The arrangement was a favorable one to De Candolle, both financially and scientifically. The new edition was of course an entirely new work, one particularly adapted to De Candolle's genius, and which gave him at once a wide reputation. Indirectly this work gave origin to the botanical explorations of the provinces of France, under the auspices of the government, which engaged much of De Candolle's attention from the summer of 1806 until he ceased to be a French subject.

And now, the death of old Adanson left a vacancy in the botanical section of the Institute, which De Candolle might hope to fill. But parties and personal dislikes, as it appears, were not unknown nor uninfluential in the Paris of half a century ago. Indeed De Candolle (let us hope without sufficient grounds) roundly charges lamentable weakness to Lamarck, and less creditable motives to Fourcroy and even to Jussieu, in respect to the nomination and canvass; while of the Abbé Hauy he relates, to his credit, that, upon being approached with the suggestion that his conscience should prevent his voting for a protestant, he replied that he was very glad of an opportunity to show that he never mixed up religious opinions with scientific judgments. Palisot de Beauvois, the rival candidate, was elected, in spite of the hearty support De Candolle received from his comrades of the " Bulletin Philomathique " and his eminent associates of the Société d'Arcueil, Berthollet, Chaptal, La Place, Cuvier, etc., — to say nothing of his scientific superiority over his rival, which De Candolle naturally regarded as very great. At that time, according to De Candolle, Beauvois had produced, "ni la ' Flore d'Oware,' ni le ' Prodrome de l'Ethéogamie,' ni en un mot aucun de ses ouvrages qui," etc. But in this De Candolle's memory was perhaps at fault; for, while this election took place in the autumn of 1806, the latter of these works of Beauvois, according to Pritzel, was published in 1805, and the first volume of the former in 1804.

Evidently the disappointment was keenly felt. Membership in the Institute secured not only an assured position but also a comfortable little annuity. This, and the prospective needs of an increasing family, disposed De Candolle to look elsewhere, and to accept, after some hesitation, the botanical chair at the University of Montpellier, which in 1807 became vacant by the death of Broussonet. Hardly was he established there when the death of Ventenat, in the autumn of 1808, made him again a candidate for a seat in the Institute; — again an unsuccessful one, but now chiefly because a considerable number of his particular friends in the Institute required a promise that if chosen he would reside at Paris, which he could not with propriety give. So they voted for Mirbel; — and De Candolle took root at Montpellier, where he flourished from 1808 to the year 1816.

That De Candolle, full of ambition and with a good opinion of his abilities, should have disliked to give up Paris is natural; but he himself afterwards records the opinion (which we share) that his removal from the metropolis was the best thing for him, as enabling him to accomplish more for botany. And as to the honors of the Institute, his disappointments were more than made up to him in the sequel by his election as one of the eight foreign associates of the Academy of Sciences.

At Montpellier, De Candolle was heartily welcomed by his

colleagues, by the official personages, and by the protestant society of the city, — in those days there was little social intercourse between catholics and protestants in the south of France, — and he gave himself with ardor and success to his new duties. He renovated the botanic garden, — the oldest in France, founded by Henry IV., — and secured additional funds for its support. He built up the botanical school, and developed peculiar talents as an instructor, — with results perhaps up to the average as respects the making of botanists; but Dunal, one of his earliest pupils, was about the only one at Montpellier who achieved a general reputation, and his fell much below expectations. He continued and extended his official botanical explorations of the provinces of France, making annual reports to the Minister of the Interior, and planning a very comprehensive work on the " Statique Végétale de la France," which, however, owing to political and other changes, was never written. He wrote and published the " Théorie Elémentaire " which made his reputation as a theoretical botanist, and well exemplifies the characteristics of his genius in this regard, — constructive rather than critical, — quick and ingenious in seizing analogies and in framing hypotheses, rather than sagacious in testing their validity,— content with an hypothesis which neatly connects observed facts, but not so solicitous to prove it actually true, nor urgent to follow it out to ultimate conclusions, — a lucid expositor, and a happy diviner within a certain reach, rather than a profound investigator, — in short, a generalizer rather than an analyzer.[1]

At Montpellier, also, De Candolle planned his " Systema Vegetabilium " — a systematic and detailed account of all known plants, arranged under their natural families, — and

---

[1] It is curious that De Candolle, who early took to the ideas of Geoffroy in anatomy who founded his morphology of the flower upon the idea of symmetry, and recognized the homology of the floral organs with leaves, and who could have got from the writings of his townsman, Bonnet, enough of phyllotaxy for the purpose, seems never to have thought of connecting the one with the other, nor to have asked himself why a flower is symmetrical.

he there prepared the first volume of this work; thus, with characteristic ardor and courage, but without calculating its immensity, entering upon the grand and most important undertaking of his life, and into that field of labor in systematic and descriptive botany for which he was eminently adapted, by his enterprising disposition and unflagging industry, his capacity for sustained labor, his excellent memory, his spirit of order and method, his quickness of eye, and his great aptitude for generalization.

The overthrow of the Empire, the Restoration, the Hundred Days, and the final fall of Napoleon supervened. De Candolle's life at Montpellier was troubled and his prospects precarious. He naturally turned to his native Geneva, where he had kept up intimate social relations; and when he had ascertained that a place would be provided for him, he exchanged the comparatively ample emoluments of the chair at Montpellier for the very humble salary of one at Geneva, encumbered with the duty of lecturing upon zoölogy as well as botany.

Pending the change he made a visit to England, in 1816, of which a detailed account is given, with reminiscences of the botanists and others whose personal acquaintance he then made. We regret that we have no room left for further extracts: his account of Brown is expressive of the great respect he entertained for him, and that of Salisbury and of Lambert is amusing.

Settled now at Geneva, at the good working age of thirty-eight, the narrative of his steadily industrious and prosperous life, and of his happy surroundings, flows on for nearly two hundred pages, down to the sad overthrow of his health by an overdose of iodine in 1836, his partial convalescence and resumption of botanical work in 1837, and ends with the record of the death of his only brother, at the beginning of the year 1841, only eight months before his own.

These twenty-five years witnessed the publication of the two volumes of the "Systema"; the change of plan to a "Species Plantarum" in a restricted form, more nearly within the limits of a mortal's life and powers; the publica-

tion of the " Organographie " and of the " Physiologie Végétale," and — not to mention a hundred other botanical and
sundry miscellaneous writings, of greater or smaller extent —
of seven out of the present fifteen volumes of the " Prodromus." Only one botanist of the present century — and one
happily who still survives — has accomplished an equal
amount of work, and good work, in systematic botany.

Our account has run on to such a length that we cannot touch
upon De Candolle's social and domestic life — of which the
memoirs reveal pleasant glimpses, nor of his useful and honorable life as a Genevan and Swiss citizen. Nor can we now
venture to gather interesting anecdotes from his notices of
friends, visitors, pupils, and collaborators ; nor notice his
methods of working, and his capital arrangements for securing and classifying details and economizing time.

It is not for us to pronounce upon De Candolle's relative
rank in the hierarchy of naturalists. He incidentally once
speaks of Brown and himself as rivals for the botanical
sceptre. It is natural that they should be compared, or
rather contrasted ; for they were the complements of each
other in almost every respect. The fusion of the two would
have made a perfect botanist. But De Candolle's facility for
generalization, zeal, and industry were as much above, as
his depth of insight and analytical power were below Brown's.
The one longed, the other loathed, to bring forth all he knew.
The editor compares De Candolle's traits of character with
those of Linnæus, as delineated by Fabricius, and finds much
resemblance. But his impress upon the science, however
broad and good, can hardly be compared with that of Linnæus.

BENJAMIN D. GREENE.[1]

BENJAMIN D. GREENE, Esq., of Boston, died on the 14th of October last, at the age of sixty-nine years. He was born in 1793, and graduated at Harvard University in the year 1812. He first pursued legal studies, partly in the then celebrated school at Litchfield, Connecticut, and was duly admitted to the bar in Boston. He then took up the study of medicine, and completed his medical course in the medical schools of Scotland and Paris, taking his medical degree at Edinburgh in the year 1821. The large advantages of such a training having been enjoyed, Mr. Greene did not engage in the practice of either profession. An ample inheritance, which rendered professional exertion unnecessary, conspiring with a remarkably quiet and contemplative disposition and a refined taste, led him to devote his time to literary culture and to scientific pursuits. His fondness for botany, which early developed, was stimulated by personal intercourse with various European botanists, and especially with his surviving friend, the now venerable Sir William Hooker, then professor at the University of Glasgow, to whom he naturally became much attached, and by whom he was highly appreciated.

In botany, as in everything else, Mr. Greene sought to be silently useful. He never himself published any of his discoveries or observations. The few species to which his name is annexed were given to the world at second hand. But his collections were extensive, his original observations numerous and accurate, and both were freely placed at the disposal of working botanists. He early saw that the great obstacles to the advantageous prosecution of botanical investigation in this country, and especially in New England, were the want of books and the want of authentic collections; and these

[1] American Journal of Science and Arts, 2 ser., xxxv. 449. (1863.)

desiderata he endeavored, so far as he could, to supply. He gathered a choice botanical library, he encouraged explorations, and he subscribed to all the large purchasable North American collections, — beginning with those of Drummond in the southern United States and in the then Mexican province of Texas. These being distributed under numbers, among the principal herbaria of the world, and named or referred to in monographs or other botanical works, were of prime importance as standards of comparison. Such collections and such books as Mr. Greene brought together were just the apparatus most needed at that time in this country ; and now when our wants are somewhat better supplied, we should not forget the essential service which they have rendered, nor the disinterested kindness with which their most amiable and excellent owner always placed them at the disposal of those who could advantageously use them. Mr. Greene's botanical library and collections have been, by gift and bequest, consigned to the Boston Society of Natural History, of which he was one of the founders and the first president, and by which they will be preserved for the benefit of future New England botanists, by whom his memory should ever be gratefully cherished. The genus Greenea, established by Wight and Arnott upon two rare Rubiaceous shrubs of India, barely anticipated a similar dedication by his old friend Mr. Nuttall, of a curious Grass of Arkansas and Texas, and will perpetuate his name in the annals of the science which he lovingly cultivated.

CHARLES WILKINS SHORT.[1]

DR. CHARLES WILKINS SHORT died at his residence at Louisville, Kentucky, on the 7th of March last, in the sixty-ninth year of his age. He was born in Woodford County of that State, on October 6, 1794, was educated in the school of Mr. Joshua Frye, near Danville, — a distinguished teacher of those days, — pursued his professional studies mainly in Philadelphia, where he took the degree of M. D. from the University of Pennsylvania in the year 1815. For ten years he devoted himself to the practice of medicine, until in the year 1825 he was called to the chair of Materia Medica and Medical Botany in the Transylvania University at Lexington, where he contributed to the reputation of that celebrated school. Relinquishing medical practice, for which he had no liking, he devoted his powers with zeal and success to the more congenial duties of his professorship, and to the cultivation of botany, the favorite pursuit of his life. At the close of the year 1838 he removed, along with some of his distinguished colleagues, to Louisville, filling the same chair in the University of that city until 1849, when he retired from public functions. The remainder of his honorable life was passed at Hayfield, at his tasteful residence near Louisville, in the bosom of his family; in the exercise of kindly but unostentatious hospitality and of all good offices ; in quietly enjoying and in causing others to enjoy the blessings of a handsome fortune, to which by inheritance, combined with the fruits of his own industry, he now attained, and in the cultivation of the "amiable science" to which he was devotedly attached.

Dr. Short's botanical publications were neither large nor many. They were chiefly articles contributed to the "Tran-

[1] American Journal of Science and Arts, 2 ser., xxxv. 451. (1863.)

sylvania Journal of Medicine," etc., of which he was for some time one of the editors. The most important is his Catalogue of the plants of his native state (which he widely and assiduously explored), and several supplements; with well considered characters of some new species, and acute and discriminating notes upon several imperfectly known plants. These and the copious manuscript observations which he was for many years accustomed to communicate to his botanical correspondents, showed what he was capable of accomplishing, had not a most retiring and unambitious disposition unduly limited his exertion. It was not activity or persevering labor, but publicity, that he shrunk from. He was a very industrious botanist, and an effectual promoter of our science in this country. His great usefulness in this field was mainly owing to the extent and the particular excellence of his personal collections, and to the generous profusion with which he distributed them far and wide among his fellow-laborers in this and other lands. He and the late Mr. Oakes — the one in the west and the other in the east, but independently — were the first in this country to prepare on an ample scale dried specimens of uniform and superlative excellence and beauty, and in lavish abundance for the purpose of supplying all who could need them. Dr. Short's disinterested activity in these respects has enriched almost every considerable herbarium both at home and abroad, and set an example which has produced large and good results among us. The vast improvement in the character of the dried specimens now generally made by our botanists may be mainly traced to the example and influence of Dr. Short and Mr. Oakes. As might be expected, Dr. Short's own herbarium is a model of taste and neatness. It is also large and important. To one himself so solicitous " to do good and to communicate," contributions from numerous sources naturally flowed abundantly. He, moreover, subscribed to all the North American distributed collections within his reach, and he set on foot or efficiently furthered several distant or difficult botanical explorations. He purchased, at a liberal price, the important botanical collections of Texas and northern Mexico, left by Berlandier, which Lieutenant (now General) Couch acquired of his widow and sent on to Washington ; and, retaining one set for his own herbarium, he caused the rest to be distributed among the botanists to whom they would be most useful, — especially including two Swiss botanists who had contributed to send out Berlandier to Mexico as a collector, but from whom (apparently through Berlandier's dishonesty) they had failed to receive any adequate return. It is understood that Dr. Short's rich herbarium — to which a lifetime of thoughtful attention and much expense were lovingly devoted — is now offered, by a wise bequest, to the custody of the Smithsonian Institution, under instructions that it shall be permanently well cared for and always open to be consulted by botanists. It will there form an excellent and conspicuous nucleus for a collection of American herbaria, such as our science needs, and the country ought to possess.

The natural effects upon his scientific career of a fastidious taste, an unwarrantable diffidence, and a too retiring disposition, were enhanced by a constitutional tendency to depression of spirits. But this never obscured the native kindness of his heart, nor the real though so quiet geniality of his disposition, nor checked an unobtrusive and considerate benevolence. With an uncompromising sense of right and justice, and a keen hatred of everything mean and unworthy, he was never harsh or even cynical. All who knew him well, and also his more intimate correspondents who never enjoyed the privilege of a personal acquaintance, can testify to the nobility and Christian excellence of his character. An appreciative tribute to his memory from the pen of a former colleague will be found in the " Louisville Journal," issued a few days after Dr. Short's lamented death.

Two or three species of Kentucky plants commemorate the name of Dr. Short as their discoverer. Also a new genus Shortia, inhabiting the Alleghany Mountains, was dedicated by him to the present writer. But, alas ! too like the botanist for whom it was named, it is so retiring in its habits that it is not known as it ought to be, but lives as yet unseen, except by a single botanist of a former generation, in some secluded recess of the Black Mountain of North Carolina. It will some day be found again and appreciated.

FRANCIS BOOTT.[1]

FRANCIS BOOTT, M. D., died at his residence in London on Christmas morning, in the seventy-first year of his age. He was born in Boston, on the 26th of September, 1792. His father, Kirk Boott, came to this country early in life, from Derbyshire, England, became a successful merchant in Boston, was one of the pioneers of manufacturing enterprise here, and one of the founders of Lowell, — the type, if not wholly the original of manufacturing towns. His Boston residence was on the site now occupied by the Revere House, of which the Boott mansion forms a part. Francis Boott entered Harvard University in the year 1806, and took his bachelor's degree in 1810. A year after, being then in his nineteenth year, namely, in the summer of 1811, he sailed for England, intending to enter a counting-room in Liverpool, as a preparation for mercantile life. This plan, however, was soon relinquished; and the three succeeding years were mainly spent with his relatives and their friends near Derby, where he made the acquaintance of Mrs. Hardcastle, his future mother-in-law, who was something of a botanist, and where he formed both the scientific and social attachments which determined the aims and secured the happiness of his whole after life. Returning to Boston at the close of the year 1814, he engaged with enthusiasm in botanical pursuits, and amassed a good collection of New England plants. In the summer of 1816 he took a leading part in a botanical exploration of the mountains of New England, ascending in the course of one journey, Wachusett, Monadnock, Ascutney, and Mount Washington ; and later in the season Dr. Boott with his brother visited and ascended Moosehillock. His companions in the extended and then formidable tour which culmi-

[1] American Journal of Science and Arts, 2 ser., xxxvii. 288. (1864.)

nated in the White Mountains — then to be reached only by a laborious journey of two days on foot — were Francis C. Gray, Judge Shaw, Nathaniel Tucker, and Dr. Jacob Bigelow, the Nestor of New England botany, now the sole surviving member of the party. An interesting account of the ascent of Mount Washington, written by Dr. Bigelow, was published at the time in the fifth volume of the "New England Journal of Medicine and Surgery."

In the year 1820 Dr. Boott crossed the Atlantic for the last time, and proceeding to London entered upon the study of medicine, under the direction of the late Dr. Armstrong. He continued his medical studies at the University of Edinburgh, where he took the degree of M. D. in 1824. The next year he established himself in London, we believe in the very house in Gower Street where he resided until the day of his death. He was soon associated with his near friend and former teacher in the work of instruction, becoming Lecturer on Botany in the Webb Street School of Medicine, where Dr. Armstrong was Professor of Materia Medica.

"His lectures are said to have been admirable, both in matter and style, and to have excited much enthusiasm ; whilst his untiring efforts to promote the welfare of his pupils in other ways were so deeply and generally felt, that, on the eve of his too early withdrawal from the lectureship, they in one day raised a large subscription to present to their friend and teacher ; — a tribute which, with the characteristic modesty and consideration, was declined as soon as heard of. He was, however, afterwards persuaded to accept a collection of books instead, in remembrance of their grateful feelings and good will."

The early death of Dr. Armstrong, cutting short a distinguished career, imposed upon his friend the duties of a biographer and expositor. Accordingly, after much preparation, Dr. Boott, in the year 1834, published two octavo volumes, entitled, "Memoir of the Life and Medical Opinions of John Armstrong, M. D. ; to which is added an Inquiry into the facts connected with those forms of Fever attributed to Malaria and Marsh Effluvium." He published, besides, in the year 1827, two introductory Lectures on Materia Medica, which gave a good idea of his excellence as a teacher. Although he did not continue in this career, his interest in medical and scientific education never abated. He was an active promoter of the establishment of London University (now University College), and was for more than a quarter of a century an influential member of its senate and council. He was successfully engaged for some time in medical practice, and was for many years Physician to the American Embassy ; but he gradually withdrew from professional cares and toils to more congenial literary and scientific pursuits. As early as the year 1819 he had become a Fellow of the Linnæan Society of London ; and afterwards, for the last twenty-five years, he gave it continuous and invaluable service as secretary, treasurer, or vice-president.

At one time it was thought that Dr. Boott might be recalled to his native country and to an active scientific life. Nearly thirty years ago he was offered the chair of natural history in Harvard University, — a chair which had remained vacant since the death of Professor Peck in 1822, although its duties were for several years fulfilled by the late Mr. Nuttall. After Nuttall left Cambridge to explore Oregon and California, arrangements were made to endow the vacant professorship properly in case Dr. Boott would accept the place. Although the offer was declined, we have been told that he intimated a willingness to accept it if the chair were simply that of botany ; and when informed that he might practically make it so, although the title was unchangeable, he insisted that he would not be called a professor of natural history, while he could pretend to a knowledge only of botany.

Nearly thirty years ago, Dr. Boott began seriously to devote his energies to the special work upon which his scientific reputation mainly rests, namely, to the study of the vast and intricate genus Carex. The first result of these studies appeared in his elaboration of the Carices of British North America in Sir William Hooker's " Flora Boreali-Americana," published in 1840. Other papers upon Carices were contributed to the " Transactions of the Linnæan Society," the " Journal of the Boston Society of Natural History," etc. As it had always been the greatest pleasure, we might say the business of his life to assist others, so now friends and correspondents from all parts of the world hastened to place in his hands the fullest sets of their collections in this difficult genus; and he was able to study, in the unrivaled caricological collection he thus formed, and the various public and private herbaria to which he had access, almost all of the six hundred or more species which the genus was computed by him to comprise, to compare them in numerous specimens of their various forms, and to examine them, group after group, with untiring and closest scrutiny. At length, early in the year 1858, he gave to the world (literally gave to the world) the first volume of his great work, entitled " Illustrations of the Genus Carex," a folio volume with two hundred plates, admirably representing about that number of species. A very large proportion of them were North American species, in which he naturally always took a special interest. In the letter of dedication of this work to his friend, John Amory Lowell, Esq., of Boston, Dr. Boott states that his original design " was limited to the Carices of North America," but that the large collections brought by Dr. Hooker from the East Indies, and placed in his hands for study, caused him to extend his plan, and to endeavor to illustrate the genus at large. With characteristic modesty he makes no allusion to the years of labor and the large amount of money (savings from a moderate income by a simple mode of life) which the volume had cost him; the drawings, engravings, and letterpress having been produced at his sole individual expense, and the larger part of the copies freely given away. Nor did he put forth any promise to continue the work. But in 1860 Part Second quietly appeared, without a word of preface. This contains 110 plates. Two years after, this was followed by Part Third, with 100 plates, making 410 in all; and it is understood that the materials of a fourth volume are left in such forwardness that they may perhaps be published by his surviving family.

Our own estimate of this work has been recorded in the pages of this Journal, as the successive volumes were received.

This motto, which the author placed upon his title-pages, —
    "The man who labors and digests things most,
    Will be much apter to despair than boast,"
is felicitously expressive both of the endless difficulties of the subject, and of his undervaluation of his endeavors to overcome them. A most competent judge briefly declares that, —
    "This work is certainly one of the most munificent contributions ever made to scientific botany, besides being one of the most accurate; on which account it certainly entitles its author to take a much higher place amongst botanists than that of an amateur, which was all that his modesty would allow him to lay claim to."

Dr. Boott's health, which had long been delicate, was much shattered in the winter of 1839–40 by a dangerous attack of pneumonia. "From this time he had repeated slight attacks; but no alarming symptoms occurred till June, 1863, when the remaining lung gave way, and from that time he never fairly rallied. He died at his residence, 24 Gower Street, on Christmas day, retaining to the last his faculties and all the characteristics of his most admirable life."

Dr. Boott was a man of singular purity, delicacy, and goodness of character, and of the most affectionate disposition. Few men of his ardent temperament and extreme sense of justice ever made fewer enemies or more friends. To the latter he attached himself with entire devotion. If there were any of the former, probably no man ever heard him speak ill of them. His published works suffice to place his name imperishably upon the records of science. But only his contemporaries and friends will know how much he has done to help others, and how disinterestedly and gracefully that aid was ever rendered. He took with him to England, upon his return in the year 1820, a valuable herbarium of New England plants, especially those of the White Mountains, which were then rare and little known. He must have valued this collection highly, and have expected to use it. But he presented the whole of it to Dr., now Sir William Hooker, when he saw how serviceable it would be to him in the preparation of the "Flora Boreali-Americana." His British herbarium was long ago similarly given to a then young American botanist.

Another, who, twenty-five years ago, called to take leave of him upon returning to this country, found, as he left, the seat of his cab loaded with choice botanical books, which Dr. Boott had at the moment sent there from the shelves of his own library, where they were not duplicates. We know of one or two instances where he had commenced a critical study of a particular genus with a view to publication, but, upon learning that another person had taken up the same subject, he dispatched to him his own notes and other materials. The Linnæan Society of London owes no little of its present prosperity to his long and faithful services and his wise counsels. He kept up an active correspondence with his friends in this country; and for more than thirty years our young professional men, naturalists, and others who have visited Europe, have experienced cordial welcome and thoughtful kindness at his hands. The following gives a good idea of the man : —

" When practicing as a physician he discarded the customary black coat, knee-breeches, and silk stockings, for the very good reason that sombre colors could not but suggest gloomy ideas to the sick; and he was one of the first who adopted the custom now universal in the profession, of dressing in the ordinary costume. In doing this, Dr. Boott adopted the blue coat, gilt buttons, and buff vest of the period, which he continued to wear to the last, and with which dress his casual acquaintance, no less than his personal friends, will ever associate him. In person he was so tall and thin as almost to suggest ill-health ; and the refinement of his manners, his expression, address, and bearing were in perfect keeping with his polished mind and many accomplishments."

The preceding extracts are all from an excellent article in the "Gardeners' Chronicle" for January 16, to which we are much indebted. In the first volume of the late Dr. Wallich's splendid " Plantæ Asiaticæ Rariores," published in the year 1830, is the figure of a handsome and curious Butomaceous plant, *Boottia cordata*, a genus dedicated " in honorem Francisci Boott, Americani, botanici ardentissimi et peritissimi, amici dilectissimi, non minus animi probitate quam scientiarum cultu, et morum suavitate egregii."

WILLIAM JACKSON HOOKER.[1]

Sir William Jackson Hooker died at Kew, after a short illness, on the 12th of August last, in the eighty-first year of his age.

Seldom, if ever before, has the death of a botanist been so widely felt as a personal sorrow, so extended were his relations, and so strongly did he attach himself to all who knew him. By the cultivators of botany in our own country, at least, this statement will not be thought exaggerated. Although few of our botanists ever had the privilege of personally knowing him, there are none who are not much indebted to him, either directly or indirectly. It is fitting, therefore, that some record of his life and tribute to his memory should appear upon the pages of the "American Journal of Science."

The incidents of his life are soon told. He was born on the 6th of July, 1785, at Norwich, England, where his father — who survived to even a greater age than his distinguished and only son — was at that period confidential clerk in a large business establishment. He was descended from the same family with "the Judicious Hooker," author of the "Ecclesiastical Polity." The name William Jackson was that of our botanist's cousin and god-father, who died young, and was soon followed by both his parents; in consequence of which their estate of Sea-salter, near Canterbury, came to young Hooker while yet a lad at the Norwich High School. He could therefore indulge the taste which he early developed for natural history, at this time mainly for ornithology. But the chance discovery of that rare and curious moss, *Buxbaumia aphylla*, which he took to his eminent townsman Sir James Edward Smith, directed his attention to botany, and fixed the bent of his long and active life. He now made ex-

[1] American Journal of Science and Arts, 2 ser., xli. 1. (1866.)

tensive botanical tours through the wildest parts of Scotland, the Hebrides, and the Orkneys, which his lithe and athletic frame and great activity fitted him keenly to enjoy. Coming up to London, he made the acquaintance of Sir Joseph Banks and of the botanists he had drawn around him, Dryander, Solander, and Robert Brown.

In 1809 he went to Iceland, to explore that then little-known island. The exploration was most successful; but the ship in which he embarked, with all his collections, notes, and drawings, was fired and destroyed, and everything was lost, he himself narrowly escaping with his life. Hooker's earliest work, the "Journal of a Tour in Iceland," in two octavo volumes, published at Yarmouth in 1811, and republished at London two years afterwards, gives an interesting account of his explorations and adventures, along with the history of a singular attempt at the time to revolutionize the island, — with which the disaster to the vessel he returned in was in some way connected, we forget how. Not disheartened by these losses, he now turned from a polar to an equatorial region, and made extensive preparations for going to Ceylon, with Sir Robert Brownrigg, then appointed governor. But the disturbances which broke out in that island, more serious than those which attended the close of his Iceland tour, again frustrated his endeavors.

The strong disposition for travel and distant exploration, frustrated in his own case, came to fruit abundantly in the next generation, in the world-wide explorations of his son. He himself made no more distant journey than to Switzerland, Italy, and France, in 1814, becoming personally acquainted with the principal botanists of the day, and laying the foundation of his wide correspondence and great botanical collections. In 1815 he married the eldest daughter of the late Dawson Turner, of Yarmouth, and established his residence at Halesworth, in Suffolk. The next year, in 1816, besides publishing some of the *Musci* and *Hepaticæ* of Humboldt and Bonpland's collection, he brought to completion his first great botanical work, "The British Jungermanniæ," with colored figures of each species, and microscopical analyses, in eighty-four plates, all from his own ready pencil, — a work which took rank as a model both for description and illustration. In 1828 he brought out, in conjunction with Dr. Taylor, the well-known "Muscologia Britannica," the second edition of which, issued in 1827, is only recently superseded. The "Musci Exotici," with 176 admirable plates, appeared, the first volume in 1818, the second in 1820. These were his principal works upon Mosses and the like, — an excellent subject for the training of a botanist, and one in which Hooker, with quick eye, skilled hand, and intuitive judgment, was not only to excel but to lay the foundation of high excellence in general descriptive botany.

When arranging for a prolonged visit to Ceylon, it appears that he sold his landed property, and that his investment of the proceeds was unfortunate ; so that the demands of an increasing family and of his enlarging collections, for which he always lavishly provided, made it needful for him to seek some remunerative scientific employment. Botanical instruction in Great Britain was then, more than now, nearly restricted to medical classes ; the botanical chairs in the universities therefore belonged to the medical faculty, and were filled by members of the profession. But, through the influence of Sir Joseph Banks, as is understood, the Regius Professorship of Botany in the University of Glasgow was offered to Hooker, and was accepted by him. He removed to Glasgow in the year 1820, and assumed the duties of this position. Here, for twenty years — the most productive years of his life — he was not only the most active and conspicuous working botanist of his country and time, but one of the best and most zealous of teachers. The fixed salary was then only fifty pounds ; and the class fees at first scarcely exceeded that sum. But his lecture-room was soon thronged with ardent and attached pupils, and the emoluments rose to a considerable sum, enabling him to build up his unrivaled herbarium, to patronize explorers and collectors in almost every accessible region, and to carry on his numerous expensive publications, very few of which could be at all remunerative.

The first publication of these busy years was the "Flora

Scotica," brought out in 1821. The next year but one brought the first of the three volumes of the " Exotic Flora," containing figures and descriptions of new, rare, or otherwise interesting exotic plants, admirably delineated, chiefly from those cultivated in the Glasgow and Edinburgh Botanic Gardens. Here first is manifested the interest in the flora of our own country, which has since identified the name of Hooker with North American botany, a considerable number of our choicest plants, especially of the Orchis family, having been here illustrated by his pencil.

The " Icones Filicum " (in which he was associated with Dr. Greville), in two large folio volumes, with two hundred and forty plates, begun in 1829 and finished in 1831, was his introduction to the great family of Ferns, to which he in later years devoted his chief attention.

In 1830 began, with the " Botanical Miscellany," that series of periodical publications which, continued for almost thirty years, stimulated the activity and facilitated the intercourse of botanists in no ordinary degree. The " Miscellany," in royal octavo, with many plates, closed with its third volume, in 1833. The " Journal of Botany," a continuation of the " Miscellany " in a cheaper form (in ordinary 8vo, issued monthly), took its place in 1834, but was itself superseded during the years 1835 and 1836 by the " Companion to the Botanical Magazine " (2 vols. imp. 8vo). In 1840 (after an interval in which the editor took charge of the botanical portion of Taylor's " Annals of Natural History "), the Journal was resumed and carried on to the fourth volume in 1842. Then, changed in title and enlarged, it appears as the " London Journal of Botany " for seven years, until 1848, and finally, as the " Journal of Botany and Kew Garden Miscellany," for nine years more, or to the close of 1857. The whole was carried on entirely at the editor's cost, he furnishing the MSS. for the letter-press, the drawings, etc., without charge, " so that it may be supposed that his expenses were heavy, while his profits were, as he always anticipated, literally nil."

The plates of the Journal being too few to contain a tithe

of the species in his herbarium which it was desirable to figure, an outlet for these was made by the "Icones Plantarum, or Figures, with brief descriptive Characters and Remarks of New or Rare Plants selected from the Author's Herbarium." Ten volumes of the work were published, with a thousand plates (in octavo), at the author's sole expense, and with no remuneration, between the years 1837 and 1854, the drawings of the earlier volumes by his own hand, of the later, by Mr. Fitch, whom he had trained to the work.

Botanists do not need to be told how rich these journals are in materials illustrative of North American botany, containing as they do accounts of collections made by Scouler, Drummond, Douglas, Geyer, etc. Equally important for the botany of our western coast, especially of California, is "The Botany of Captain Beech's Voyage" (4to), in the elaboration of which Sir William Hooker was associated with Professor Walker-Arnott. But his greatest contribution to North American botany — for which our lasting gratitude is due — was his "Flora Boreali-Americana" (2 vols. 4to, with 238 plates), of which the first part was issued in 1833, the last in 1840. Although denominated "the botany of the Northern parts of British America," it embraced the whole continent from Canada and Newfoundland, and on the Pacific from the borders of California, northward to the Arctic sea. Collections made in the British arctic voyages had early come into his hands, as afterwards did all those made in the northern land expeditions by the late Sir John Richardson, Drummond, etc., and the great western collections of Douglas, Scouler, Tolmie, and others, while his devoted correspondents in the United States contributed everything they could furnish from this region. So that this work marks an epoch in North American botany, which now could be treated as a whole.

We should not neglect to notice that, from the year 1827 down to his death, he conducted that vast repertory of figures of the ornamental plants cultivated in Great Britain, the "Botanical Magazine" (contributing over 2500 plates and descriptions); a work always as important to the botanist as to the cultivator, and under his editorship essential to both.

For the use of students at home, in 1830 he produced the "British Flora," which ran through five or six editions before it was consigned to his successor in the chair at Glasgow, Professor Arnott, who has edited two or three more.

We have enumerated the principal works published before he returned to England, including those which were reëdited or (as the periodicals) continued later. After twenty years' service in the Scotch University, Dr., now Sir William Hooker, K. H. (for in 1836 he accepted from William IV. — the last British sovereign who could bestow it — the honor of Knight of the Hanoverian Order), was appointed by government to take the direction of the Royal Gardens at Kew, until then in the private occupation of the crown, but now to be developed into a national scientific establishment.

Even since the death of Banks and Dryander, and while Aiton, the director, grew old and lost any scientific ambition he may once have had, Kew Gardens had declined in botanical importance. The little they preserved, indeed, was chiefly owing to the scientific spirit and unaided exertions of Mr. John Smith, then a foreman, afterwards for many years the superintending gardener (and well known to botanists for his writings upon Ferns), who, retired from his labors, still survives to rejoice in the changed scene.

The idea of converting Kew Gardens into a great national botanical establishment is thought to have originated either with Sir William Hooker himself, or with his powerful friend, and excellent patron of botany and horticulture, John, Duke of Bedford, the father of the present British Premier. Lord John Russell was in the ministry under Lord Melbourne when this project was pressed upon the authorities, and recommended to Parliament by the report of a scientific commission, and, succeeding to the Premiership, he had the honor of carrying it into execution at the propitious moment, and in the year 1841, of appointing Sir William Hooker to the direction of the new establishment. The choice could hardly have been different, even without such influential political support; indeed his patron and friend, the Duke of Bedford, died two years before the appointment was made; but Hooker's special fitness for the place was manifest, and his claims were heartily seconded by the only other botanist who could have come into competition with him in this respect. We refer of course to Dr. Lindley. The office, moreover, was no pecuniary prize ; the salary being only three hundred pounds a year (less, we believe, than the retiring pension of his unscientific superannuated predecessor), " with two hundred pounds to enable him to rent such a house as should accommodate his herbarium and library, by this time of immense extent, and essential, we need not say, to the working of the establishment, whether in a scientific or economic point of view." The salary, if we mistake not, has since been increased in some moderate proportion to the enlarged responsibilities and cares of the vast concern ; but up to his death, so important an auxiliary as his unrivaled herbarium, and the greatest scientific attraction of the institution, was left to be supported (excepting some incidental aid) out of the Director's own private means.

Such record as needs here be made of Sir William Hooker as Director of Kew Gardens can be best and most briefly given mainly in the words of a writer in the " Gardeners' Chronicle " for September 2d, to whose ripe judgment and experience we may defer.

" Sir William entered upon his duties in command of unusual resources for the development of the gardens, such as had never been combined in any other person. Single in purpose and straightforward in action, enthusiastic in manner, and at the same time prepared to advance by degrees, he at once won the confidence of that branch of the government under which he worked. . . . To those in office above him, he imparted much of the zeal and interest he himself felt, which was proved by constant visits to the gardens, resulting in invariable approval of what he was doing, and promises of aid for the future. Another means at his disposal, and which he at once brought to bear on the work in hand, was his extensive foreign and colonial correspondence, including especially that with a large number of students whom he had imbued with a love of botany, and who were scattered over the most remote countries of the globe, and several of whom, indeed, remained in more or less active correspondence with the Director up to the day of his death. His views were further greatly facilitated by his friendly intercourse with the foreign and colonial offices, the admiralty, and the East India Company, to all of whom he had the means of rendering services, by the recommendation of former pupils to posts in their employment, and by publishing the botanical results of the expeditions they sent out. . . .

" At the time of Sir William's taking office, the gardens consisted of eleven acres, with a most imperfect and generally dilapidated series of ten hot-houses and conservatories. Most of these have since been gradually pulled down, and, with the exception of the great orangery (now used as a museum for woods) and the large architectural house near the garden gates, which has just previously been removed from Buckingham Palace, not one now remains. They have been replaced by twenty-five structures (in most cases of much larger dimensions) exclusive of the Palm-stove and the hitherto unfinished great conservatory in the pleasure grounds.

" To describe the various improvements which have resulted in the present establishment, — including, as it does, a botanic garden of seventy-five acres, a pleasure ground or arboretum of two hundred and seventy acres, three museums, stored with many thousand specimens of vegetable products, and a magnificent library and herbarium, the finest in Europe, placed in the late king of Hanover's house on one side of Kew Green, near the gardens, — would rather be to give a history of the gardens than the life of their director." . . .

" It might be supposed that the twenty-four years of Sir William's life spent at Kew in the above public improvements, added to the daily correspondence and superintendence of the gardens, would have left little time and energy for scientific pursuits. Such, however, was far from being the case. By keeping up the active habits of his early life, he was enabled to get through a greater amount of scientific work than any other botanist of his age."

From this period his contributions to scientific botany, if we except the journals and illustrated works (contained until

lately, and some of them to the last), were mainly restricted
to his old favorites, the Ferns. Some years before he removed to Kew, he found the veteran Francis Bauer, then an
octogenarian, or near it, employed in drawing under the
microscope admirable and faithful illustrations of the fructi-
fication of Ferns. He arranged immediately for their publication, drew up the letter-press, and so brought out, between
1838 and 1842, the well-known work entitled, " Genera Filicum, or illustrations of the Ferns and other allied genera."
His large quarto, " Filices Exoticæ, in colored figures and
descriptions of exotic Ferns, chiefly of such as are cultivated
in the Royal Gardens of Kew," (100 plates,) appeared in
1859 ; — the drawings of these, as of nearly all his illustrated
works for the last thirty years, by Walter Fitch, his indefatigable coadjutor, whom he had trained in Scotland, and who
soon became " the most distinguished botanical artist in Europe." " A Second Century of Ferns " (imp. 8vo) was published in 1860 and 1861, the First Century being the tenth
and closing volume of the " Icones Plantarum."

But the principal systematic work of these later years was
his " Species Filicum, being descriptions of the known Ferns,
. . . accompanied with numerous figures," in five volumes,
8vo. The first volume of this work appeared in 1846, and
the last only a year and a half ago.

The crowd of new Ferns and the new knowledge which has
accumulated in the interval of seventeen or eighteen years,
demanded large revision and augmentation of the earlier volumes to bring them up to the level of the later ones. Moreover, a compendious work on this favorite class of plants was
much needed. Both objects might be well accomplished by
a synopsis of known Ferns in a single volume, to be for our
day what Swartz's " Synopsis Filicum " was just sixty years
ago. To this Sir William Hooker, upon the verge of fourscore, undauntedly turned, as soon as the last sheets of the
" Species Filicum " passed from his hands, devoting to it
the time that remained after attending to his administrative
duties. Upon it he steadily labored, with unabated zeal and
with powers almost unimpaired, conscientiously diligent and

constitutionally buoyant to the last. He had made no small progress in the work, and had carried the sheets of the initial number through the press, when an attack of diphtheria, then epidemic at Kew, suddenly closed his long, honored and most useful life.

Our survey of what Sir William Hooker did for science would be incomplete indeed, if it were confined to his published works — numerous and important as they are — and the wise and efficient administration through which, in a short space of twenty-four years, a Queen's flower and kitchen - garden and pleasure-grounds have been transformed into an imperial botanical establishment of unrivaled interest and value. Account should be taken of the spirit in which he worked, of the researches and explorations he promoted, of the aid and encouragement he extended to his fellow-laborers, especially to young and rising botanists, and of the means and appliances he gathered for their use no less than for his own.

The single-mindedness with which he gave himself to his scientific work, and the conscientiousness with which he lived for science while he lived by it, were above all praise. Eminently fitted to shine in society, remarkably good-looking and of the most pleasing address, frank, cordial, and withal of a very genial disposition, he never dissipated his time and energies in the round of fashionable life, but ever avoided the social prominence and worldly distinctions which some sedulously seek. So that, however it may or ought to be regarded in a country where court honors and government rewards have a factitious importance, we count it a high compliment to his sense and modesty that no such distinctions were ever conferred upon him in recognition of all that he accomplished at Kew.

Nor was there in him, while standing in a position like that occupied by Banks and Smith in his early days, the least manifestation of a tendency to overshadow the science with his own importance, or of indifference to its general advancement. Far from monopolizing even the choicest botanical materials which large expenditure of time and toil and money brought into his hands, he delighted in setting other botanists to work on whatever portion they wished to elaborate ; not only imparting freely, even to young and untried men of promise, the multitude of specimens he could distribute, and giving to all comers full access to his whole herbarium, but sending portions of it to distant investigators, so long as this could be done without too great detriment or inconvenience. He not only watched for opportunities for attaching botanists to government expeditions and voyages, and secured the publication of their results, but also largely assisted many private collectors, whose fullest sets are among the treasures of far the richest herbarium ever accumulated in one man's lifetime, if not the amplest anywhere in existence.

One of the later and not least important services which Sir William Hooker has rendered to botany is the inauguration, through his recommendation and influence, of a plan for the publication, under government patronage, of the Floras of the different British colonies and possessions, scattered over every part of the world. Some of these (that of Hongkong and that of the British West Indies) are already completed ; others (like that of Australia, and the Cape Flora of Harvey and Sonder, adopted into the series) are in course of publication ; and still others are ready to be commenced.

The free and cordial way in which Hooker worked in conjunction with others is partly seen in the various names which are associated with his authorship. This came in part from the wide range of subjects over which his survey extended, a range which must have contributed much to the breadth of his views and the sureness of his judgment. Invaluable as such extent of study is, in the present state and prospects of our science we can hardly expect to see again a botanist so widely and so well acquainted both with Cryptogamic and Phanerogamic botany, or one capable of doing so much for the advancement and illustration of both.

Our narrative of Sir William Hooker's scientific career and our estimate of his influence has, we trust, clearly, though incidentally, informed our readers what manner of man he was. To the wide circle of botanists, in which he has long filled so conspicuous a place, to his surviving American friends and correspondents, some of whom have known him long and well, — and "none knew him but to love him, or named him but to praise," — it is superfluous to say that Sir William Hooker was one of the most admirable of men, a model Christian gentleman.

There could really be no question as to the succession to the charge of the great botanical establishment at Kew. But we may add, for the information of many of our readers, that the directorship vacated by Sir William's death has been filled by the appointment of his only surviving son, Dr. Joseph Dalton Hooker, whose well-established scientific fame and ability, no less than his lineage, may assure the continued equally successful administration of this most interesting and important trust.

JOHN LINDLEY.[1]

JOHN LINDLEY, one of the most renowned botanists of the age, died at his residence near London, on the 1st of November last, at the age of sixty-six years. He was born at Catton, near Norwich, where his father was a nurseryman, on February 5, 1799; and was educated at the Norwich Grammar School, as was his friend and earliest scientific acquaintance, Sir William Hooker. It was at the house of the latter, soon after his removal to Halesworth, that young Lindley began his career of authorship by translating Richard's "Analyse du Fruit," which was published in 1819. He appears already to have devoted himself to botanical and horticultural pursuits, and, it is said, had arranged to visit Sumatra and the Malayan Islands; but for some reason, perhaps connected with his father's reverses in business, the project was abandoned. At this juncture he was introduced by his friend Hooker to Sir Joseph Banks, who employed him as his assistant librarian. Sir Joseph recommended him to Mr. Cattley, for whom he edited the folio " Collectanea Botanica," illustrating some of the new and curious plants cultivated in Mr. Cattley's collection. He had already published his " Monograph of Roses " (1820) and his " Monograph of Digitalis " (1821), the latter illustrated by plates from Ferdinand Bauer's drawings. The next year (1822) began his connection with the Horticultural Society, as garden assistant secretary, when he took charge of the laying out of the garden at Chiswick. In 1826 he became sole assistant secretary, Mr. Sabine being honorary secretary until 1830, and then Mr. Bentham until 1841 ; nearly the whole active charge of the establishment falling upon Dr. Lindley. Then, as vice-secretary he conducted the operations of this great and pros-

[1] American Journal of Science and Arts, 2 ser., xli. 265. (1866.)

perous society, with almost undivided responsibility until
1858, when, dropping the laboring oar, he became secretary
and member of the council, and took a leading part in the
reorganization of the society, until, fairly broken down by
overwork, in 1862 he was obliged to retire from the management. Besides his work in the Horticultural Society, sufficient in itself to task any ordinary powers, Dr. Lindley was
professor of botany in University College from 1829 to 1861,
giving elaborate courses of lectures every year, and also lecturer at the Apothecaries Garden at Chelsea for nearly the
same period. He conducted the " Botanical Register " from
about 1823 (although his name does not appear upon the
title-page until somewhat later) down to its close in 1847;
he did the principal botanical work in Loudon's " Encyclo-
pedia of Plants," and wrote the botanical articles for the
" Penny Cyclopedia," down to the letter R; contributed to
the Transactions and Proceedings of the Horticultural Society,
and edited its Journal; prepared the later volumes of Sib-
thorp's magnificent " Flora Græca," etc., etc.; besides writing
and often reëditing his numerous classical botanical works,
which, with his lectures to successive classes of pupils inspired
by his own ardor, have made his name so famous wherever
botany is cultivated. Of these numerous works we can mention only the principal. His " Synopsis of the British Flora "
arranged according to the Natural Orders, first issued in 1829,
has only a local and historical interest, as a part of his successful endeavors to introduce and popularize the natural
system in England, where it had peculiar obstacles to contend
against. His " Genera and Species of Orchidaceous Plants,"
with his " Sertum Orchidaceum," and, later, his " Folia
Orchidacea " (which he was able only to commence), embody
a portion of his labors upon an important and curious family
of plants, upon which he became the paramount authority.
His " Introduction to Botany," which ran through four editions, his outlines of the " First Principles of Botany," at
length expanded into his " Elements of Botany," and his
" School Botany," form a series of introductory works which
have done much more for botanical instruction than any others

in the English language. By his "Flora Medica" he supplied
to medical students a good botanical account of all the more
important plants used in medicine. By his "Theory of Horticulture," explaining the principal operations of gardening
upon physiological principles, in connection with his articles
upon the subject in the "Gardeners' Chronicle," he may
almost be said to have raised this branch of knowledge "from
the conditions of an empirical art to that of a developed
science." And, finally, in his "Introduction to the Natural
System of Botany," the first edition of which, published in
1830, was the earliest systematic exposition of the natural
system in the English language, or fairly available to English
and American students, and his further development of this
work into his classical "Vegetable Kingdom," — the one book
which may take the place of a botanical library, — Dr. Lindley made his most important contributions to the advancement
of systematic botany. The coming generation of botanists
cannot be expected to appreciate the vast influence exerted
by the earlier of these works in its day; the latter, however
open to adverse criticism in particulars, is still unrivaled and
is probably "that by which his name will be best known to
posterity." Physiologist, morphologist, and systematist, he
displayed equal genius in all these departments of the science,
but he worked too rapidly to do himself full justice in any of
them. "His power of work was indeed astonishing; whatever he undertook (and his undertakings were wonderful in
amount and variety) he did with the utmost conscientiousness,
never flagging until he had done it; and he was a splendid
example of what can be accomplished by a man of strong will,
habitually acting up to his oft-repeated saying, that to method,
zeal, and perseverance nothing is impossible." "Until he had
passed fifty years of age," it is stated that "he never knew
what it was to feel tired either in body or mind." Such persons are sure to be overtasked. The Great Exhibition of 1851,
adding protracted and onerous duties to his ordinary work,
prostrated him with serious illness; the Second Exhibition,
in 1862, in which he took charge of the whole colonial department, fatally injured his bodily and mental powers, and

cut short his scientific career. He was able, however, to enjoy the society of his immediate friends and to keep up an interest in his favorite pursuits quite to the close, which occurred from an apoplectic attack, on the morning of the 1st of November last.

Dr. Lindley was a man of marked character. While his biographer declared that "he was hot in temper and impatient of opposition," he no less truly adds that, "on the other hand, he had the warmest of hearts and the most generous of dispositions." He seemed as incapable of cherishing a resentment, as of repressing the expression of indignation for what he thought wrong ; and if at times he made enemies, he was almost sure in time to convert his enemies into friends.

WILLIAM HENRY HARVEY.[1]

William Henry Harvey was born at Summerville, near Limerick, Ireland, on the 5th of February, 1811. His father, Joseph M. Harvey, was a highly respected merchant in that city and a member of the Society of Friends. William Henry was, we believe, the youngest of several children. He received a good education at Ballitore school, — an institution of the Friends, — and on leaving it was engaged for a time in his father's counting-room, devoting, however, all his spare time to natural history, his favorite pursuit even from boyhood. He made considerable attainments in entomology and conchology, and in botany he early turned his attention to Mosses and *Algæ*. To the study of the latter, in which he became pre-eminent, he was attracted from the first by the opportunities which he enjoyed on the productive western coast of Ireland, the family usually spending a good part of the summer at the seaside, mostly on the bold and picturesque shore of Clare. As the late Sir William Hooker's bent for botany was fixed by his accidental discovery of a rare moss, which he took to Sir J. E. Smith, so in turn was Harvey's, by his discovery of two new habitats of another rare moss, the *Hookeria lætevirens*, which led to a correspondence with Hooker, and to a life-long mutual attachment of these most excellent men. Encouraged by his illustrious friend and patron, Harvey sought some position in which he might devote himself to science; and it would appear was selected by Mr. Spring Rice (the late Lord Monteagle) for the post of colonial treasurer at the Cape of Good Hope; that by some accident the appointment was made out in the name of an elder brother, and an inopportune change of ministry frustrated all attempts at rectification. There was no other way but for the brother

[1] American Journal of Science and Arts, 2 ser., xlii. 273. (1866.)

to accept the undesigned appointment, and to take the young botanist with him to the Cape as his assistant. This was done, and the brothers sailed for that colony in the year 1835. But the health of the elder brother suddenly and hopelessly failed within a year, and he died in 1836 on the passage home. William Harvey's appointment to succeed his brother had been sent to the Cape while he was on his homeward voyage : he immediately returned to his post, and fulfilled its duties for three years, devoting his mornings to collecting and his nights to botanical investigation, with such assiduity that his health also gave way, and he was compelled to return home in 1839. The summer of the next year found him reëstablished and on his way to the Cape for the third time. But he could not long endure the sultry climate and the intense application ; with broken health he came back in 1841 and gave up the appointment.

After two years of prostration and seclusion he was well again ; and in 1844, on the death of Dr. Coulter, he was appointed keeper of the herbarium of Trinity College, Dublin. The most important portion of the herbarium then consisted of the collections, yet unassorted, made by Coulter in northwestern Mexico and California. Harvey generously added his own large collections, for which he was allowed fifty pounds a year, in addition to a slender salary, and he proceeded to build up the herbarium into a first-class establishment. The professorship of botany in the college, which was pretty well endowed, fell vacant about this time; and the college authorities, wishing to elect Harvey to the chair and so to combine the two offices, conferred upon him the necessary degree of M. D. But it was contended that an honorary degree did not meet the requirements, and so Dr. Allman, the present distinguished professor of natural history at Edinburgh, carried the election.

Except for the slenderness of his salary, Dr. Harvey was now well placed for scientific work, the object to which he wished to devote his life ; and he entered upon and pursued his distinguished career henceforth with an entire and well-directed energy that never flagged until he was prostrated by mortal disease.

He had already published, at the Cape, in 1838, his "Genera of South African Plants," hastily prepared, solely for local use, but no unworthy beginning of his work in Phænogamous botany; and in his favorite department of the science he had brought out, in 1841, his "Manual of British Algæ," which he reëdited in 1849. He now commenced the first of the series of his greatest works, illustrated by his facile pencil, for he drew admirably. The first (monthly) part of his excellent and beautiful "Phycologia Britannica, a History of British Seaweeds," containing colored figures of all the species inhabiting the shores of the British Islands, appeared in January, 1846; and the undertaking was completed in 1851 in three (or four) volumes, with three hundred and sixty plates, all drawn on stone by his own hand. A similar but less extended work, the "Nereis Australis, or Algæ of the Southern Ocean," which was begun in 1847, was carried only to fifty plates of selected and beautiful species.

In 1848, Dr. Harvey succeeded Dr. Litton as professor of botany in the Royal Dublin Society, to which belonged the botanic garden at Glasnevin; this required him to deliver short courses of lectures annually in Dublin or some other Irish town, and provided a welcome addition to his income.

In 1848, at the request of his friend Van Voorst, the publisher, he wrote his charming little volume, "The Seaside Book," the unsurpassed model of that class of popular scientific books; it was published in 1849, and has passed through several editions. In July of that year, having arranged a visit to this country, and having been invited to deliver a course of lectures before the Lowell Institute, he took steamer for Halifax and Boston, passed the summer and autumn in exploring the shores of the northern States, and in the society of his friends and relatives; for the late Mr. Jacob Harvey, still well and pleasantly remembered in New York, who married the daughter of Dr. Hosack, was his elder brother. In the autumn he gave an admirable course of lectures upon Cryptogamic botany before the Lowell Institute in Boston, and afterwards a shorter course at the Smithsonian Institution at Washington. He then traveled in the southern Atlantic States, continuing the exploration of our *Algæ* down to Florida and the Keys ; and in May, 1850, he returned to Ireland. Under the wise and liberal arrangements made by Professor Henry in behalf of the Smithsonian Institution, and with his own large collections augmented by the contributions which every student or lover of *Algæ* was glad to place in such worthy hands, Professor Harvey now prepared his " Nereis Boreali-Americana, or Contributions to a History of the Marine Algæ of North America." The work is a systematic account of all the known Marine *Algæ* of North America, but with figures only of the leading species. It was issued in three parts : the first part, the *Melanospermeæ*, in 1852, in the third volume of the Smithsonian Contributions to Knowledge ; the second, *Rhodospermeæ*, in the fifth volume ; and the third, or *Chlorospermeæ*, in the tenth volume of the series, published in 1858 ; and the three parts, collected for separate issue, compose a thick imperial quarto volume, of five hundred and fifty pages of letter-press and fifty plates. The work remains the principal if not the only guide to the American student of *Algæ*, and one of the most popular as well as useful of the various contributions to knowledge which the well-managed bequest of Smithson has given to the world.

Before the last part of the " Nereis Boreali-Americana " was published, Professor Harvey had sought a wider field of scientific labor and observation. Obtaining a long leave of absence, and some assistance from the University in addition to the continuance of his salary, he left England in August, 1853, by the overland route for Australia, stopping at Aden and Ceylon to collect : he visited the east, south, and west coasts of Australia, as well as Tasmania. Taking advantage of a missionary ship, which was to cruise among the South Sea Islands, and which offered him unexpected facilities, he visited the Fiji, Navigators', and Friendly Islands, touching also at New Zealand. Returning to Sydney, he sailed to Valparaiso, which he reached much prostrated through over-exertion in a warm climate ; and when recuperated he returned home by way of the Isthmus, arriving in October,

1856. The algological collections of these three laborious years, or the Australian portion of them, formed the subject of Professor Harvey's third great illustrated work, and one of the most exquisite of the kind, the "Phycologia Australica," the serial publication of which began in 1858 and was concluded in 1863, in five imperial octavo volumes, each of sixty colored plates. All but the last century of plates were put upon stone by the author.

Upon Dr. Harvey's return in 1856 from his long expedition he found the chair of botany in the University of Dublin vacated by the appointment of Dr. Allman to that of natural history in the University of Edinburgh; and he was at once preferred to the position which he had sought when younger and freer, and which he now occupied till his death. The exhausting duties of this chair, and of that which he still held in the Royal Dublin Society, undiminished by the transference to the government Museum of Irish Industry, did not prevent Professor Harvey from entering with unabated ardor upon an undertaking of greater magnitude than any preceding one. This was the "Flora Capensis," a full systematic account of all the plants of the Cape Colony and the adjacent provinces of Caffraria and Natal, — in which he was associated with Dr. Sonder of Hamburg. Three thick octavo volumes of this work have appeared, the last in 1865, including the *Compositæ.* Along with this Dr. Harvey — learning for the purpose another form of lithographic drawing — brought out, between the years 1859 and 1864, two volumes of his "Thesaurus Capensis, or Illustrations of the South African Flora," comprising two hundred plates of interesting phænogamous plants. A complete list of his publications would include several contributions to scientific periodicals, mainly to "Hooker's Journal of Botany," and a few miscellaneous writings.

In April, 1861, Dr. Harvey married Miss Phelps of Limerick. If not robust, he was apparently in good health, in the full maturity of his powers, and it was hoped only at the noonday of his allotted course of usefulness. But ere the lecture season of that summer was over, an attack of hemorrhage from the lungs gave notice of a serious pulmonary disease. Yet he seemed to recover from this almost completely : he resumed his stated work, and gave his lectures as usual in 1863, and also in the spring of the following year, but with some difficulty. The winter and spring of 1864–65 were spent in the south of France, with only transient benefit. Returning to his home and his herbarium, he worked on still at the Cape Flora, with cheerful spirit but feeble hands, until he could work no longer. Last spring he sought in Devonshire a milder air, and found a peaceful rest. " On Tuesday, the 15th of May, 1866, at the age of fifty-five years, he quietly breathed his last, at the residence of Lady Hooker, the widow of his long-attached friend Sir William J. Hooker, surrounded by kind and anxious relatives and friends, and was buried in the cemetery at Torquay on Saturday, the 19th of May."

Mr. Harvey was one of the few botanists of our day who excelled both in Phænogamic and Cryptogamic botany. In Algology, his favorite branch, he left probably no superior ; in systematic botany generally he had won an eminent position. He was a keen observer and a capital describer. He investigated accurately, worked readily and easily with microscope, pencil, and pen, wrote perspicuously, and where the subject permitted, with captivating grace, affording, in his lighter productions, mere glimpses of the warm and poetical imagination, delicate humor, refined feeling, and sincere goodness which were charmingly revealed in intimate intercourse and correspondence, and which won the admiration and the love of all who knew him well. Handsome in person, gentle and fascinating in manners, genial and warm-hearted, but of very retiring disposition, simple in his tastes and unaffectedly devout, it is not surprising that he attracted friends wherever he went, so that his death will be sensibly felt on every continent and in the islands of the sea.

HENRY P. SARTWELL.[1]

DR. HENRY P. SARTWELL died at Penn Yan, New York, on the 15th of November, at the age of seventy-five years. He was, we believe, indigenous to the western part of the state of New York, and when a medical student resided at Gorham, Ontario County. More than forty years ago he was established as a physician in the neighboring town of Penn Yan, where he passed his honorable and useful life, engaged to the last in the practice of his profession. It is said that the illness of which he died was brought on by over-exertion in attendance at the bedside of a sick friend. He was fond of all branches of natural history, and a diligent observer and collector in more than one; but in botany he has secured an enduring reputation. He was in his way a model local botanist. He thoroughly explored the district within his reach; he prepared admirable specimens in great numbers, and distributed them with a free hand. Few botanists in this country have contributed to so many herbaria, home or foreign, none more disinterestedly and generously. He accumulated a large herbarium, specially rich and attractive in plants of western New York, in Carices and Ferns. Desirous of insuring its preservation and future usefulness, and needing in his old age the very moderate sum which it would bring him, he a few years ago transferred his herbarium to Hamilton College, where it is valued and well cared for. Most local botanists, when they have nearly exhausted the resources of the district they are confined to, are liable to sink into inactivity. Dr. Sartwell avoided this destiny, and prolonged to the last his enjoyment and usefulness, by making a specialty of the great genus Carex. Dewey, Torrey, Tuckerman, Carey, Boott, all who have published within the last thirty years

[1] American Journal of Science and Arts. 2 ser., xlv. 121. (1868.)

upon this genus, have had frequent occasion to acknowledge their obligations to him; and he erected for himself a monument of his zeal and devotion, and a testimony to his powers of observation, in his " Carices Americæ Septentrionalis Exsiccatæ," the first part of which was issued in the year 1848, the second in 1850; while his interest and activity in the study continued with little abatement down to the last year or two. He published nothing else, we believe, excepting a " Catalogue of the Plants growing in the vicinity of Penn Yan "; but he contributed no little to the value of the publications of others, especially to those of the writer of this notice, and to the "Flora of the State of New York." The Carex which had been dedicated to Dr. Sartwell has proved to be identical with an old European species, but a very distinct and peculiar genus of *Compositæ* found in southwestern Texas, keeps up the name of Sartwellia. His most intimate associate in caricological study survived him only one month.

CHESTER DEWEY.[1]

PROFESSOR CHESTER DEWEY died at Rochester, New York, December 13, having completed the eighty-third year of his age. " He was born at Sheffield, Massachusetts, October 25, 1784; was graduated at Williams College in 1806; studied for the ministry; was licensed to preach in 1808, and during the latter half of that year officiated in Tyringham in western Massachusetts. The same year he accepted a tutorship in Williams College, and in 1810 was appointed professor of mathematics and natural philosophy, an office which he discharged for seventeen years. During his connection with the college he did much to advance the standard of scholarship and enlarge the course of study in his own and kindred departments. Between 1827 and 1836 he was principal of the 'Gymnasium,' a high school for boys at Pittsfield, Massachusetts. In the latter year he removed to this city (Rochester), and became principal of the Rochester Collegiate Institute, which post he held until 1850, when he was elected professor of chemistry and natural philosophy in the Rochester University. He was actively engaged in the duties of that position till 1860, when he retired at the age of seventy-seven, though he continued to teach to some extent till his eightieth year. The last four years he has passed in easy and dignified retirement, happy in the society of his family and friends, beloved and respected by all, and occupying himself still with his scientific studies and with meteorological observations, which he conducted with great care and regularity."

Dr. Dewey was an early and a frequent contributor to this Journal, upon several subjects, but especially upon that with which his name is inseparably connected, the Carices of North America. His " Caricography," commenced in 1824,

[1] American Journal of Science and Arts, 2 ser., xlv. 122. (1868.)

was continued year after year with few breaks, down to the close of 1866, when it terminated with a general Index to Species. It is not for us to speak particularly of the merits of this elaborate monograph, patiently prosecuted through more than forty years. This and the monograph of Schweinitz and Torrey laid the foundation and insured the popularity of the study of Sedges in this country. But while the latter systematic arrangement was published as a whole in 1825, Dr. Dewey's, carried on without particular order, extended through a lifetime, and represents both the earlier and the later knowledge. What is needed to render these stores of observation and their permanent results most available, is a systematic digest or synopsis, something like that which the author contributed to Wood's Botany, in the article Carex, but with all the more important references. Hopes were entertained that he might be able to crown his life's work in this way. But at past fourscore and ten this could not be expected. Beyond this favorite genus, Professor Dewey's botanical writings were few; the most considerable was his " History of the Herbaceous Plants of Massachusetts," published under the authority of the State, being the companion volume to the better known " Report on the Trees and Shrubs," by Mr. Emerson. Botany was one of the occupations of Professor Dewey's leisure hours; his long life was mainly devoted to education. Turning his attention to a special yet almost inexhaustible subject, however, and laboring perseveringly and faithfully, although under many disadvantages, he has permanently and honorably impressed his name upon the science in which the Californian Umbelliferous genus Deweya records his services. He was an excellent, simple-hearted, devout man, a fine specimen of the western New Englander of the old school.

The lovers of Carex, so numerous in this country, will cherish the memory of these two venerable men, Sartwell and Dewey, long associated in congenial pursuits, and gone to their rest together. May the turf of the Sedges they loved, and which cover or ought to cover the low mounds under which their dust reposes, keep them perennially green, and adorn them each returning spring with their sober blossoms!

GEORGE A. WALKER-ARNOTT.[1]

GEORGE A. WALKER-ARNOTT, Professor of Botany in the University at Glasgow, died on the 17th of June last, in the seventieth year of his age. He was born in Edinburgh, February 6, 1799, educated at the celebrated high school of that city, and at the university, where he took high rank as a scholar, especially in the mathematics, — publishing two papers in Tilloch's "Philosophical Magazine" in 1817 and 1818, while yet a student in arts, — and then, turning to law studies, he was called to the bar as a member of the faculty of advocates in the year 1821. He hardly entered, however, upon the duties of his profession, his taste for natural history having been early developed under the lectures of Professor Jameson and of Mr. Stewart, — the latter a well-known teacher of botany at that time, and his patrimonial estate of Arlary in Kinros-shire suffering for his support, so that he could devote himself to botany, as he did, with unsurpassed ardor and success. His earliest botanical paper, upon some Brazilian Mosses, was written in France, and published in a journal at Paris in 1823. In 1826 and 1827 he contributed to the "Edinburgh New Philosophical Journal" a lively narrative of a botanical tour to the south of France and the Pyrenees. He resided for some time at Montpellier and in Paris, examining the principal herbaria there, also that of De Candolle at Geneva, and in 1828 the herbaria at St. Petersburg. In 1831 he married and established himself with his collections at Arlary, where he resided until, in 1845, he accepted the professorship of botany in the University of Glasgow. It was during these fourteen years that the vast amount of scientific work he was able to accomplish was mainly done. He wrote the article "Botany" in the seventh edition of the "Encyclo-

[1] American Journal of Science and Arts, 2 ser., xlvii. 140. (1869.)

pedia Britannica," — the best treatise of the kind of its day in the English language, and one of the most influential. In conjunction with his early friend, Sir William Hooker, he wrote the "Contributions to the Flora of South America," etc., which form a long series of articles in the "Botanical Miscellany," "Journal of Botany," and other similar periodical or serial publications edited by Sir William. He took a similar part in the "Botany of Beechey's Voyage"; in connection with Dr. Wight he brought out the first volume of the "Prodromus Floræ Peninsulæ Indiæ Orientalis"; and made numerous contributions to various periodicals. Up to 1845 or somewhat later Dr. Arnott was one of the foremost botanists of the time, one of the most zealous and sagacious, versed alike in European and exotic botany. But upon assuming the duties of his chair at Glasgow he appears soon to have abandoned the field in which he had won the highest honors, and in which much more was justly expected. He assumed, however, the joint authorship of Hooker's "British Flora," taking, we believe, the whole charge and responsibility of the later editions. As he began with Mosses, so for the last fifteen or twenty years of his life he devoted himself principally to the *Diatomaceæ*, bringing to their investigation all the ardor of his nature and the keenest powers of observation, combined with indomitable patience and unwearied care. So that he became in this department of microscopical research one of the highest authorities, and amassed one of the richest collections extant. As a professor he was greatly esteemed and respected, although he may be thought to have come almost too late in life to the professor's chair. In his later years he was much withdrawn from general botanical intercourse; but his surviving correspondents and friends on this side of the ocean cherish very pleasant memories of him.

NATHANIEL BAGSHAW WARD.[1]

NATHANIEL BAGSHAW WARD, Fellow of the Royal and Linnæan societies, after whom, as its inventor, the Wardian case is named, died at the ripe age of seventy-seven years, on the 4th of June last. He was born in the east end of London, where his father was a medical practitioner of repute, and where for the greater part of his busy and most useful life he laboriously devoted himself to the same profession. About twenty years ago he exchanged the smoke-charged atmosphere and dingy dwellings of Wellclose Square for the pleasant and airy suburb of Clapham Rise, but still actively engaged almost to the last in professional practice, and in his various official duties, mainly in connection with the Apothecaries Society, filling in succession nearly all its important offices. The renovation and even the maintenance of the celebrated Apothecaries Garden at Chelsea — the oldest botanical establishment of the country — is probably mainly due to his counsels and exertions. We cannot here enter into the interesting history of the now familiar Wardian case, — a discovery which grew out of Mr. Ward's persistent endeavors to cultivate the plants he delighted in under the smoke and soot of the dingiest part of London, and which resulted in providing for the poor as well as the rich denizens of the smoky towns of the old world the inexpensive but invaluable luxury or comfort of being surrounded at all seasons with growing plants and fresh flowers. Nor is the invention less applicable to house-culture, especially of Ferns, under the clearer and purer air of our own country rendered arid by the cold of winter, as hundreds could testify who have enjoyed the benefit, perhaps without knowing even the name of their benefactor. Equally important is the application of the Wardian case to the conveyance of living plants between distant countries. The writer well remembers the

[1] American Journal of Science and Arts, 2 ser., xlvii. 141. (1869.)

first case of growing plants sent to New York thirty-five years ago, which arrived as fresh and healthy as when they left London; and the transmission was quite successful between England and Australia, when the voyage, confined to sailing-ships, was far longer than now. So useful has this contrivance proved to be in this respect, that the director of Kew Gardens "feels safe in saying that a large proportion of the most valuable economic and other tropical plants now culti-vated in England would, but for these cases, not yet have been introduced." The earliest published account of the War-dian case was given by Mr. Ward in the form of a letter to his near friend, the late Sir William Hooker, and was printed in the "Companion to the Botanical Magazine" for May, 1836. His volume "On the Growth of Plants in Closely Glazed Cases" appeared in the year 1842, and a second edi-tion, considerably enlarged and suitably illustrated, was pub-lished a few years later. These were, we believe, Mr. Ward's only scientific publications, excepting reports of communica-tions to various societies with which he was connected, several of them relating to a subject near to his heart: the improve-ment of the dwellings of the poor in England, and the amel-ioration, in other respects, of their hard condition. A most enthusiastic and, in some departments, a learned botanist, his contributions to his favorite avocation were not in the form of authorship, to which he seemed averse: a man "given to hos-pitality" indeed, but as unpretending as it was cordial and unlimited. The coming generation will hardly appreciate the extent of the influence he exerted and the strength of the attachment he inspired so widely among the cultivators of natural science, nor understand, perhaps, how it could be said of him, and without exaggeration, that "for very many years his hospitable house, first in Wellclose Square, and lat-terly at Clapham Rise, was the most frequented metropolitan resort of naturalists from all quarters of the globe of any since Sir Joseph Banks' day." But while any survive of those who have had the privilege of knowing him personally, or in the friendly correspondence he delighted in, Mr. Ward will be remembered as "one of the gentlest, kindest, and purest," and in the highest sense one of the best of men.

MOSES ASHLEY CURTIS.[1]

MOSES ASHLEY CURTIS was born in Stockbridge, Massa chusetts, on the 11th of May, 1808. His father was the Rev Jared Curtis, of Stockbridge, afterward for many years chap lain of the state prison at Charlestown. His mother was a daughter of General Moses Ashley. He was fitted for col lege chiefly under his father's tuition, and was graduated at Williams in the class of 1827. Three years afterward he went to Wilmington, North Carolina, as a tutor in the family of Governor Dudley, while at the same time he studied divin ity. There he resided until the year 1841, with the exception of a year and a half passed with his father in Charlestown. In the autumn of 1834 he married Miss De Rosset, of Wil mington, who survives him. He took holy orders at Rich mond, Virginia, in the summer of 1835 ; became rector of the Protestant Episcopal Church at Hillsborough, North Carolina, in 1841, and fulfilled the duties of this station for the re mainder of his life, with the exception of ten years, from 1847 to 1857, during which he had the pastoral charge of a parish at Society Hill, South Carolina. The degree of Doctor of Divinity was conferred on him by the University of North Carolina, at Chapel Hill. His health for a few years past was sensibly impaired ; but he was able to perform his pro fessional duties, and, in a measure, to prosecute his scientific studies, until the 10th of April last, when he died suddenly, probably of heart-disease.

Dr. Curtis's attention must have early been attracted to botany, and his predilection fixed by his residence at Wil mington, one of the richest and most remarkable botanical stations in the United States. For it was in the year 1843, after only three years' residence there, that he communicated

[1] American Journal of Science and Arts, 3 ser., v. 391. (1873.)

to the Boston Society of Natural History his first botanical work, namely, his "Enumeration of Plants growing spontaneously around Wilmington, North Carolina, with remarks on some New and Obscure Species." This was printed in the first volume and second number of that society's Journal; but the original impression having been mainly destroyed by fire, important additions and emendations were made in the subsequent reprint. The author's powers of observation and aptitude for research are well shown in this publication, and it is one of the first of the kind in this country in which the names are accented. In his note upon the structure of Dionæa, or Venus's Fly-trap, — a plant found only in the district around Wilmington, — Dr. Curtis corrected the account of the mode of its wonderful action which had prevailed since the time of Linnæus, and confirmed the statement and inferences of the first scientific describer, Ellis, namely, that this plant not only captures insects, but consumes them, enveloping them in a mucilaginous fluid which appears to act as a solvent. Extending his botanical observations to the western borders of his adopted State, Dr. Curtis was among the first to retrace the steps and rediscover the plants found and published by the elder Michaux, in the higher Alleghany Mountains. But for the last twenty-five years his scientific studies were mainly given to mycology, in which he became a proficient, and the highest American authority. His papers upon *Fungi*, some of which are large, and all are important, were mainly published by the American Philosophical Society, and by the Linnæan Society of London. Several of them are the joint productions of Dr. Curtis and the able English mycologist Mr. Berkley.

His other published writings mainly are " A Commentary on the Natural History of Dr. Hawks's ' History of North Carolina,' " — a good specimen of his appreciation of exact research and of sharpness of wit without acerbity; two papers in Silliman's Journal on " New and Rare Plants of the Carolinas "; and the botanical portion of the " Geological and Natural History Survey of North Carolina," in two parts; — the first a popular account of the trees and shrubs, issued in

1860 ; the other, a catalogue of all the plants of the State, in 1867. This includes the lower Cryptogamia, especially the *Fungi,* of which he enumerates almost 2400 species, while the phænogamous plants are less than 1,900. All our associate's work was marked by ability and conscientiousness. With a just appreciation both of the needs of the science and of what he could best do under the circumstances, when he had exhausted the limited field in Phænogamous botany within his reach, he entered upon the inexhaustible ground of mycology, which had been neglected in this country since the time of Schweinitz. In this difficult department he investigated and published a large number of new species, as well as determined the old ones, and amassed an ample collection, the preservation of which is most important, comprising as it does the specimens, drawings, and original notes which are to authenticate his work. By his unremitting and well-directed labors, filling the intervals of an honored and faithful professional life, he has richly earned the gratitude of the present and ensuing generations of botanists. Several years ago he prepared drawings of the edible *Fungi* of the country, with a view to making them better known in an accessible and popular publication ; but he was unable to find a publisher. He was much impressed with their importance as a source of food. During the hardships of the Rebellion he turned his knowledge of them to useful account for his family and neighborhood ; and he declared that he could have supported a regiment upon excellent and delicious food which was wasting in the fields and woods around him.

HUGO VON MOHL.[1]

HUGO VON MOHL, the acknowledged chief of the vegetable anatomists of this generation, died on the first day of April last. He was born at Stuttgart, April 8, 1805, the youngest of four brothers who all became men of mark in political and scientific life; Julius the orientalist and Hugo the botanist being the most distinguished. The latter was educated at the Stuttgart Gymnasium and Tübingen University, where he studied medicine as well as natural history and physics. His first publication, while a student, in the year 1827, was his "Essay on the Structure and Coiling of Tendrils and Twiners," written in response to a prize-question offered by the Tübingen Medical Faculty. In it he divined the real nature of the movements which coiling stems and tendrils execute, as has recently been clearly made out. In the following year appeared his inaugural dissertation on the " Pores of the Cellular Tissue of Plants," in which his later views and discoveries, respecting the structure, growth, and component parts of cells, as subsequently developed, are already foreshadowed. About this time his choice was made for a scientific rather than a medical career; and he went to Munich to prosecute more advantageously his favorite studies. Here the late Von Martius and Zuccarni were his botanical masters, and Agassiz, Karl Schimper, Braun, and Engelmann his fellow-students. Here he made those researches upon the anatomy of Ferns, Cycads, and especially of Palms, — the latter a most important contribution to Martius's great work upon Palms, the former also contributed to another work by Martius, — which first displayed his remarkable talents for histological investigation, to which his subsequent scientific life was mainly devoted. His merits were promptly recognized by a call to the Imperial

[1] American Journal of Science and Arts, 3 ser., v. 393. (1873.)

Botanic Garden of St. Petersburg, as assistant to its director, Dr. Fischer, and to the chair of physiology in the Academy of Berne. He accepted the latter in 1832, and occupied it until 1835. Then, upon the death of Schubler, he returned to Tübingen, accepted the professorship of botany in its high school, in which chair and in that of Tübingen the rest of his life was passed. Invitations to more prominent and lucrative positions, as, for example, to the botanical chair at Berlin University, when vacated by the death of the veteran Link, were unhesitatingly declined. Although he published numerous (about ninety) special papers or articles, most of them important and timely, and some of great pith and moment, he resolutely declined to bring out any general work. His " Mikrographie " (1846) and his " Principles of the Anatomy and Physiology of the Vegetable Cell " are his only writings which may claim to be such. The latter, an admirable and still invaluable treatise, appeared as an article in Rudolf Wagner's " Cyclopædia of Physiology," but is best known to English readers in its separate form, in a translation made by the late Professor Henfrey, with the author's sanction, issued by Van Voorst in 1852. A year or two later it was for a time understood, to the great satisfaction of botanists, that Mohl had agreed to take a prominent part in the production of a general manual of the anatomy and physiology of plants; but his promise was soon withdrawn. For thirty years he was one of the editors of the " Botanische Zeitung " ; but the editorial labor must have devolved mainly upon Schlechtendal and his successor, although occasional articles from Mohl's pen appeared as late as the year 1871. During that year his health became seriously impaired; yet as the new year advanced, apprehension disappeared. Upon Easter Monday he was apparently well, and so retired to nightly rest; in the morning he was found to have died in sleep.

ROBERT WIGHT.[1]

ROBERT WIGHT, M. D., died at his residence, near Reading, England, May 26, 1872, at the age of seventy-six years. He was born in East Lothian, Scotland, educated at the Edinburgh High School, and professionally at Edinburgh University, where he took his medical degree in 1816. He went to India, the field of his botanical career and most useful administrative activity for forty years, in 1819. He was first assistant surgeon and afterward full surgeon of a native regiment in the East India Company's service, but was soon transferred to the charge of the Botanic Garden at Madras, and finally to that of the important cotton plantations at Coimbatoor. His earliest botanical contributions occupy a conspicuous place in Hooker's "Botanical Miscellany," commencing in 1830, and in the continuation of that work under other names and firms. In 1834, after a temporary sojourn in his native city, appeared the first volume of a model flora, the "Prodromus Floræ Peninsulæ Indiæ Orientalis," by Dr. Wight and Mr. (afterwards Professor) Arnott, of which their successors in the field remarked, that it is the most able and valuable contribution to Indian botany which has ever appeared, and one which has few rivals in the whole domain of botanical literature. Dr. Wight returned to India immediately after the publication of this initial volume of the work, which was never continued. In India, assisted by native artists whom he had trained, he brought out two quarto volumes of "Illustrations of Indian Botany," with one hundred and eighty-two colored plates ; his "Spicilegium Nielgherrense," of similar character ; and finally his " Icones Plantarum Indiæ Orientalis," in six volumes, with 2101 uncolored lithographic plates, and elaborate analysis, of unequal merit, many of them

[1] American Journal of Science and Arts, 3 ser., v. 395. (1873.)

truly excellent, but all wonderful, under the circumstances of their production. When he returned to England, nearly twenty years ago, his productive season, as it proved, was nearly over. But he distributed his collections with a liberal hand, as indeed he had always done, and in spite of a failing health enjoyed in a serene and happy old age the quiet country residence to which he retired.

FREDERIK WELWITSCH.[1]

FREDERIK WELWITSCH, M. D., died in London, on the 20th of October last, in the sixty-ninth year of his age. He was a native of Carinthia; was educated at Vienna; was commissioned by the Würtemberg Unio Itineraria to collect the plants of the Azores and Cape Verde Islands; but on reaching Lisbon and finding good employment there, he made Portugal the field of his investigations, until, in 1850, he was sent by the Portuguese government to explore the natural history of its possessions on the west coast of Africa. His exploration of Angola and Benguela was rewarded by the discovery of more highly curious plants, probably, than any other that has been undertaken since Australia was opened to botanists; among them, and strangest of all, the genus which commemorates the discoverer, *Welwitschia mirabilis*, which Dr. Hooker, who described and illustrated it, does "not hesitate to consider the most wonderful, in a botanical point of view, that has been brought to light during the present century." Perhaps the limitation in the latter clause of the sentence is needless. This inhabits a most arid waste. In another district, under almost opposite conditions, Welwitsch had the good fortune to find the only Cactaceous plant indigenous out of America, namely, *Rhipsalis Cassytha*, and in a lake a new and most remote habitat of our *Brasenia peltata!* In his "Sertum Angolense," a splendid memoir published by the Linnæan Society, with twenty-five plates, some of his most interesting discoveries are described; but the still unpublished portions of his collections must furnish most important contributions to the "Flora of Tropical Africa," now in progress under the orders of the British Colonial department and the editorship of Professor Oliver of Kew. It is to be hoped that they are to be more fully available for this flora than they have thus far been.

[1] American Journal of Science and Arts, 3 ser., v. 396. (1873.)

JOHN TORREY.[1]

JOHN TORREY, M. D., LL. D., died at New York, on the 10th of March, 1873, in the seventy-seventh year of his age. He has long been the chief of American botanists, and was at his death the oldest, with the exception of the venerable ex-President of the American Academy (Dr. Bigelow), who entered the botanical field several years earlier, but left it to gather the highest honors and more lucrative rewards of the medical profession, about the time when Dr. Torrey determined to devote his life to scientific pursuits.

The latter was of an old New England stock, being, it is thought, a descendant of William Torrey, who emigrated from Combe St. Nicholas, near Chard, in Somersetshire, and settled at Weymouth, Massachusetts, about the year 1640.[2]

His grandfather, John Torrey, with his son William, removed from Boston to Montreal at the time of the enforce-

[1] Proceedings American Academy of Arts and Science, ix. 262. (1873.)

[2] In some notes furnished by a member of the family, the descent is endeavored to be traced through the eldest of the five sons who survived their parent, namely, Samuel, who came with him from England, became a minister of the Gospel, and had the unprecedented honor of preaching three election sermons (in 1674, 1683, and 1695), as well as of having three times declined the presidency of Harvard College (after Hoar, after Oakes, and after Rogers). Although educated at the college, he was not a graduate, because he left it in 1650, after three years' residence, just when the term for the A. B. degree was lengthened to four years. The tradition has it, that, " at the prayer-meetings of the students, he was generally invited to make the concluding prayer," — for which an obvious reason suggests itself, — for, " such was his devotion of spirit that, after praying for two hours, the regret was that he did not continue longer." Students of the present day are probably less exacting.

The desire to claim a descent through so eminent a member of the family is natural. But our late venerable associate, Mr. Savage, in his Dictionary of early New England families, states that he could not ascertain that Samuel had any children.

ment of the Boston Port Bill. But neither of them was disposed to be a refugee. For the son, then a lad of seventeen years, ran away from Canada to New York, joined his uncle, Joseph Torrey, a major of one of the two light infantry regiments of regulars (called Congress's Own) which were raised in that city; was made an ensign, and was in the rearguard of his regiment on the retreat to White Plains; served in it throughout the war with honor, and until at the close he reentered the city upon Evacuation Day, when he retired with the rank of captain. Moreover, the father soon followed the son, and became quartermaster of the regiment. Captain Torrey, in 1791, married Margaret Nichols, of New York.

The subject of this biographical notice was the second of the issue of this marriage, and the oldest child who survived to manhood. He was born in New York, on the 15th of August, 1796. He received such education only as the public schools of his native city then afforded, and was also sent for a year to a school in Boston. When he was fifteen or sixteen years old his father was appointed fiscal agent of the state prison at Greenwich, then a suburban village, to which the family removed.

At this early age he chanced to attract the attention of Amos Eaton, who soon afterwards became a well-known pioneer of natural science, and with whom it may be said that popular instruction in natural history in this country began. He taught young Torrey the structure of flowers and the rudiments of botany, and thus awakened a taste and kindled a zeal which were extinguished only with his pupil's life. This fondness soon extended to mineralogy and chemistry, and probably determined the choice of a profession. In the year 1815, Torrey began the study of medicine in the office of the eminent Dr. Wright Post, and in the College of Physicians and Surgeons, in which the then famous Dr. Mitchill and Dr. Hosack were professors of scientific repute; he took his medical degree in 1818; opened an office in his native city, and engaged in the practice of medicine with moderate success, turning the while his abundant leisure to scientific pursuits, especially to botany. In 1817, while yet a medical student, he reported to the Lyceum of Natural History — of which he was one of the founders — his " Catalogue of the Plants growing spontaneously within Thirty Miles of the City of New York," which was published two years later ; and he was already, or very soon after, in correspondence with Kurt Sprengel and Sir James Edward Smith abroad, as well as with Elliott, Nuttall, Schweinitz, and other American botanists. Two mineralogical articles were contributed by him to the very first volume of the " American Journal of Science and Arts " (1818–1819), and several others appeared a few years later, in this and in other journals.

Elliott's " Sketch of the Botany of South Carolina and Georgia " was at this time in course of publication, and Dr. Torrey planned a counterpart systematic work upon the botany of the northern States. The result of this was his " Flora of the Northern and Middle Sections of the United States," *i. e.*, north of Virginia, — which was issued in parts, and the first volume concluded in the summer of 1824. In this work Dr. Torrey first developed his remarkable aptitude for descriptive botany, and for the kind of investigation and discrimination, the tact and acumen, which it calls for. Only those few — now, alas, very few — surviving botanists who used this book through the following years can at all appreciate its value and influence. It was the fruit of those few but precious years which, seasoned with pecuniary privation, are in this country not rarely vouchsafed to an investigator, in which to prove his quality before he is haply overwhelmed with professional or professorial labors and duties.

In 1824, the year in which the first volume (or nearly half) of his Flora was published, he married Miss Eliza Robinson Shaw, of New York, and was established at West Point, having been chosen professor of chemistry, mineralogy, and geology in the United States Military Academy. Three years later he exchanged this chair for that of chemistry and botany (practically that of chemistry only, for botany had already been allowed to fall out of the medical curriculum in this country) in the College of Physicians and Surgeons, New York, then in Barclay Street. The Flora of the Northern

States was never carried further; although a Compendium, a pocket volume for the field, containing brief characters of the species which were to have been described in the second volume, along with an abridgment of the contents of the first, was issued in 1826. Moreover, long before Dr. Torrey could find time to go on with the work, he foresaw that the natural system was not much longer to remain, here and in England, an esoteric doctrine, confined to profound botanists, but was destined to come into general use and to change the character of botanical instruction. He was himself the first to apply it in this country in any considerable publication.

The opportunity for this, and for extending his investigations to the Great Plains and the Rocky Mountains on their western boundary, was furnished by the collections placed in Dr. Torrey's hands by Dr. Edwin James, the botanist of Major Long's expedition in 1820. This expedition skirted the Rocky Mountains belonging to what is now called Colorado Territory, where Dr. James, first and alone, reached the charming alpine vegetation, scaling one of the very highest summits, which from that time and for many years afterward was appropriately named James's Peak; although it is now called Pike's Peak, in honor of General Pike, who long before had probably seen, but had not reached it.

As early as the year 1823, Dr. Torrey communicated to the Lyceum of Natural History descriptions of some new species of James's collection, and in 1826 an extended account of all the plants collected, arranged under their natural orders. This is the earliest treatise of the sort in this country, arranged upon the natural system; and with it begins the history of the botany of the Rocky Mountains, if we except a few plants collected early in the century by Lewis and Clark, where they crossed them many degrees farther north, and which are recorded in Pursh's Flora. The next step in the direction he was aiming was made in the year 1831, when he superintended an American reprint of the first edition of Lindley's "Introduction to the Natural System of Botany," and appended a catalogue of the North American genera arranged according to it.

Dr. Torrey took an early and prominent part in the investigation of the United States species of the vast genus Carex, which has ever since been a favorite study in this country. His friend, Von Schweinitz, of Bethlehem, Pennsylvania, placed in his hands and desired him to edit, during the author's absence in Europe, his "Monograph of North American Carices." It was published in the "Annals of the New York Lyceum," in 1825, much extended, indeed almost wholly rewritten, and so much to Schweinitz's satisfaction that he insisted that this classical monograph "should be considered and quoted in all respects as the joint production of Dr. Torrey and himself." Ten or eleven years later, in the succeeding volume of the "Annals of the New York Lyceum," appeared Dr. Torrey's elaborate Monograph of the other North American *Cyperaceæ*, with an appended revision of the Carices, which meanwhile had been immensely increased by the collections of Richardson, Drummond, etc., in British and arctic America. A full set of these was consigned to his hands for study (along with other important collections), by his friend Sir William Hooker, upon the occasion of a visit which he made to Europe in 1833. But Dr. Torrey generously turned over the Carices to the late Professor Dewey, whose rival Caricography is scattered through forty or fifty volumes of the "American Journal of Science and Arts"; and so had only to sum up the results in this regard, and add a few southern species at the close of his own monograph of the order.

About this time, namely, in the year 1836, upon the organization of a geological survey of the State of New York upon an extensive plan, Dr. Torrey was appointed botanist, and was required to prepare a Flora of the State. A laborious undertaking it proved to be, involving a heavy sacrifice of time, and postponing the realization of long-cherished plans. But in 1843, after much discouragement, the "Flora of the State of New York," the largest if by no means the most important of Dr. Torrey's works, was completed and published, in two large quarto volumes, with one hundred and sixty-one plates. No other State of the Union has produced a Flora to compare with this. The only thing to be regretted is that it interrupted, at a critical period, the prosecution of a far more important work.

Early in his career Dr. Torrey had resolved to undertake a general Flora of North America, or at least of the United States, arranged upon the natural system, and had asked Mr. Nuttall to join him, who, however, did not consent. At that time, when little was known of the regions west of the valley of the Mississippi, the ground to be covered and the materials at hand were of comparatively moderate compass ; and in aid of the northern part of it, Sir William Hooker's Flora of British America — founded upon the rich collections of the arctic explorers, of the Hudson's Bay Company's intelligent officers, and of such hardy and enterprising pioneers as Drummond and Douglas — was already in progress. At the actual inception of the enterprise, the botany of eastern Texas was opened by Drummond's collections, as well as that of the coast of California by those of Douglas, and afterwards those of Nuttall. As they clearly belonged to our own phyto-geographical province, Texas and California were accordingly annexed botanically before they became so politically.

While the field of botanical operations was thus enlarging, the time which could be devoted to it was restricted. In addition to his chair in the Medical College, Dr. Torrey had felt obliged to accept a similar one at Princeton College, and to all was now added, as we have seen, the onerous post of state botanist. It was in the year 1836 or 1837 that he invited the writer of this · notice — then pursuing botanical studies under his auspices and direction — to become his associate in the Flora of North America. In July and in October, 1838, the first two parts, making half of the first volume, were published. The great need of a full study of the sources and originals of the earlier published species was now apparent ; so, during the following year, his associate occupied himself with this work in the principal herbaria of Europe. The remaining half of the first volume appeared in June, 1840. The first part of the second volume followed in 1841; the second, in the spring of 1842 ; and in February, 1843, came the third and the last; for Dr. Torrey's associate was now also immersed in professorial duties and in the consequent preparation of the works and collections which were necessary to their prosecution.

From that time to the present the scientific exploration of the vast interior of the continent has been actively carried on, and in consequence new plants have poured in year by year in such numbers as to overtask the powers of the few working botanists of the country, nearly all of them weighted with professional engagements. The most they could do has been to put collections into order in special reports, revise here and there a family or a genus monographically, and incorporate new materials into older parts of the fabric, or rough-hew them for portions of the edifice yet to be constructed. In all this Dr. Torrey took a prominent part down almost to the last days of his life. Passing by various detached and scattered articles upon curious new genera and the like, but not forgetting three admirable papers published in the "Smithsonian Contributions to Knowledge" (Plantæ Fremontianæ, and those on Batis and Darlingtonia), there is a long series of important, and some of them very extensive, contributions to the reports of government explorations of the western country, — from that of Long's expedition, already referred to, in which he first developed his powers, through those of Nicollet, Fremont, and Emory, Sitgreaves, Stansbury, and Marcy, and those contained in the ampler volumes of the Surveys for Pacific Railroad routes, down to that of the Mexican Boundary, the botany of which forms a bulky quarto volume, of much interest. Even at the last, when he rallied transiently from the fatal attack, he took in hand the manuscript of an elaborate report on the plants collected along our Pacific coast in Admiral Wilkes's celebrated expedition, which he had prepared fully a dozen years ago, and which (except as to the plates) remains still unpublished through no fault of his. There would have been more to add, perhaps of equal importance, if Dr. Torrey had been as ready to complete and publish, as he was to investigate, annotate, and sketch. Through undue diffidence and a constant desire for a greater perfection than was at the time attainable, many interesting observations have from time to time been anticipated by other botanists.

All this botanical work, it may be observed, has reference to the Flora of North America, in which, it was hoped, the diverse and separate materials and component parts, which he and others had wrought upon, might some day be brought together in a completed system of American botany.

It remains to be seen whether his surviving associate of nearly forty years will be able to complete the edifice. To do this will be to supply the most pressing want of the science, and to raise the fittest monument to Dr. Torrey's memory.

In the estimate of Dr. Torrey's botanical work, it must not be forgotten that it was nearly all done in the intervals of a busy professional life; that he was for more than thirty years an active and distinguished teacher, mainly of chemistry, and in more than one institution at the same time; that he devoted much time and remarkable skill and judgment to the practical applications of chemistry, in which his counsels were constantly sought and too generously given; that when, in 1857, he exchanged a portion, and a few years later the whole, of his professional duties for the office of United States Assayer, these requisitions upon his time became more numerous and urgent.[1] In addition to the ordinary duties of his office, which he fulfilled to the end with punctilious faithfulness (signing the last of his daily reports upon the very day of his death, and quietly telling his son and assistant that it would not be necessary to bring him any more), he was frequently requested by the head of the Treasury Department to undertake the solution of difficult problems, especially those relating to counterfeiting, or to take charge of some delicate or confidential commission, the utmost reliance being placed upon his skill, wisdom, and probity.

[1] It ought to be added, that, when the government Assay Office at New York was established, the Secretary of the Treasury selected Dr. Torrey to be its superintendent, — which would have given to the establishment the advantage of a scientific head. But Dr. Torrey resolutely declined the less laborious and better paid post, and took in preference one the emoluments of which were much below his worth and the valuable extraneous services he rendered to the government, — simply because he was unwilling to accept the care and responsibility of treasure.

In two instances these commissions were made personally gratifying, not by pecuniary payment, which, beyond his simple expenses, he did not receive, but by the opportunity they afforded to recruit failing health and to gather floral treasures. Eight years ago he was sent by the Treasury Department to California, by way of the Isthmus ; and last summer he went again across the continent, and in both cases enjoyed the rare pleasure of viewing in their native soil, and plucking with his own hands, many a flower which he had himself named and described from dried specimens in the herbarium, and in which he felt a kind of paternal interest. Perhaps this interest culminated last summer, when he stood on the flank of the lofty and beautiful snow-clad peak to which a grateful former pupil and ardent explorer, ten years before, gave his name, and gathered charming alpine plants which he had himself named forty years before, when the botany of the Colorado Rocky Mountains was first opened. That age and fast-failing strength had not dimmed his enjoyment, may be inferred from his remark when, on his return from Florida the previous spring, with a grievous cough allayed, he was rallied for having gone to seek Ponce de Leon's Fountain of Youth. "No," said he, "give me the fountain of old age. The longer I live, the more I enjoy life." He evidently did so. If never robust, he was rarely ill, and his last sickness brought little suffering, and no diminution of his characteristic cheerfulness. To him, indeed, never came the "evil days" of which he could say, "I have no pleasure in them."

Evincing in age much of the ardor and all of the ingenuousness of youth, he enjoyed the society of young men and students, and was helpful to them long after he ceased to teach, — if, indeed, he ever did cease. For, as Emeritus Professor in Columbia College (with which his old Medical School was united), he not only opened his herbarium, but gave some lectures almost every year, and as a trustee of the college for many years he rendered faithful and important service. His large and truly invaluable herbarium, along with a choice botanical library, he several years ago made over to Columbia College, which charges itself with its safe preservation and maintenance.

Dr. Torrey leaves three daughters, a son, who has been appointed United States assayer in his father's place, and a grandson.

This sketch of Dr. Torrey's public life and works, which it is our main duty to exhibit, would fall short of its object if it did not convey, however briefly and incidentally, some just idea of what manner of man he was. That he was earnest, indefatigable, and able, it is needless to say. His gifts as a teacher were largely proved and are widely known through a long generation of pupils. As an investigator he was characterized by a scrupulous accuracy, a remarkable fertility of mind, especially as shown in devising ways and means of research, and perhaps by some excess of caution.

Other biographers will doubtless dwell upon the more personal aspects and characteristics of our distinguished and lamented associate. To them, indeed, may fittingly be left the full delineation and illustration of the traits of a singularly transparent, genial, delicate, and conscientious, unselfish character, which beautified and fructified a most industrious and useful life, and won the affection of all who knew him. For one thing, they cannot fail to notice his thorough love of truth for its own sake, and his entire confidence that the legitimate results of scientific inquiry would never be inimical to the Christian religion, which he held with an untroubled faith, and illustrated, most naturally and unpretendingly, in all his life and conversation. In this, as well as in the simplicity of his character, he much resembled Faraday.

Dr. Torrey was an honorary or corresponding member of a goodly number of the scientific societies of Europe, and was naturally connected with all prominent institutions of the kind in this country. He was chosen into the American Academy in the year 1841. He was one of the corporate members of the National Academy at Washington. He presided in his turn over the American Association for the Advancement of Science; and he was twice, for considerable periods, president of the New York Lyceum of Natural History, which was in those days one of the foremost of our scientific societies. It has been said of him that the sole distinction on which he prided himself was his membership in the order of the Cincinnati, the only honor in this country which comes by inheritance.

As to the customary testimonial which the botanist receives from his fellows, it is fortunate that the first attempts were nugatory. Almost in his youth a genus was dedicated to him by his correspondent, Sprengel: this proved to be a Clerodendron, misunderstood. A second, proposed by Rafinesque, was founded on an artificial dismemberment of Cyperus. The ground was clear, therefore, when, thirty or forty years ago, a new and remarkable evergreen tree was discovered in our own southern States, which it was at once determined should bear Dr. Torrey's name. More recently a congener was found in the noble forests of California. Another species had already been recognized in Japan, and lately a fourth in the mountains of northern China. All four of them have been introduced, and are greatly prized as ornamental trees in Europe. So that, all round the world, *Torreya taxifolia, Torreya Californica, Torreya nucifera,* and *Torreya grandis* — as well as his own important contributions to botany, of which they are a memorial — should keep our associate's memory as green as their own perpetual verdure.

WILLIAM STARLING SULLIVANT.[1]

WILLIAM STARLING SULLIVANT, LL. D., died at his residence in Columbus, Ohio, on the 30th of April, ultimo. In him we lose the most accomplished bryologist which this country has produced; and it can hardly be said that he leaves behind him anywhere a superior.

He was born, January 15, 1803, at the little village of Franklinton, then a frontier settlement in the midst of primitive forest, near the site of the present city of Columbus. His father, a Virginian, and a man of marked character, was appointed by government to survey the lands of that district of the Northwestern Territory which became the central part of the now populous State of Ohio; and he early purchased a large tract of land, bordering on the Scioto River, near by, if not including, the locality which was afterwards fixed upon for the state capital.

William, his eldest son, in his boyhood, if he endured some of the privations, yet enjoyed the advantages of this frontier life, in the way of physical training and early self-reliance. But he was sent to school in Kentucky; he received the rudiments of his classical education at the so-called Ohio University at Athens, upon the opening of that seminary; and was afterward transferred to Yale College, where he was graduated in the year 1823. His plans for studying a profession were frustrated by the death of his father in that year. This required him to occupy himself with the care of the family property, then mainly in lands, mills, etc., and demanding much and varied attention. He became surveyor and practical engineer, and indeed took an active part in business down to a recent period. Leisure is hardly to be had in a

[1] Proceedings of the American Academy of Arts and Sciences, ix. 271. (1873.)

newly settled country, and least of all by those who have possessions. Mr. Sullivant must have reached the age of nearly
thirty years, and, having married early,[1] was established in
his suburban residence in a rich floral district, before his taste
for natural history was at all developed. His brother Joseph,
next in age, was already somewhat proficient in botany as well
as in conchology and ornithology; and when in some way his
own interest in the subject was at length excited, he took it
up with characteristic determination to know well whatever
he undertook to know at all. He collected and carefully
studied the plants of the central part of Ohio, made neat
sketches of the minuter parts of many of them, especially of
the Grasses and Sedges, entered into communication with the
leading botanists of the country, and in 1840 he published
" A Catalogue of Plants, Native or Naturalized, in the vicinity
of Columbus, Ohio," (pp. 63,) to which he added a few pages
of valuable notes. His only other direct publication in Phænogamous botany is a short article upon three new plants which
he had discovered in that district, contributed to the " American Journal of Science and Arts," in the year 1842. The
observations which he continued to make were communicated
to his correspondents and friends, the authors of the Flora of
North America, then in progress. As soon as the flowering
plants of his district had ceased to afford him novelty, he
turned to the Mosses, in which he found abundant scientific
occupation, of a kind well suited to his bent for patient and
close observation, scrupulous accuracy, and nice discrimination. His first publication in his chosen department, the
" Musci Alleghanienses," was accompanied by the specimens
themselves of Mosses and *Hepaticæ* collected in a botanical
expedition through the Alleghany Mountains from Maryland
to Georgia, in the summer of 1843, the writer of this notice
being his companion. The specimens were not only critically determined, but exquisitely prepared and mounted, and
with letter-press of great perfection; the whole forming two
quarto volumes, which well deserve the encomium bestowed

[1] His first wife, Jane Marshall of Kentucky, was a niece of Chief
Justice Marshall. She died a few years after marriage.

by Pritzel in his Thesaurus.[1] It was not put on sale, but fifty copies were distributed with a free hand among bryologists and others who would appreciate it.[2]

In 1846, Mr. Sullivant communicated to the American Academy the first part, and in 1849, the second part of his "Contributions to the Bryology and Hepaticology of North America," which appeared, one in the third, the other in the fourth volume (new series) of the Academy's Memoirs, each with five plates, from the author's own admirable drawings. These plates were engraved at his own expense, and were generously given to the Academy.

When the second edition of Gray's "Manual of the Botany of the Northern United States" was in preparation, Mr. Sullivant was asked to contribute to it a compendious account of the *Musci* and *Hepaticæ* of the region; which he did, in the space of about one hundred pages, generously adding, at his sole charge, eight copperplates crowded with illustrations of the details of the genera, — thus enhancing vastly the value of his friend's work, and laying a foundation for the general study of bryology in the United States, which then and thus began.

So excellent are these illustrations, both in plan and execution, that Schimper, then the leading bryologist of the Old World, and a most competent judge, since he has published hundreds of figures in his "Bryologia Europæa," not only adopted the same plan in his "Synopsis of the European Mosses," but also the very figures themselves (a few of which were, however, originally his own), whenever they would serve his purpose, as was the case with most of them. A separate edition was published of this portion of the Manual, under the title of "The Musci and Hepaticæ of the United States east

---

[1] "Huic splendidæ impressæ 292 specierum enumerationi accedit elegantissima speciminum omnium exsiccatorum collectio."

[2] A tribute is justly due to the memory of the second Mrs. (Eliza G. Wheeler) Sullivant, a lady of rare accomplishments, and, not least, a zealous and acute bryologist, her husband's efficient associate in all his scientific work until her death, of cholera, in 1850 or 1851. Her botanical services are commemorated in *Hypnum Sullivantiæ* of Schimper, a new Moss of Ohio.

of the Mississippi River " (New York, 1856, imperial 8vo), upon thick paper, and with proof-impressions directly from the copperplates. This exquisite volume was placed on sale at far less than its cost, and copies are now of great rarity and value. It was with regret that the author of the Manual omitted this Cryptogamic portion from the ensuing editions, and only with the understanding that a separate " Species Muscorum," or Manual for the Mosses of the whole United States, should replace it. This most needful work Mr. Sullivant was just about to prepare for the press.

About the same time that Mr. Sullivant thus gave to American students a text-book for our Mosses, he provided an unequaled series of named specimens for illustrating them. The ample stores which he had collected or acquired, supplemented by those collected by M. Lesquereux (who was associated with him from the year 1848) in a journey through the mountainous parts of the southern States under his auspices, after critical determination were divided into fifty sets, each of about three hundred and sixty species or varieties, with printed tickets, title, index, etc., and all except a few copies for gratuitous distribution were generously made over, to be sold at less than cost, for his esteemed associate s benefit, and still more for that of the botanists and institutions who could thus acquire them. The title of this classical work and collection is " Musci Boreali Americani quorum specimina exsiccati ediderunt W. S. Sullivant et L. Lesquereux; 1856." Naturally enough the edition was immediately taken up.

In 1865 it was followed by a new one, or rather a new work, of between five and six hundred numbers, many of them Californian species, the first - fruits of Dr. Bolander's researches in that country. The sets of this unequaled collection were disposed of with the same unequaled liberality, and with the sole view of advancing the knowledge of his favorite science. This second edition being exhausted, he recently and in the same spirit aided his friend Mr. Austin, both in the study and in the publication of his extensive " Musci Appalachiani."

To complete here the account of Mr. Sullivant's bryological labors illustrated by "exsiccati," we may mention his " Musci

Cubenses," named, and the new species described in 1861, from Charles Wright's earlier collections in Cuba, and distributed in sets by the collector. His researches upon later and more extensive collections by Mr. Wright remain in the form of notes and pencil sketches, in which many new species are indicated. The same may be said of an earlier still unpublished collection, made by Fendler in Venezuela. Another collection, of great extent and interest, which was long ago elaborately prepared for publication, and illustrated by very many exquisite drawings, rests in his portfolios, through delays over which Mr. Sullivant had no control; namely, the Bryology of Rodgers's United States North Pacific Exploring Expedition, of which Charles Wright was botanist. Brief characters of the principal new species were, however, duly published in this as in other departments of the botany of that expedition. It is much to be regretted that the drawings which illustrate them have not yet been engraved and given to the scientific world.

This has fortunately been done in the case of the South Pacific Exploring Expedition, under Commodore Wilkes. For, although the volume containing the Mosses has not even yet been issued by government, Mr. Sullivant's portion of it was published in a separate edition in the year 1859. It forms a sumptuous imperial folio, the letter-press having been made up into large pages, and printed on paper which matches the plates, twenty-six in number.

One volume of the Pacific Railroad Reports, *i. e.* the fourth, contains a paper by Mr. Sullivant, being his account of the Mosses collected in Whipple's Exploration. It consists of only a dozen pages of letter-press, but is illustrated by ten admirable plates of new species.

The "Icones Muscorum," however, is Mr. Sullivant's crowning work. It consists, as the title indicates, of " Figures and Descriptions of most of those Mosses peculiar to eastern North America which have not been heretofore figured," and forms an imperial octavo volume, with one hundred and twenty-nine copperplates, published in 1864. The letter-press and the plates (upon which last alone several

WILLIAM STARLING SULLIVANT.

thousand dollars and immense pains were expended) are simply exquisite and wholly unrivaled ; and the scientific character is acknowledged to be worthy of the setting. Within the last few years most of the time which Mr. Sullivant could devote to science has been given to the preparation of a second or supplementary volume of the Icones. The plates, it is understood, are completed, the descriptions in a good degree written out, and the vernal months in which his mortal life closed were to have been devoted to the printing. The Manual of North American Mosses was speedily to follow.

He was remarkably young for his years, so that the hopes and expectations in which we were indulging seemed reasonable. But in January, not far from his seventieth birthday, he was prostrated by pneumonia, from the consequences of which, after some seeming convalescence, he died upon the last day of April. He leaves a wife, Mrs. Caroline E. (Sutton) Sullivant, children, grandchildren, and great-grandchildren, to inherit a stainless and honored name, and to cherish a noble memory.

In personal appearance and carriage, no less than in all the traits of an unselfish and well-balanced character, Mr. Sullivant was a fine specimen of a man. He had excellent business talents, and was an exemplary citizen ; he had a refined and sure taste, and was an accomplished draughtsman. But after having illustrated his earlier productions with his own pencil, he found that valuable time was to be gained by employing a trained artist. He discovered in Mr. A. Schrader a hopeful draughtsman, and he educated him to the work, with what excellent results the plates of the Icones and of his other works abundantly show. As an investigator he worked deliberately, slowly indeed and not continuously, but perseveringly. Having chosen his particular department, he gave himself undeviatingly to its advancement. His works have laid such a broad and complete foundation for the study of bryology in this country, and are of such recognized importance everywhere, that they must always be of classical authority ; in fact, they are likely to remain for a long time unrivaled. Wherever Mosses are studied, his name will be honorably remembered ; in this country it should long be remembered with peculiar gratitude.

In accordance with his wishes, his bryological books and his exceedingly rich and important collections and preparations of Mosses are to be consigned to the Gray Herbarium of Harvard University, with a view to their preservation and long-continued usefulness. The remainder of his botanical library, his choice microscopes, and other collections are bequeathed to the State Scientific and Agricultural College, just established at Columbus, and to the Starling Medical College, founded by his uncle, of which he was himself the senior trustee.

Mr. Sullivant was chosen into the American Academy in the year 1845. He received the honorary degree of Doctor of Laws from Gambier College, in his native State. His oldest botanical associates long ago enjoyed the pleasure of bestowing the name of *Sullivantia Ohionis* upon a very rare and interesting but modest and neat Saxifragaceous plant, which he himself discovered in his native State, on the secluded banks of a tributary of the river which flows by the place where he was born, and where his remains now repose.

JEFFRIES WYMAN.[1]

WHEN we think of the associate and friend whose death
this Society now deplores, and remember how modest and retiring he was, how averse to laudation and reticent of words,
we feel it becoming to speak of him now that he is gone,
with much of the reserve which would be imposed upon us if
he were living. Yet his own perfect truthfulness and nice
sense of justice, and the benefit to be derived from the con-
templation of such a character by way of example, may be
our warrant for reasonable freedom in the expression of our
judgments and our sentiments, taking care to avoid all exaggeration.

Appropriate and sincere eulogies and expressions of loss,
both official and personal, have, however, already been pronounced or published; and among them one from the governors of that institution to which, together with our own
Society, most of Professor Wyman's official life and services
were devoted, — which appears to me to delineate in the few-
est words the truest outlines of his character. In it the President and Fellows of Harvard University " recall with affectionate respect and admiration the sagacity, patience, and
rectitude which characterized all his scientific work, his clearness, accuracy, and conciseness as a writer and teacher, and
the industry and zeal with which he labored upon the two admirable collections which remain as monuments of his rare
knowledge, method, and skill. They commend to the young
men of the University this signal example of a character
modest, tranquil, dignified, and independent, and of a life
simple, contented, and honored."

What more can be or need be said ? It is left for me, in

[1] An address delivered at a memorial meeting of the Boston Society of
Natural History, October 7, 1874.

compliance with your invitation, Mr. President, to say something of what he was to us, and has done for us, and to put upon record, for the use of those who come after us, some account of his uneventful life, some notice, however imperfect, of his work and his writings. I could not do this without the help of friends who knew him well in early life, and of some of you who are much more conversant than I am with most of his researches. Such aid, promptly rendered, has been thankfully accepted and freely used.

Our associate's father, Dr. Rufus Wyman, — born in Woburn, graduated at Harvard College in 1799, and in the latter part of his life physician to the McLean Asylum for the Insane, — was a man of marked ability and ingenuity. Called to the charge of this earliest institution of the kind in New England at its beginning, he organized the plan of treatment and devised excellent mechanical arrangements, which have since been developed, and introduced into other establishments of the kind. His mother was Ann Morill, daughter of James Morill, a Boston merchant. This name is continued, and is familiar to us, in that of our associate's elder brother.

Jeffries Wyman, the third son, derived his baptismal name from the distinguished Dr. John Jeffries of Boston, under whom his father studied medicine. He was born on the 11th of August, 1814, at Chelmsford, a township of a few hundred inhabitants in Middlesex County, Massachusetts, not far from the present city of Lowell. As his father took up his residence at the McLean Asylum in 1818, when Jeffries was only four years old, he received the rudiments of his education at Charlestown, in a private school, but afterwards went to the Academy at Chelmsford, and in 1826 to Philips Exeter Academy, where, under the instruction of Dr. Abbot, he was prepared for college. He entered Harvard College in 1829, the year in which Josiah Quincy took the presidency, and was graduated in 1833, in a class of fifty-six, six of whom became professors in the University. He was not remarkable for general scholarship, but was fond of chemistry, and his preference for anatomical studies was already developed. Some of his classmates remember the interest which was excited among them by a skeleton which he made of a mammoth bull-frog from Fresh Pond, probably one which is still preserved in his museum of comparative anatomy. His skill and taste in drawing, which he turned to such excellent account in his investigations and in the lecture room, as well as his habit of close observation of natural objects met with in his strolls, were manifested even in boyhood.

An attack of pneumonia during his senior year in college caused much anxiety, and perhaps laid the foundation of the pulmonary affection which burdened and finally shortened his life. To recover from the effects of the attack, and to guard against its return, he made, in the winter of 1833–34, the first of those pilgrimages to the coast of the southern States, which in later years were so often repeated. Returning with strength renewed in the course of the following spring, he began the study of medicine under Dr. John C. Dalton, who had succeeded to his father's practice at Chelmsford, but who soon removed to the adjacent and thriving town of Lowell. Here, and with his father at the McLean Asylum and at the Medical College in Boston, he passed two years of profitable study. At the commencement of the third year he was elected house-student in the medical department, at the Massachusetts General Hospital, — then under the charge of Drs. James Jackson, John Ware, and Walter Channing, — a responsible position, not only most advantageous for the study of disease, but well adapted to sharpen a young man's power of observation.

In 1837, after receiving the degree of Doctor of Medicine, he cast about among the larger country towns for a field in which to practise his profession. Fortunately for science he found no opening to his mind; so he took an office in Boston, on Washington Street, and accepted the honorable but far from lucrative post of demonstrator of anatomy under Dr. John C. Warren, the Hersey professor. His means were very slender, and his life abstemious to the verge of privation; for he was unwilling to burden his father, who indeed had done all he could in providing for the education of two sons.

It may be interesting to know, that, to eke out his subsistence, he became at this time a member of the Boston Fire Department, under an appointment of Samuel A. Eliot, Mayor, dated September 1, 1838. He was assigned to Engine No. 18. The rule was that the first-comer to the engine-house should bear the lantern, and be absolved from other work. Wyman lived near by, and his promptitude generally saved him from all severer labor than that of enlightening his company.

The turning-point in his life, *i. e.*, an opportunity which he could seize of devoting it to science, came when Mr. John A. Lowell offered him the curatorship of the Lowell Institute, just brought into operation, and a course of lectures in it. He delivered his course of twelve lectures upon comparative anatomy and physiology in the winter of 1840–41 ; and with the money earned by this first essay in instructing others, he went to Europe to seek further instruction for himself. He reached Paris in May, 1841, and gave his time at once to human anatomy at the School of Medicine, and comparative anatomy and natural history at the Garden of Plants, attending the lectures of Flourens, Majendie, and Longet on physiology, and of De Blainville, Isidore St. Hilaire, Valenciennes, Duméril, and Milne-Edwards on zoölogy and comparative anatomy. In the summer, when the lectures were over, he made a pedestrian journey along the banks of the Loire, and another along the Rhine, returning through Belgium, and by steamer to London. There, while engaged in the study of the Hunterian collections at the Royal College of Surgeons, he received information of the alarming illness of his father ; he immediately turned his face homeward, but on reaching Halifax he learned that his father was no more.

He resumed his residence in Boston, and devoted himself mainly to scientific work, under circumstances of no small discouragement. But in 1843 the means of a modest professional livelihood came to him in the offer of the chair of anatomy and physiology in the medical department of Hampden-Sidney College, established at Richmond, Virginia. One advantage of this position was that it did not interrupt his

residence in Boston except for the winter and spring; and
during these months the milder climate of Richmond was even
then desirable. He discharged the duties of the chair most
acceptably for five sessions, until in 1847 he was appointed
to succeed Dr. Warren as Hersey professor of anatomy in
Harvard College, the Parkman professorship in the Medical
School in Boston being filled by the present incumbent, Dr.
Holmes. Thus commenced Professor Wyman's most useful
and honorable connection as a teacher with the University, of
which the President and Fellows speak in the terms I have
already recited. He began his work in Holden Chapel, the
upper floor being the lecture room, the lower containing the
dissecting room and the anatomical museum of the college,
with which he combined his own collections and preparations,
which from that time forward increased rapidly in number
and value under his industrious and skillful hands. At length
Boylston Hall was built for the anatomical and the chemical
departments, and the museum, lecture and working rooms
were established commodiously in their present quarters; and
Professor Wyman's department assumed the rank and importance which it deserved. Both human and comparative
anatomy were taught to special pupils, some of whom have
proved themselves worthy of their honored master, while the
annual courses of lectures and lessons on anatomy, physiology, and for a time the principles of zoölogy, imparted
highly valued instruction to undergraduates and others.

In the formation and perfecting of his museum — the first
of the kind in the country, arranged upon a plan both physiological and morphological — no pains and labors were spared,
and long and arduous journeys and voyages were made to contribute to its riches. In the summer of 1849, — having re-
plenished his frugal means with the proceeds of a second
course of lectures before the Lowell Institute (namely, upon
comparative physiology, a good condensed shorthand report of
which was published at the time), — he accompanied Captain
Atwood of Provincetown, in a small sloop, upon a fishing voyage high up the coast of Labrador. In the winter of 1852,
going to Florida for his health, he began his fruitful series of

explorations and collections in that interesting district. In 1854, accompanied by his wife, he traveled extensively in Europe, and visited all the museums within his reach. In the spring of 1856, with his pupils Green and Bancroft as companions and assistants, he sailed to Surinam, penetrated far into the interior in canoes, made important researches upon the ground, and enriched his museum with some of its most interesting collections. These came near being too dearly bought, as he and his companions took the fever of the country, from which he suffered severely, and recovered slowly. Again, in 1858-9, accepting the thoughtful and generous invitation of Captain J. M. Forbes, he made a voyage to the La Plata, and ascended the Uruguay and the Parana in a small iron steamer which Captain Forbes brought upon the deck of his vessel ; then, with his friend George Augustus Peabody as a companion, he crossed the pampas to Mendosa, and the Cordilleras to Santiago and Valparaiso, whence he came home by way of the Peruvian coast and the Isthmus.

By such expeditions many of the choice materials of his museum and of his researches were gathered, at his own expense, to be carefully prepared and elaborated by his own unaided hands. A vast neighboring museum is a splendid example of what munificence, called forth by personal enthusiasm, may accomplish. In Dr. Wyman's we have an example of what one man may do, unaided, with feeble health and feebler means, by persistent and well directed industry, without eclat, and almost without observation. While we duly honor those who of their abundance cast their gifts into the treasury of science, let us not — now that he cannot be pained by our praise — forget to honor one who in silence and penury cast in more than they all.

Of penury in a literal sense we may not speak; for although Professor Wyman's salary, derived from the Hersey endowment, was slender indeed, he adapted his wants to his means, foregoing neither his independence nor his scientific work ; and I suppose no one ever heard him complain. In 1856 came unexpected and honorable aid from two old friends of his father, who appreciated the son, and wished him to go on with his scientific work without distraction. One of them, the late Dr. William J. Walker, sent him ten thousand dollars outright; the other, the late Thomas Lee, who had helped in his early education, supplemented the endowment of the Hersey professorship with an equal sum, stipulating that the income thereof should be paid to Professor Wyman during life, whether he held the chair or not. Seldom, if ever, has a moderate sum produced a greater benefit.

Throughout the later years of Professor Wyman's life a new museum has claimed his interest and care, and is indebted to him for much of its value and promise. In 1866, when failing strength demanded a respite from oral teaching, and required him to pass most of the season for it in a milder climate, he was named by the late George Peabody one of the seven trustees of the museum and professorship of American archæology and ethnology, which this philanthropist proceeded to found in Harvard University; and his associates called upon him to take charge of the establishment. For this he was peculiarly fitted by all his previous studies, and by his predilection for ethnological inquiries. These had already engaged his attention, and to this class of subjects he was thereafter mainly devoted, — with what sagacity, consummate skill, untiring diligence and success, his seventh annual report, — the last published just before he died, — his elaborate memoir on shell-heaps, now printing, and especially the Archæological Museum in Boylston Hall, abundantly testify. If this museum be a worthy memorial of the founder's liberality and foresight, it is no less a monument to Wyman's rare ability and devotion. Whenever the enduring building which is to receive it shall be erected, surely the name of its first curator and organizer should be inscribed, along with that of the founder, over its portal.

Of Professor Wyman's domestic life, let it here suffice to record, that in December, 1850, he married Adeline Wheelwright, who died in June, 1855, leaving two daughters; that in August, 1861, he married Anna Williams Whitney, who died in February, 1864, shortly after the birth of an only and a surviving son.

Of his later days, of the slow yet all too rapid progress of fatal pulmonary disease, it is needless to protract the story. Winter after winter, as he exchanged our bleak climate for that of Florida, we could only hope that he might return. Spring after spring he came back to us invigorated, thanks to the bland air and to open life in boat and tent, which acted like a charm; — thanks, too, to the watchful care of his attached friend, Mr. Peabody, his constant companion in Florida life. One winter was passed in Europe, partly in reference to the Archæological Museum, partly in hope of better health; but no benefit was received. The past winter in Florida produced the usual amelioration, and the amount of work which Dr. Wyman undertook and accomplished last summer might have tasked a robust man. There were important accessions to the archæological collections, upon which much labor, very trying to ordinary patience, had to be expended. And in the last interview I had with him, he told me that he had gone through his own museum of comparative anatomy, which had somewhat suffered in consequence of the alterations in Boylston Hall, and had put the whole into perfect order. It was late in August when he left Cambridge for his usual visit to the White Mountain region, by which he avoided the autumnal catarrh; and there, at Bethlehem, New Hampshire, on the 4th of September, a severe hemorrhage from the lungs suddenly closed his valuable life.

Let us turn to his relations with this society. He entered it in October, 1837, just thirty-seven years ago, and shortly after he had taken his degree of Doctor in Medicine. He was recording secretary from 1839 to 1841; curator of ichthyology and herpetology from 1841 to 1847; of herpetology from 1847 to 1845; of comparative anatomy from 1855 to 1874. While in these later years his duties may have been almost nominal, it should be remembered that in the earlier days a curator not only took charge of his portion of the museum, but in a great degree created it. Then for fourteen years, from 1856 to 1870, he was the president of this society, as assiduous in all its duties as he was wise in council; and he resigned the chair which he so long adorned and dignified, only when the increasing delicacy of his health, to which night exposure was prejudicial, made it unsafe for him any longer to undertake its duties. The record shows that he has made here one hundred and five scientific communications, several of them very important papers, every one of some positive value; for you all know that Professor Wyman never spoke or wrote except to a direct purpose, and because there was something which it was worth while to communicate. He bore his part also in the American Academy of Arts and Sciences, of which he was a Fellow from the year 1843, and for many years a councilor. To it he made a good number of communications; among them one of the longest and ablest of his memoirs.

Then he was from the first a member of the faculty of the Museum of Comparative Zoölogy, where his services and his advice were highly valued. He was chosen president of the American Association for the Advancement of Science for the year 1857, but did not assume the duties of the office.

Some notice — brief and cursory though it must be — of such portion of Dr. Wyman's scientific work as is recorded in his published papers, should form a part of this account of his life.

His earliest publication, so far as we know, was an article in the "Boston Medical and Surgical Journal," in 1837, signed only with the initials of his name. It is upon "The indistinctness of images formed from oblique rays of light," and the cause of it. The handling of the subject is as characteristic as that of any later paper. In January, 1841, we find his first recorded communication of this society, "On the Cranium of a Seal." The first to the American Academy is the account of his dissection of the electrical organs of a new species of Torpedo, in 1843, part of a paper by his friend Dr. Storer, published in Silliman's Journal. In the course of that year, eleven communications were made to our society, beside the annual address, which he delivered on the 17th of May. The most important of these was the memoir, by Dr. Savage and himself, on the Black Orang or Chimpanzee of Africa, *Troglodytes niger*, published in full in the Journal of this society, the anatomical part by Professor Wyman. Two other papers of that early year, on the anatomy of two Mollusca, *Tebennophorus Carolinensis* and *Glandina truncata,* published in the fourth volume of the society's Journal, each with a copper plate, are noteworthy, as showing that he possessed from the first that happy faculty of clear, terse, and closely relevant exposition, and that skill and neatness of illustration with his pencil, which characterize all his work, both of research and instruction.

Another paper of that year, "On the microscopic structure of the teeth of the *Lepidostei,* and their analogies with those of the *Labyrinthodonts,*" read to this society in August, and published in Silliman's Journal in October, 1843, was important and timely. In it he demonstrated that the Labyrinthine structure of the teeth, considered at the time to be peculiar to certain sauroid reptiles, equally belong to gar-fishes, and consequently that many fossil teeth which had been referred by the evidence of this character alone to a group of reptiles founded upon this peculiarity, might as well belong to ancient sauroid fishes.

Although not of any importance now to remember, I may here mention his report to this society on the *Hydrachos Sillimani* of Koch, a factitious Saurian of huge length, successfully exhibited in New York and elsewhere under high auspices, and I think also in Germany, but which Dr. Wyman exposed at sight, showing that it was made up of an indefinite number of various cetaceous vertebræ, belonging to many individuals, which (as was afterward ascertained) were collected from several localities.

But the memoir by which Professor Wyman assured his position among the higher comparative anatomists was that communicated to and published by this society in the summer of 1847, in which the Gorilla was first named and introduced to the scientific world, and the distinctive structure and affinities of the animal so thoroughly made out from the study of the skeleton, that there was, as the great English anatomist remarked, "very little left to add, and nothing to correct." In this memoir the "Description of the habits of *Troglodytes*

Gorilla " is by Dr. Thomas S. Savage, to whom, along with Dr. Wilson, " belongs the credit of the discovery ; " the Osteology of the same and the introductory history are by Dr. Wyman. Indeed, nearly all since made known of the Gorilla's structure, and of the affinities soundly deduced therefrom, has come from our associate's subsequent papers, founded on additional crania brought to him in 1849, by Dr. George A. Perkins of Salem ; on a nearly entire male skeleton of unusual size, received in 1852 from the Rev. William Walker, and now in Wyman's museum ; and on a large collection of skins and skeletons placed at his disposal in 1859, by Du Chaillu, along with a young Gorilla in spirits, which he dissected. It is in the account of this dissection that Professor Wyman brings out the curious fact that the skull of the young Gorilla and Chimpanzee bears closer resemblance to the adult than to the infantile human cranium.

In Professor Wyman's library, bound up with a quarto copy of the Memoir by Dr. Savage and himself, is a terse but complete history of this subject, in his neat and clear handwriting, and with copies of the letters of Dr. Savage, Professor Owen, Mr. Walker, and M. du Chaillu.

In the introductory part of the Memoir, Professor Wyman states that " the specific name, Gorilla, has been adopted, a term used by Hanno in describing wild men found on the coast of Africa, probably one of the species of the Orang." The name *Troglodytes Gorilla* is no doubt to be cited as of Savage and Wyman, and it was happily chosen by Professor Wyman, after consultation with his friend, the late Dr. A. A. Gould, for the reason just stated. But it is interesting to see, in the correspondence before me, how strenuously each of the joint authors deferred to the other the honor of nomenclature. Dr. Savage from first to last insists, in repeated and emphatic terms, that the scientific name shall be given by Dr. Wyman as the scientific describer, and that he could not himself honestly appropriate it. Professor Wyman in his MS. account, after mentioning what his portion of the Memoir was, and that " the determination of the differential characters on which the establishment of the species rests was prepared by me," briefly and characteristically adds : " In view of this last fact, Dr. Savage thought, as will be seen in his letter, that the species should stand in my name; but this I declined."

This Memoir was read before this society on the 18th of August, 1847, and was published before the close of the year. But it had not, as it appears, come to Professor Owen's knowledge when the latter presented to the London Zoölogical Society, on the 22d of February, 1848, a memoir founded on three skulls of the same species, just received from Africa through Captain Wagstaff. When Professor Owen received the earlier memoir, he wrote to compliment Professor Wyman upon it, substituted in a supplementary note the specific name imposed by Savage and Wyman, and reprinted in an appendix the osteological characters set forth by the latter. "It does not appear, however [adds Dr. Wyman], either in the Proceedings or the Transactions of the [Zoölogical] Society, at what time our Memoir was published, nor that we had anticipated him in our description."

It is safe to assert that in this and the subsidiary papers of Dr. Wyman may be found the substance of all that has since been brought forward, bearing upon the osteological resemblances and differences between men and apes. After summing up the evidence he concludes : —

" The organization of the anthropoid Quadrumana justifies the naturalist in placing them at the head of the brute creation and placing them in a position in which they, of all the animal series, shall be nearest to man. Any anatomist, however, who will take the trouble to compare the skeletons of the Negro and Orang, cannot fail to be struck at sight with the wide gap which separates them. The difference between the cranium, the pelvis, and the conformation of the upper extremities of the Negro and Caucasian, sinks into comparative insignificance when compared with the vast difference which exists between the conformation of the same parts in the Negro and the Orang. Yet it cannot be denied, however wide the separation, that the Negro and Orang do afford the points where man and the brute, when the totality of their organization is considered, most nearly approach each other."

Selecting now for further comment only some of the more noticeable contributions to science, we should not pass by his investigations of the anatomy of the blind fish of the Mammoth Cave. The series began in that prolific year, 1843, with a paper published in Silliman's Journal, and closed with an article in the same Journal in 1854. Although Dr. Fellkamph had preceded him in ascertaining the existence of rudimentary eyes and the special development of the fifth pair of nerves, yet for the whole details of the subject, and the minute anatomy, we are indebted to Professor Wyman. Many of the details, however, as well as the admirable drawings illustrating them, remained unpublished until 1872, when he placed them at Mr. Putnam's disposal, and they were brought out in his elaborate article in the "American Naturalist." Here the extraordinary development of tactile sense, taking the place of vision, and perfectly adapting the animal to its subterranean life, is completely demonstrated.

If Professor Wyman's first piece of anatomical work was the preparation of a skeleton of a bull-frog, in his undergraduate days, his most elaborate memoir is that on the anatomy of the nervous system of the same animal (*Rana pipiens*), published in the "Smithsonian Contributions," in 1852.

Anything like an analysis of this capital investigation and exposition would much overpass our limits. For, although the special task he assigns to himself is the description of the nervous system of a single Batrachian, chiefly of its peripheral portion, and of the changes undergone during metamorphosis, he is led on to the consideration of several abstruse or controverted questions; — such, for instance, as the attempts that have been made to homologize the nervous system of Articulates with that of Vertebrates, upon which he has some acute criticism; — the theories that have been propounded respecting the functions of the cerebellum and its relation to locomotion, which he tests in a characteristic way by a direct appeal to facts; — the supposition of Cuvier that the special enlargements of the spinal cord are in proportion to the force of the respective limbs supplied therefrom; which he controverts

decisively by similar appeal, an extract from which I beg
leave to append in a note.[1]

So, in describing the structure of the optic nerves in the
frog, and the development of the eye and optic lobes, he proceeds to remark : —

" The instances of Proteus and Amblyopsis naturally suggest the questions, whether one and the same part may not
combine functions wholly different in different animals, and
whether the same may not hold true with regard to the cere-
bral organs which is known to obtain with regard to the
skeleton, the teeth, the tongue, and the nose, that identical or
homologous parts in different animals may perform functions
wholly distinct. If the doctrine here suggested can be admitted
(and if this were the place, facts could be cited in support of
it), may we not find in it an explanation of many inconsis-
tencies which now exist between the results of comparative
anatomy and of physiology ? "

Then, in his chapter on the philosophical anatomy of the
cranial nerves and skull, after showing that there are but three
pairs of cranio-spinal nerves, he takes up the controverted
question as to the number of vertebræ which compose the

---

[1] " If by force is meant the muscular energy and development of the
limbs, this statement does not appear to be sustained in the present instance, nor in many other instances brought to notice by comparative
anatomy. In man the brachial enlargement is always larger than the
crural, though the legs are so much more powerfully developed than the
arms, and the same is true of the greater number of mammals. In frogs
there is a still greater disproportion between legs and arms, yet there is
not a corresponding difference in the size of the bulgings. They cannot,
therefore, be said to be in proportion to the muscular force only of the
limbs, but correspond far more nearly to the acuteness of the sense of
touch, which in man and mammals is more delicate in the hands and arms
than in the legs and feet. In bats, it is true that the muscular force of
the arms is greater than that of the legs, and that the brachial far surpasses the crural enlargement ; but, at the same time, the sense of touch
in the membranes of the wings is exalted to a most extraordinary degree.
In birds the posterior bulging is almost universally the largest, though
this condition is in part dependent upon the presence of the rhomboidal
sinus. In these animals, while the muscular energy of the wings is the
most developed, the sensibility of the feet is the more acute."

skull, and supports the opinion that they also are only three in a characteristic manner.[1]

Of this whole memoir it is thought that, notwithstanding the great advance which has been made in comparative anatomy during the twenty-five years which have elapsed since it was published, its importance to the student has not at all diminished.

Next to this in extent and value may be ranked Professor Wyman's paper on the development of the common skate of our waters (*Raia Batis*), communicated to the American Academy in 1864, and published among its Memoirs. It gives an account of the peculiar egg-case of the Selachians, and of the several stages of the development of the embryo skate, expressed in the concise and clear language — as little technical as possible — for which he was distinguished, and leading up to not a few problems in comparative anatomy, morphology, or systematic zoölogy, — problems which Professor Wyman never evaded when they came directly in his way, and seldom handled without making some real contribution to their elucidation. For instance, in describing the external branchial fringes of the young skate, he notes the agreement of this character in the Batrachians; and in studying the seven branchial fissures of the embryo, he is

[1] " The conclusions which have been drawn from the statements made above are as follows : that in frogs the vagus comprises the glosso-pharyngeal and accessory nerves ; that the trigeminus comprises the facial, the abducens and in the salamanders the patheticus and portions of the motor communis ; that other evidence sustains the hypothesis, that the whole of the motor communis is a dependence of the trigeminus ; if to these we add the hypoglossus (which in frogs is exceptionally a spinal nerve), we shall have three pairs of cranial nerves, each having all the characters of a common spinal nerve, namely, motor and sensitive roots and a ganglion ; that there are no nerves to indicate a fourth vertebra, unless the special sense nerves are considered ; if these are admitted as indications, then we must presuppose either two pairs of nerves to each vertebra, or the existence of six vertebræ, which is a larger number than can be accounted for on an osteological basis. The functions and mode of development of the special sense nerves we have taken as affording sufficient grounds for considering them as of a peculiar order, and not to be classified with common spinal nerves."

brought into contact with the view of Huxley, that the formation of the external ear is by involution of the integument.
After confirming the contrary observations of Reichert on the
embryo pig, he concludes that " the first of the seven branchial fissures of the embryo skate is converted into the spiracle,
which is the homologue of the Eustachian tube and the outer
ear-canal." After a full discussion of the homology of the
upper jaw in sharks and skates, under the light afforded by
his investigation of the embryo skate, he suggests that the
cartilage which extends from the olfactory fossæ towards the
pectoral fin is the probable homologue of a maxillary bone,
and that in the lobe, the homologue of an intermaxillary;
that, if so, the skates and proteiform reptiles agree in having
the nostrils open in front of the dental arch; that while in
all Batrachians the nasal groove becomes closed, in the skate
it remains permanently open; and finally, that this view, if
confirmed, " will add another feature which justifies Owen,
Agassiz, and others, in dissenting from Cuvier so far as to
give the Selachians a place in the zoölogical series higher than
that of the bony fishes. But at the same time, it will give
corroborative proof of the correctness of Cuvier's view, that
' the rudiments of the maxillaries and intermaxillaries . . .
are evident in the skeleton.' "

In attempting these analyses, I am drifting into a fault
which Professor Wyman never committed, that of being too
long. So I must leave many of his papers unmentioned, and
barely refer to two or three others which cannot be passed
over. The most noteworthy of the shorter papers, however,
are upon less technical or more generally interesting topics,
so that we have need only to be reminded of them. Among
them are his " Observations on the Development of the Surinam Toad," the paper on *Anableps Gronovii*, and that " On
some unusual Modes of Gestation." The importance of these
papers lies, not in being accounts of some of the most striking
curiosities of the animal world, but in the sagacity and quickness with which he discerned, and the clearness with which
he taught the lessons to be learned from them. Any good
zoölogist, with the same excellent opportunities, would have

worked out all the details of the development of the Surinam toads in the skin of the back of their mother, and would
equally have noted the morphological significance of the
branchiæ and tail, that are never to know anything of the
element they are adapted for; but Dr. Wyman remarks upon
the development of the limbs independently of the vertebral
axis, as showing that whatever view be taken of their homology, they are something superadded to it, and not evolved
from it; he notes how the whole yelk mass is moulded into
a spiral intestine; and that the embryo at the end of incu-
bation forms a larger and heavier mass than existed in the
egg when it commenced, — showing that there was an absorp-
tion of material furnished by the dermal sac of the mother,
— " a solitary instance among Batrachians, if not among reptiles generally, in which the embryo is nourished at the expense of materials derived from the parent." From this he
is led (in the last paper above mentioned) to infer the probability that the developed larvæ of *Hylodes lineatus* — carried about inland upon the back of their mother, and destitute
of limbs adapted to terrestrial locomotion — may depend
upon a secretion from the body for needful sustenance, — an
interesting and rudimentary foreshadowing of mammalian
life, of which he discerned the bearings.

His "Description of a Double Fœtus" (in the "Boston
Medical and Surgical Journal," March, 1866) gives him the
opportunity of briefly recording some of the results of his
studies of the development of double monsters, and to bring
out his view, that "the force, whatever it be, which regulates
the symmetrical distribution of matter in a normal or abnormal embryo, has its analogy, if anywhere, in those known
as polar forces;" that, "studying the subject in the most
general manner, there are striking resemblances between the
distribution of matter capable of assuming a polar condition,
and free to move around a magnet, and the distribution of
matter around the nervous axis of an embryo." That this is
not one of those vague conceptions by which many speculators
set about to explain that of which they know little by means
of that of which they know less, but that he had striking

parallelisms to adduce, the close of this striking paper shows.

The subject of fore and hind symmetry, thus brought directly under notice, had been broached by Dr. Wyman several years before. He returned to it the year following, in his very important morphological paper, "On Symmetry and Homology in Limbs," read to this society in June, 1867, and published in the Proceedings of that date. It is interesting to observe with what caution and restraint he handled this doctrine of "reversed repetitions," which has since been freely developed by one of his pupils who has a special predilection for speculative morphology, Professor Burt Wilder.

Professor Wyman's "Notes on the Cells of the Bee," in the "Proceedings of the American Academy" for January, 1866, is a characteristic specimen of his way of coming directly down to the facts, and making them tell their own story. I could not recapitulate his results much more briefly than he records them in his paper. I need not recall to you how neatly he made this investigation, and represented some of the results, filling the comb with plaster-of-paris and then cutting it across midway, so that the observations might be made and the cells measured just where they are most nearly perfect; and then printing impressions of the comb upon the woodblock, he reproduces on the pages of his article the exact outlines of the cells, with all their irregularities and imperfections. But I cannot refrain from citing a portion of his remarks at the close : —

"Here, as is so often the case elsewhere in nature, the typeform is an ideal one ; and with this, real forms seldom or never coincide. . . . An assertion, like that of Lord Brougham, that there is in the cell of the bee 'perfect agreement' between theory and observation, in view of the analogies of nature, is more likely to be wrong than right; and his assertion in the case before us is certainly wrong. Much error would have been avoided if those who have discussed the structure of the bee's cell had adopted the plan followed by Mr. Darwin, and studied the habits of the cell-making insects comparatively, beginning with the cells of the humble-

bee, following with those of wasps and hornets, then with those
of the Mexican bees (Melipona), and finally with those of the
common hive-bee. In this way, while they would have found
that there is a constant approach to the perfect form, they
would at the same time have been prepared for the fact, that
even in the cell of the hive-bee perfection is not reached. The
isolated study of anything in natural history is a fruitful
source of error."

Let me add to this important aphorism its fellow, which I
have from him, but know not if he ever printed it. "*No
single experiment in physiology is worth anything.*"

The spirit of these aphorisms directed all his work. It is
well exemplified in his experimental researches — the last
which I can here refer to — upon "The formation of Infusoria in boiled solutions of organic matter, inclosed in hermet-
ically sealed vessels and supplied with pure air," and its supplement, "Observations and Experiments on Living Organisms in Heated Water," published in the "American Journal
of Science and Arts," the first in the year 1862, the other in
1867. Milne-Edwards could not have known the man, when
he questioned the accuracy of the first series because they do
not agree with those of Pasteur, and thought the difference
in the results depended upon a defective mode of conducting
the experiments. As Dr. Wyman remarks, in a note to the
second series, "the recent experiments of Dr. Child of Oxford, and those reported in this communication, are sufficient
answer to the criticisms of M. Edwards." Then, as to his
thoroughness, most persons would have rested on the results
of his thirty-three well-devised experiments proving "that the
boiled solutions of organic matter made use of, exposed only to
air which has passed through tubes heated to redness, became
the seat of infusorial life;" but all would not have concluded
that, after all, they "throw but little light on the immediate
source from which the organisms have been derived," nor
would many have closed an impartial summary of the opposing views in this judicial way: —

"If, on the one hand, it is urged that all organisms, in so
far as the early history of them is known, are derived from

ova, and therefore from analogy we must ascribe a similar origin to these minute beings the early history of which we do not know, it may be urged with equal force, on the other hand, that all ova and spores, in so far as we know anything about them, are destroyed by prolonged boiling ; therefore from analogy we are equally bound to infer that Vibrios, Bacterians, etc., could not have been derived from ova, since these would have all been destroyed by the conditions to which they have been subjected. The argument from analogy is as strong in the one case as in the other."

Returning to the subject again a few years later, with a critical series of twenty experiments, each of three, five, ten, fifteen, or even twenty flasks, used by way of checks and comparisons, — a rigorous experimenter would have been satisfied when he had proved that sealed solutions subjected to a heat of at least 212° for from one to four hours, became the seat of infusorial life, at least of such as Vibrios, Bacterians, and Monads, while all infusoria having the faculty of locomotion were shown by a special series of experiments to lose this at a temperature of 120° or at most 134° Fahr. But Professor Wyman carried the boiling up to five hours, and in these flasks no infusoria of any kind appeared. The question of abiogenesis stands to-day very much where Professor Wyman left it seven years ago.

I must omit all notice of the ethnological work which has occupied his later years, merely referring to the seven annual "Reports of the Trustees of the Peabody Museum of American Archæology and Ethnology," of which he was curator. The last of these, issued just before the writer's death, contain the principal results of his investigations of the human remains he collected in the shell-heaps of east Florida, and convincing evidence of the cannibalism of those who made them. A fuller memoir, embodying all his observations of the last six winters upon the Florida shell-mounds, was sent to the printer just before he died.

The thought that fills our minds upon a survey even so incomplete as this is : How much he did, how well he did it all, and how simply and quietly ! We know that our associate,

though never hurried, was never idle, and that his great repose of manner covered a sustained energy; but I suspect
that none of us, without searching out and collecting his published papers, had adequately estimated their number and
their value. There is nothing forth-putting about them,
nothing adventitious, never even a phrase to herald a matter
which he deemed important.

His work as a teacher was of the same quality. He was
one of the best lecturers I ever heard, although, and partly
because, he was most unpretending. You never thought of the
speaker, nor of the gifts and acquisitions which such clear
exposition were calling forth, — only of what he was simply
telling and showing you. Then to those who, like his pupils
and friends, were in personal contact with him, there was the
added charm of a most serene and sweet temper. He was
truthful and conscientious to the very core. His perfect freedom, in lectures as well as in writing, and no less so in daily
conversation, from all exaggeration, false perspective, and
factitious adornment, was the natural expression of his innate
modesty and refined taste, and also of his reverence for the
exact truth.

It has been a pleasure to learn, from former college students,
who hardly ever saw him except in the lecture room, that he
gave to them much the same impression of his gifts and
graces and sterling worth, that he gave us who knew him
intimately — so transparent was he and natural.

With all his quick sense of justice, and no lack of occasion
for controversy, it seemed to cost him no effort to avoid it
altogether. He made no enemies, and was surrounded by
troops of life-long friends. When he first went abroad, in
1841, he was told by some near friends, who recognized his
promise, that a chair of natural history in his alma mater
would soon have to be filled, and that he should be presented
as a candidate. In the winter following. the present incumbent, responding to an invitation to visit Boston, which he
had never seen, and to consider if he would be a candidate,
then first heard of Wyman's name and his friends' expectations or hopes; whereupon he dismissed the subject from his

mind. Probably he felt more surprise than did Dr. Wyman when notified, a few months afterward, of the choice of the corporation. The exigencies of the botanic garden probably overbore other considerations. I doubt if Dr. Wyman ever had an envious feeling. Certain it is that no one welcomed the new professor with truer cordiality, or proved himself a more constant friend.

In these days it is sure to be asked how an anatomist, physiologist, and morphologist like Professor Wyman regarded the most remarkable scientific movement of his time, the revival and apparent prevalence of doctrines of evolution. As might be expected, he was neither an advocate or an opponent. He was not one of those persons who quickly make up their minds, and announce their opinions, with a confidence inversely proportionate to their knowledge. He could consider long, and hold his judgment in suspense. How well he could do this appears from an early, and so far as I know, his only published presentation of the topic, in a short review of Owen's " Monograph of the Aye-Aye " (in Am. Journ. Science, Sept., 1863) — the paper in which Professor Owen's acceptance of evolution, but not of natural selection, was promulgated. Dr. Wyman compares Owen's view with that of Darwin (to whom he had already communicated interesting and novel illustrations of the play of natural selection) ; and he adds some acute remarks upon a rather earlier speculation by Mr. Agassiz, in which the latter suggests that the species of animals might have been created as eggs rather than as adults. He states the case between the two general views with perfect impartiality, and the bent of his own mind is barely discernible. In due time he satisfied himself as to which of them was the more probable, or, in any case, the more fertile hypothesis. As to this I may venture to take the liberty to repeat the substance of a conversation which I had with him some time after the death of the lamented Agassiz, and not long before his own. I report the substance only, not the words.

Agassiz repeated to me, he said, a remark made to him by Humboldt, to the effect that Cuvier made a great mistake, and missed a great opportunity, when he took the side he did in the famous controversy with Geoffroy St. Hilaire; he should have accepted the doctrines of morphology, and brought his vast knowledge of comparative anatomy and zoölogy, and his unequaled powers, to their illustration. Had he done so, instead of gaining by his superior knowledge some temporary and doubtful victories in a lost cause, his preëminence for all our time would have been assured and complete. I thought, continued Wyman, that there was a parallel case before me, that if Agassiz had brought his vast stores of knowledge in zoölogy, embryology, and palæontology, his genius for morphology, and all his quickness of apprehension and fertility in illustration, to the elucidation and support of the doctrine of the progressive development of species, science in our day would have gained much, some grave misunderstandings been earlier rectified, and the permanent fame of Agassiz been placed on a broader and higher basis even than it is now.

Upon one point Wyman was clear from the beginning. He did not wait until evolutionary doctrines were about to prevail, before he judged them to be essentially philosophical and healthful, "in accordance with the order of nature as commonly manifested in her works," and that they need not disturb the foundations of natural theology.

Perhaps none of us can be trusted to judge of such a question impartially, upon the bare merits of the case; but Wyman's judgment was as free from bias as that of any one I ever knew. Not at all, however, in this case from indifference or unconcern. He was not only, philosophically, a convinced theist, in all hours and under all "variations of mood and tense," but personally a devout man, an habitual and reverent attendant upon Christian worship and ministrations.

Those of us who attended his funeral must have felt the appropriateness for the occasion of the words which were there read from the Psalmist: —

"The heavens declare the glory of God, and the firmament showeth his handy-work. . . . O Lord, how manifold are thy works! In wisdom hast thou made them all; the earth is full of thy riches; so is this great and wide sea, wherein are things creeping innumerable, both great and small beasts. Thou sendest forth thy spirit, they are created, and thou renewest the face of the earth."

These are the works which our associate loved to investigate, and this the spirit in which he contemplated them. Not less apposite were the Beatitudes that followed:

" Blessed are the meek ; blessed are the peace-makers ; blessed are the merciful ; blessed are the pure in heart."

Those who knew him best, best know how well he exemplified them.

DANIEL HANBURY.[1]

NOT long ago we called attention to a most valuable book, the "Pharmacographia, a History of Drugs," by Professor Flückiger of Strasburg and Daniel Hanbury of London, the first-fruits of much investigation, the precursor, as was hoped, of more extended similar works by the English author. We have now sadly to record the decease of Mr. Hanbury, of enteric typhoid, on the 24th of March, at his residence on Clapham Common, in the fiftieth year of his age. The obituary and biographical notices which have appeared in the London scientific journals and in the Proceedings of the learned societies, as well as loving individual tributes to an endeared memory, have given expression to the loss which has been sustained, and delineated the outlines of a most worthy and winning character. The loss is deplored, personally and scientifically, over wider circles and on this side of the Atlantic. The pupil and friend of Pareira and his successor in his line of work, an adept in pharmaceutical knowledge, a keen botanist, and a most assiduous and conscientious investigator, a man of simple and pure tastes, and happily of sufficient means, he had just withdrawn wholly from business in the noted house in which he had an inherited share, so that he might devote his powers and acquisitions without distraction to the natural history of drugs and useful vegetable products. He had already done much : more than sixty articles were contributed by him to a single journal, the editor of which declares that "the quality of what he did was almost faultless," that "he never wrote without having original information to impart, and his papers uniformly bear evidence of careful investigation and thorough knowledge." The Transactions and Journal of the Linnæan Society (of which he was repeat-

[1] American Journal of Science and Arts, 3 ser., ix. 476. (1875.)

edly a councilor and the treasurer at the time of his death) contain several of his papers. His first published paper, "On Turnsole," appeared in 1850; his latest, on the "Countess Chinchon and the Cinchona Genus," appeared, since his death, in "The Academy" of the 3d of April. An ardent botanist and lover of plants, he traveled much in the south of Europe, accompanied Dr. Hooker in his explorations of Lebanon, and took an active interest especially in the introduction of officinal plants and in ornamental cultivation. With one villa-garden on the shores of the Mediterranean — that of his brother at Mortola — his memory to us is indelibly associated. Although remarkably self-reliant, Mr. Hanbury was the opposite of self-asserting or ambitious; but his sterling worth was soon recognized by the principal learned societies and associations. He was early chosen into the Royal Society and served upon its council in 1873. Born and educated in the Society of Friends, he remained a devout and attached member of it, while the graces and the goodness of his character endeared him to all who knew him. With the sense of personal loss his scientific comrades mingle deep regrets that a career of unusual usefulness and promise is cut short, and that in a line in which, it is to be feared, he leaves no successor. *Hanburia Mexicana*, a striking Cucurbitaceous genus, commemorated his services to botany.

ALEXANDER BRAUN.[1]

WE announce with sorrow the death of this excellent botanist, which took place in Berlin, on the 29th of March, after a
short illness. Systematic botanists of the first class are everywhere rare, and especially in Germany, where they have gone
out of fashion, all attention being turned to histology and the
like. In Braun's earlier days there was a goodly array of
systematic botanists in Germany; at his decease there are
very few of mark, although signs of revival are apparent.
Alexander Braun was born at Ratisbon, May 10, 1805, but
was brought up at Carlsruhe, where his father became a trusted
officer in the post-office department. Fifty years ago there
was a knot of closely - allied students at the university of
Heidelberg, consisting of Braun, Carl Schimper, Agassiz, and
Engelmann. Two of them were transferred to our own soil;
the latter is now the sole survivor. Three of them went
soon to Munich, where Oken, Schelling, Dollinger, and Martius were teaching; but Braun, Agassiz, and Engelmann met
again as fellow - students at Paris in 1832. The first two
became allied afterwards by the marriage of Agassiz with
Braun's sister. About the time that Dr. Engelmann came to
the United States, Braun was made professor of botany and
zoölogy in the Polytechnic School of Carlsruhe. In 1846 he
took the chair of botany in the university of Freiburg in the
Breisgau; was transferred to Giessen in 1850; but in the
spring of 1851 was called to Berlin, as the successor to Link
and Kunth, taking charge of the botanic garden as well as of
the professorship. Although he had nearly reached the age of
seventy-two, and felt the full weight of his years, yet he was
assiduously attending to his official duties when he was suddenly prostrated by acute disease of the chest, terminating

[1] American Journal of Science and Arts, 3 ser., xiii. 471. (1877.)

fatally. Although learned in almost every department of his science, his forte, like that of Agassiz, was morphology, and his systematic work mainly among the higher and some of the lower Cryptogamia, Marsilia, Isoetes, Chara, etc. Although his communications to the scientific journals began as early as the year 1822, when he was only seventeen years old, his first contribution to science, of much extent and of high and permanent value, was his memoir on the arrangement of the scales of Pine-cones, etc., published in 1830. With this publication began the present knowledge of phyllotaxis. It is well understood that the first steps were taken by his fellow-student Carl Schimper, and that the early investigations were pursued in common by the two. But Schimper published nothing, or next to nothing, either then or since, although he lived until the year 1867. His name in connection with the subject is preserved by the favorable mention of his companions and contemporaries; but Braun's treatise was timely and fruitful, and became classical. Braun's ability for the philosophical treatment of vegetable morphology and development was manifested in his next large paper, namely, in his memoir entitled "Rejuvenescence in Nature," especially in the life and development of plants. This was first published at Freiburg in 1849–50, and again at Leipsic in 1851, and an English translation of it was published by the Ray Society in 1853. Of a similar character, and marked with equal acuteness, is his essay on "The Vegetable Individual in its relation to Species," etc., published in 1853, at Berlin, and which, in a translation by a pupil of mine, was mainly reproduced in this Journal (May and September, 1855). He reaches the conclusion — which would now be more confidently expressed — "that the individual appears in its full import only in the higher steps of the series of created beings."

In his systematical work, Braun was exceedingly laborious, persevering, and conscientious. When we add that throughout the riper and what should have been the most productive years of his life, he was overtasked with official duties and cares, we shall not wonder that much which he hoped to accomplish is left undone. His work upon Marsilia, Pilularia, and Isoetes may be essentially complete. But his prolonged studies of Chara, which began forty years ago, and the completion of which would have crowned his career, have probably not been finished, or brought into such form that the results may be fully secured.

His influence as a teacher is said to have been great; as an investigator he stood in the first rank among the botanists of our time ; as a man, his simple, earnest, and transparently truthful character won the admiration and love of all who knew him.

CHARLES PICKERING.[1]

CHARLES PICKERING, M. D., died in Boston, of pneumonia, on the 17th of March, 1878, in the seventy-third year of his age. He was of a noted New England stock, being a grandson of Colonel Timothy Pickering, a member of Washington's military family, and of his first cabinet as President; and he was elected into this Academy under the presidency of his uncle, John Pickering. He was born on Starucca Creek, on the Upper Susquehanna, in the northern part of Pennsylvania, at a settlement made on a grant of land taken up by his grandfather, who then resided there. His father, Timothy Pickering, Jr., died at the age of thirty years, leaving to the care of the mother — who lived to a good old age — the two sons, Charles and his brother Edward, who were much united in their earlier and later lives, and were not long divided in death, the subject of this notice having been for only a year the survivor.

Dr. Pickering was a member of the class of 1823 at Harvard College, but left before graduation. He studied medicine, and took the degree of M. D. at the Harvard Medical School in 1826. Living in these earlier years at Salem, he was associated with the late William Oakes in botanical exploration; and it is believed that the two first explored the White Mountains together, following in the steps of the first botanist to ascend Mount Washington, Dr. Manasseh Cutler of Essex County, and of Francis Boott and the still surviving Dr. Bigelow. His taste for natural history showed itself in boyhood, both for botany and zoölogy, and probably decided his choice of a profession. He may have intended to practise medicine for a livelihood, when, about the year 1829, he took

[1] Proceedings of the American Academy of Arts and Sciences, xiii. 414. (1878.)

CHARLES PICKERING. 407

up his residence at Philadelphia; but it is probable that he was attracted thither more by the facilities that city offered for the pursuit of natural history than by its renown as a centre of medical education. We soon find him acting as one of the curators of the Academy of Natural Sciences, and also as librarian, and with reputation established as the most erudite and sharp-sighted of all the young naturalists of that region. His knowledge then, as in mature years, was encyclopædic and minute; and his bent was toward a certain subtlety and exhaustiveness of investigation, which is characteristic of his later writings. Still, in those days in which he was looked up to as an oracle, and consulted as a dictionary by his co-workers, he had published nothing which can now be recalled, except a brief essay on the geographical distribution and leading characteristics of the United States flora, which very few of our day have ever seen.

When the United States surveying and exploring expedition to the South Seas, which sailed under command of then Lieutenant Charles Wilkes in the autumn of 1838, was first organized under Commodore T. Ap-Catesby Jones, about two years before, Dr. Pickering's reputation was such that he was at once selected as the principal zoölogist. Subsequently, as the plan expanded, others were added. Yet the scientific fame of that expedition most largely rests upon the collections and the work of Dr. Pickering and his surviving associate, Professor Dana, the latter taking, in addition to the geology, the Corals and the Crustacea, other special departments of zoölogy being otherwise provided for by the accession of Mr. Couthouy and Mr. Peale. Dr. Pickering, although retaining the ichthyology, particularly turned his attention during the three and a half years' voyage of circumnavigation to anthropology, and to the study of the geographical distribution of animals and plants; to the latter especially affected by or as evidence of the operations, movements, and diffusion of the races of man. To these, the subjects of his predilections and to investigations bearing upon them, all his remaining life was assiduously devoted. The South Pacific exploring expedition visited very various parts of the world; but it necessarily left out regions of the highest interest to the anthropological investigator, those occupied in early times by the race to which we belong, and by the peoples with which the Aryan race has been most in contact. Desirous to extend his personal observations as far as possible, Dr. Pickering, a year after the return of the expedition, and at his own charges, crossed the Atlantic, visited Egypt, Arabia, the eastern part of Africa, and western and northern India. Then, in 1848, he published his volume on "The Races of Man, and their Geographical Distribution," being the ninth volume of the "Reports of the Wilkes' Exploring Expedition." Some time afterward, he prepared, for the fifteenth volume of this series, an extensive work on the "Geographical Distribution of Animals and Plants." But, in the course of the printing, the appropriations by Congress intermitted or ceased, and the publication of the results of the celebrated expedition was suspended. Publication it could hardly be called : for Congress printed only one hundred copies, in a sumptuous form, for presentation to states and foreign courts; and then the several authors were allowed to use the types and copperplates for printing as many copies as they required, and could pay for. Under this privilege, Dr. Pickering brought out in 1854 a small edition of the first part of his essay,— perhaps the most important part,— and in 1876 a more bulky portion, "On Plants and Animals in their Wild State," which is largely a transcript of the note-book memoranda as jotted down at the time of observation or collection.

These are all his publications, excepting some short communications to scientific journals and the proceedings of learned societies to which he belonged. But he is known to have been long and laboriously engaged upon a work for which, under his exhaustive treatment, a lifetime seems hardly sufficient; a digest, in fact, of all that is known of all the animals and plants with which civilized man has had to do from the earliest period traceable by records. When Dr. Pickering died, he was carrying this work through the press at his own individual expense, had already in type five or six hundred quarto pages, and it is understood that the remainder, of about equal extent, is ready for the printer. This formidable treatise is entitled " Man's Record of his own Existence." Its character is indicated in the brief introductory sentences : —

" In the distribution of species over the globe, the order of Nature has been obscured through the interference of man. He has transported animals and plants to countries where they were previously unknown ; extirpating the forest and cultivating the soil, until at length the face of the globe itself is changed. To ascertain the amount of this interference, displaced species must be distinguished, and traced each to its original home. Detached observations have already been given in the twenty-first and succeeding chapters of my ' Races of Man ' ; but when such observations are extended to all parts of the globe, the accumulated facts require some plan of arrangement. A list will naturally assume the chronological order, beginning with Egypt, the country that contains the earliest records of the human family, and receding geographically from the same central point of reference."

Then, starting with " 4713 B. C.," and " 4491 B. C., beginning of the first Great Year in the Egyptian reckoning," he begins the list, which, under the running heading of " Chronological Arrangement of Accompanying Animals and Plants," first treats of the vegetables and animals mentioned in the book of Genesis, and of the " Commencement of Bedouin or Nomadic Life in the Desert ; " passes to the " Colonization of Egypt," and to critical notices (philological and natural-historical) of its plants and animals, as well their earliest mention as their latest known migrations ; reaches the beginning of the Christian era at about the 470th page ; and so proceeds, till our wonder at the patience and the erudition of the writer passes all bounds. We are ready to agree with a biographer who declares that our associate was " a living encyclopædia of knowledge," — that there never was a naturalist " who had made more extended and minute original explorations ; " and we fully agree that " no one ever had less a passion or a gift for display ; " " that he was engaged during a long life in the profoundest studies, asking neither fame nor money, nor any other reward, but simply the privilege of gaining knowledge and of storing it up in convenient forms for the service of others ; " that " the love of knowledge was the one passion of his life," and that " he asked no richer satisfaction than to search for it as for hidden treasure." He was singularly retiring and reticent, very dry in ordinary intercourse, but never cynical ; delicate and keen in perception and judgment ; just, upright, and exemplary in every relation ; and to those who knew him well communicative, sympathetic, and even genial. In the voyage of circumnavigation he was the soul of industry, and a hardy explorer. The published narrative of the commander shows that he took a part in every fatiguing excursion or perilous ascent. Perhaps the most singular peril (recorded in the narrative) was that in which this light-framed man once found himself on the Peruvian Andes, when he was swooped upon by a condor, evidently minded to carry off the naturalist who was contemplating the magnificent ornithological specimen.

Dr. Pickering married in the year 1851, and leaves a widow, but no children to inherit his honored name.

ELIAS MAGNUS FRIES.[1]

ELIAS MAGNUS FRIES died at Upsal on February 8, in the eighty-fourth year of his age, five months after the celebration, in which he was able to take some part, of the four hundredth anniversary of the foundation of that university, and a month after the hundredth anniversary of the death of Linnæus. Born, as was Linnæus, in Smoland, a southern province of Sweden, and like him called in middle age to the renowned Scandinavian university, he might be regarded as the most distinguished of Linnæus's successors, except for the fact that he did not occupy the chair of Linnæus ; for when, more than forty years ago, Fries, then demonstrator of botany at Lund, was called to Upsal, Wahlenberg was in the botanical chair, and Fries was made professor of practical economy. His son, however, by the retirement of Areschoug, is now botanical professor.

Fries's earliest work, the first part of his Novitiæ, appeared in the year 1814, when the author was only twenty years old. His last of any moment, a new edition of his " Hymenomycetes Europæi," was published on his eighty-first birthday, August 15, 1874. Most of the sixty intervening years are marked by some publication from his busy and careful hand. His work was wholly in systematic botany, and of the highest character of its kind. In Phænogamous botany, it related chiefly to the Scandinavian flora, in which for critical judgment he had no superior ; in Mycology, of which he was the reformator and to a good degree in Lichenology, he had no rival except as regards microscopical research. The modern microscope did not exist when he began his work, and, while showing how much can be done without it, he may too long

[1] Proceedings of the American Academy of Arts and Science, xiii. 453. (1878.)

have underrated its value. But he lived to see it confirm many conclusions which his insight foresaw, and solve riddles which he had pondered, but was unable to divine. He was the prince, Nestor, and last survivor of an excellent school of systematic botanists, whose teachers were taught by Linnæus or his contemporaries.

JACOB BIGELOW.[1]

Dr. Jacob Bigelow died at his residence in Boston, on the 10th of January last, near the close of the ninety-second year of his age.

While we would pay the tribute due to his memory as by far the most venerable of American botanists, the last survivor of a school in this country which culminated half a century ago, it should also be remembered that he was even at that time distinguished in their scientific avocations, and that from middle to old age he was among the most eminent of physicians. It is not often that we can contemplate a life so long, so richly various, and so well-rounded as his. He was born in Sudbury, Massachusetts, on February 27, 1787; and his father was the minister of the town. That almost goes without saying, most of our distinguished professional men of his and the preceding generations in New England having been the sons of country ministers. He was graduated at Harvard College in the year 1806, Alexander H. Everett and the late Dr. J. G. Cogswell being among the most notable of his classmates, all of whom he long survived. He directly took up the study of medicine, was licensed as a practitioner in 1809, and after attending one course of lectures in Philadelphia, took his degree of M. D. at Harvard in 1810, and established himself in Boston. There he was a practising physician for about sixty years, and since the death of his senior, Dr. James Jackson, probably the most eminent one. What turned his attention to botany we know not. He early showed an abiding taste for poetry. His commencement part was a poem, and he delivered a Φ. B. K. poem not long after. At about the same time, however, he gave a course of popular botanical lectures in Boston, in connection with Professor Peck, who must have been installed as natural-history professor at

[1] American Journal of Science and Arts, 3 ser., xvii. 263. (1879.)

Cambridge while Dr. Bigelow was a medical student. The latter possessed the gift of exposition which Dr. Peck lacked; and it naturally came to pass that Dr. Bigelow repeated this course of lectures alone for a year or two afterward.

In the spring of 1814 he brought out the first edition of his "Florula Bostoniensis," the book which, mainly in its second edition, has been the manual for New England herborization down to a recent day, or rather to a day which seems to us recent. The original volume, of 268 octavo pages, describes the plants which "have been collected during the two last seasons in the vicinity of Boston, within a circuit of from five to ten miles," exceeding those limits only in the case of Magnolia (from Manchester) and one or two more remarkable plants. We know of no other Flora of the kind which has been prepared so quickly and so well. The characters are short diagnoses, and in good part compiled. But the descriptive matter must have been original; and it shows that aptitude for seizing the best points of character or most available distinctions, and of indicating them in few and clear words, which has made this Manual so deservedly popular. Similar merits distinguish, on its botanical side, Dr. Bigelow's "American Medical Botany," a quarto work which was published, in three parts or volumes, between 1817 and 1821, with colored plates — at that time thought to be very good ones indeed — of the principal medicinal plants of the country. He also brought out an American edition of Sir James Edward Smith's "Introduction to Botany"; and his botanical knowledge, along with that of the Materia Medica generally and his classical scholarship, placed him at the head, or at the laboring oar, of the committee which in 1820 formed the American Pharmacopœia. The writer used this volume in his medical student days, and remembers dimly how the account of minor preparations, coming down to jams and conserves, ended with the classical "Jam satis est mihi."

The second edition of the "Florula Bostoniensis," published in 1824, while retaining its modest title, was nearly doubled in size and in the number of plants it contained, the whole area of New England included; and it became the manual of

botany for the region. What a popular and satisfactory work
it was, especially to hundreds of amateur botanists, some still
living may testify.

The third and last edition, issued in 1840, was a reprint,
with various additions and corrections, furnished mainly from
those who had learned their botany from the preceding one.
This is the last Flora or Manual of this and perhaps any
other country, arranged upon the Linnæan artificial system.
Much later in life the author contemplated a revision of the
work, brought up to the time and illustrated by chromo-
lithographic plates, such as we have lately seen turned to good
account. But after some consideration the project was abandoned. He did not propose himself to undertake the editorial
work; for he had long since passed from actual service into
the emeritus or honorary rank of botanists; and his active
professional life, already verging to its close, was diversified
or relieved by other avocations. Indeed, some of these were
taken up very early. He became Rumford professor of the
applications at Cambridge in 1816, and delivered annual
courses of lectures until 1827, when he published the substance of them in a volume entitled " Elements of Technology," here coining this apt word. During all this time, and
much longer, he was professor of Materia Medica in the medical school of Harvard University, namely from 1815 to 1855;
for many of these years one of the physicians of the Massachusetts General Hospital; through all of them, and until old
age disabled him, a leading physician of Boston. From the
year 1847 to 1863 he was president of the American Academy
of Arts and Sciences, of which body he was a member for
sixty-seven years!

We cannot here refer to Dr. Bigelow's various professional
and literary writings. They are not numerous, but are weighty.
His treatise on "Nature in Disease," which contains the famous discourse "On Self-limited Disease," is the most important of them; and an address "On the Limits of Education,"
delivered in the year 1865 before the Massachusetts Institute
of Technology, is notable. It has been said of the latter, that
never before was the depreciation of classical study or general

culture, as a preparation for technical scientific education, undertaken by so ripe a classical scholar or so wide-cultured a man. His many essays in English and Latin verse, some of which have been privately printed, ought to be collected. Dr. Bigelow lived, honored and trusted, to a good old age before infirmities touched his frame, and only toward the close was the brightness of his acute mind dimmed. The candle at length burnt down, the flame flickered awhile in the socket, and the light went out.

The name will abide in botanical nomenclature. First appeared in Rees' Cyclopedia the Bigelowia of Smith, founded on the Adelia of Michaux. But that is Forestiera. Then Sprengel, in 1821, founded a genus Bigelovia on a Brazilian plant which he took to be a Rhamnacea, but it is a species of Casearia. Again, in 1824, Sprengel gave the name to a part of Spermacoce, the Borreria of G. Meyer. Then De Candolle, in 1824, was proposing a Bigelowia on *Solea concolor*, of our own New England, as the " Prodromus " records, when he found he had to refer it to Noisettia. Lastly, in 1836, De Candolle bestowed the name of Bigelovia upon some golden-flowered *Compositæ* of the southern United States, which had borne the name of an Old World genus, Chrysocoma (Anglice, Golden-tuft), and he added the complimentary phrase : " A Chrysocoma separatum dicavi cl. J. Bigelow qui floræ Americanæ auream coronam flora Bostoniensi et medica addidit." Although this genus was founded upon only two or three species, it has been vastly extended by the exploration of the western regions of our country, where it forms a conspicuous and characteristic portion of the low shrubby vegetation. More than thirty North American species of Biglovia, besides one of Mexico and two of the Andes of South America, now commemorate our venerable late associate. Most of them were introduced to the genus by the present writer.

JOHN CAREY.[1]

JOHN CAREY — of whom few of the botanists of our day can have a personal remembrance — died at Blackheath, near London, March 26 ult., in the 83d year of his age. He came from London to the United States, in the spring of 1830, accompanied by three young and motherless children and by his brother, Samuel T. Carey, who was also addicted to botany. Both, we believe, were Fellows of the Linnæan Society, and were near friends of Thomas Bell, afterwards the president of that society, who also lived to a good old age, dying only a few weeks earlier than the subject of this notice. Samuel T. Carey remained in the city of New York, in active business, and so was only an amateur botanist. His brother John went into the country, first to Towanda, in the northern part of Pennsylvania, then to Bellows Falls, Vermont, where, giving much of his leisure to botanical pursuits, he resided until the year 1836, when he removed to New York, upon the entrance of his sons into Columbia College. He did not enter into business, but his administrative talents and great worth were so appreciated that he was at various times called to very responsible temporary positions. These positions, although unsought, were not unwelcome, for no small part of the moderate property he had brought from England had been lost in investments made through reliance upon the honor and probity of defaulting States.

From the time of his arrival in the United States down to the year of his return to England in 1852, most of his leisure was given to botany, and much of it in the companionship of the present writer, who was generously and greatly assisted by him in many critical studies. The proofs of the writer's first botanical book were revised by him, and to the first

[1] American Journal of Science and Arts, 3 ser., xix. 421. (1880.)

edition of the "Manual of Botany" Mr. Carey contributed the articles on Salix and on Carex, at that time the two most difficult parts of the work. In the year 1841 the two made a botanical journey together into the mountains of North Carolina, extending to the Grandfather and to the Roan, though a mishap upon the former mountain prevented Mr. Carey from reaching the latter. After the establishment of the writer at Cambridge, Mr. Carey was his frequent guest and invaluable companion. His botanical career may have said to have closed in the year 1852. In that year he returned to England alone, having successively lost his aged mother and his two younger sons, and seen the elder son happily established in marriage. He engaged for several years in business, in connection with a friend of his youth, whose daughter he soon married, but lost within three years, after the birth of the second of the two children, the solace and comfort and care of a serene old age, who survive to perpetuate his name, we trust, on that side of the ocean also. Mr. Carey's first herbarium was destroyed by a calamitous fire in New York, at the time of the death of his youngest son. American botanists vied with each other in the endeavor to repair this serious loss, and another large collection of United States plants was formed, critically studied, and carefully annotated. This was presented to the Kew Gardens herbarium eleven years ago. Several species of United States plants commemorate this honored name, among them a Saxifrage, which was discovered upon the excursion to the mountains of North Carolina, where the survivor of the party re-collected it last summer. The almost sole survivor of a botanical circle, of which Torrey was the centre, sadly but serenely pays the tribute of this brief note to the memory of a near and faithful friend, an accomplished botanist, a genial and warm-hearted and truly good man.

THOMAS POTTS JAMES.[1]

THOMAS POTTS JAMES died at his residence in Cambridge, February 22, 1882, in the seventy-ninth year of his age. He had been a Fellow of the Academy for only four years, most of his life having been spent in Philadelphia, in the neighborhood of which city he was born on the 1st of September, 1803. His paternal and maternal ancestors were notable persons among the earlier settlers of Pennsylvania. For forty years he was engaged in business in Philadelphia as a wholesale druggist, on the relinquishment of which he removed to Cambridge, bringing his wife and their four children to her paternal home. From his youth he was more or less devoted to botany ; but of late years, having more leisure for the indulgence of his taste, and wishing to be more than an amateur, he devoted himself exclusively and most sedulously to bryology, in which he became a proficient. After the death of Mr. Sullivant in 1873, Mr. James and our associate, Lesquereux, were looked to as the principal authorities upon Mosses in this country ; and the duty appropriately devolved upon them of preparing the systematic work upon North American Bryology which Mr. Sullivant had planned. Owing to the preoccupation of Mr. Lesquereux in vegetable palæontology, the laboring oar fell to Mr. James. He had already published some papers upon the subject in the "Transactions of the American Philosophical Society," of which he had long been an active member, and he had contributed to Mr. Watson's "Botany of Clarence King's Exploration on the Fortieth Parallel" a notable article on the Mosses of that survey. Our own Academy has also published some of the results of the joint study of these two veteran bryologists. The characters of Mosses in these days are mostly drawn

[1] Proceedings American Academy of Arts and Science, xvii. 405. (1882.)

from their minute structure. Hundreds of species and varieties in numerous specimens had to be patiently scrutinized under the compound microscope, the details sketched and collated, and the differences weighed. To this task Mr. James gave himself with single and untiring devotion. He had nearly brought this protracted labor of microscopical analysis to a conclusion, and was actually engaged in this work, when the eye suddenly was dimmed and the pencil dropped from his hand. Partial paralysis was soon followed by coma, and he died within a few hours. So very much has been done, that it is confidently hoped that his coadjutor may soon bring the work to a completion, and give to bryological students the Manual of North American Mosses which is greatly needed, and to which a vast amount of faithful research has been devoted. The name of Mr. James will thereby be inseparably associated with the advancement of an interesting branch of botany. He was not often seen at our meetings, but he is greatly missed by his associates in study, and his memory is cherished by all who in the various relations of life came to know this diligent and conscientious student of nature, and most estimable, simple-hearted, kindly, and devout man.

JOHN AMORY LOWELL.[1]

JOHN AMORY LOWELL died at his residence in Boston, on the 31st of October last, when he had almost completed the eighty-third year of his age, for he was born on the 11th of November, 1798. A few years of his boyhood — from 1803 to 1806 — were passed in Paris, where he was a spectator of some of the glorifications of the First Empire, especially on the occasion of the return from Austerlitz. He entered Harvard College in 1811, Messrs. Sparks, Parsons, and Palfrey being among his classmates, and after graduation he entered a mercantile house. He was elected into this Academy on the 10th of November, 1841, at the same time with two other Fellows assigned to the botanical section. One was William Oakes, of Ipswich, who died seven years afterward ; to the other is assigned the duty of preparing this memorial. When the Fellows of the Academy were arranged in classes and sections, the pronounced tastes inherited from his father and culti-vated by his own studies made it natural that he should belong to the small section of botany. But he might with equal pro-priety have been relegated to more than one section of the third class. For, notwithstanding his devotion to business affairs, his classical and linguistic knowledge were always well kept up, and his authority upon economical and financial questions was great.

The family has always had a marked representation in this Academy. To mention only the direct line, the subject of our notice was chosen into it very shortly after the death of his father, — the John Lowell who, after achieving distinction and a competency at the bar, retired from active practice at the age of thirty-four, to be known through his valuable writ-ings as " The Norfolk Farmer," and as a principal promoter,

[1] Proceedings American Academy of Arts and Science, xvii. 408. (1882.)

if not the founder, of scientific agriculture and horticulture in New England. John Lowell — the father of John Amory Lowell — was elected into the Academy in the year 1804, soon after the decease of his father, the Hon. John Lowell, first judge of the United States District Court of Massachusetts, under a commission from Washington. This office is now held by his great-grandson, the eldest son of our deceased associate, who has been a Fellow since the year 1877, thus continuing the line from the very foundation of the Academy, for Judge Lowell was one of the sixty-two members incorporated by the charter in 1780. In tracing the genealogy one step farther back, we come (as is almost universal in New England families of note) upon a clergyman, the Rev. John Lowell, of Newbury, a man of mark in his day.

Mr. Lowell was the fourth of his family to be a member of the Corporation of Harvard University, to which he gave a continuous and most valuable service of forty years. He was for more than fifty years one of the directors of the Suffolk Bank, which was chartered in his time, and which early established a very useful plan for the redemption of the currency of the New England banks in Boston. Not to mention other important public trusts, — as of the Athenæum, of the Massachusetts General Hospital, of the Agricultural Trustees, of the Provident Institution for Savings, to all of which he rendered assiduous and wise service, — nor to refer here to the very important part which he has taken for a lifetime in the development of the manufacturing interests of Massachusetts, especially as prosecuted in the town which was named in commemoration of similar services by his cousin, — we proceed to speak of that most important " corporation sole " founded by that cousin, the Lowell Institute. This trust was specifically consigned to our late associate and to such successor as he should appoint, — with preference to the family and the name of Lowell, — subject to no other than a formal visitatorial control, mainly for auditorship. And " to him, single and alone, it fell to shape the whole policy and take the whole direction of this great educational foundation," the history of which for almost half a century has justly been said to be a "record of his own intellectual breadth and scope, as well as of his large administrative capacity." We all know with what good judgment, with what liberality, and with what success this peculiar trust has been administered, and how on the one hand a series of most distinguished men have been attracted into its service, while on the other the efforts of younger men have been stimulated and rewarded at the period when such encouragement was most important to them. Suffice it to mention the names of Lyell and Agassiz, — the former early and also a second time brought from England for courses of lectures at the Lowell Institute, the latter a permanent acquisition to us and to our country. Through Mr. Lowell's discernment, moreover, the first encouragement to devote his life to scientific pursuits was afforded to Jeffries Wyman, by the offer of the curatorship of the Institute as well as of a lectureship. The intellectual and the financial interests of this trust have equally prospered in Mr. Lowell's hands; for while the number of lecture-courses has been doubled, and various subsidiary lines of instruction have been developed, the principal of the fund has been increased to thrice its original amount.

Mr. Lowell's fondness for botany developed shortly after he left college, and was incited by the botanical intercourse between his father and the late Dr. Francis Boott, with whom he maintained a lifelong friendship. But it was only in about the year 1844 or 1845 that he began the formation of an herbarium and botanical library; and this was actively prosecuted for several years, in evident expectation of comparative leisure which he could devote to scientific studies. He subscribed liberally to the botanical explorations in our newly-acquired or newly-opened western territories; and when in Europe, in 1850 and 1851, he added largely to his store of rare and costly botanical books. But just when he was ready to use the choice materials and appliances which had been brought together, the financial crisis of 1857 remanded him to business. The grave duties and responsibilities which he resumed he carried up nearly to the age of fourscore — carried as it were with the vigor of early manhood and the cheerful ease that attends "a real love of work for the work's own sake." And when it became evident that the comparatively unbroken attention requisite for serious botanical study was not to be secured, and as soon as a building was prepared for their reception, he presented all his botanical books which were needed to the herbarium of Harvard University; and the remainder, with his herbarium, to the Boston Society of Natural History, — not giving up the while his studious habits, but transferring his attention back to the Latin and the French classics, and in a certain degree to German and Italian literature.

As his father was one of the leading promoters of the establishment of the Botanic Garden of Harvard University, Mr. Lowell was also its most efficient supporter through its years of sorest need; and, in memory of his father, he bequeathed to it the sum of $20,000 in order to make his annual subvention perpetual. He made a legacy of equal amount to the general Library of the University, which he along with his father and grandfather had served in a most responsible trust for seventy years. He never sought or accepted any office in city or state; but few men were more sought for responsible trusts, or ever served their day and generation more devotedly, disinterestedly, and wisely. He seemed always to have a firm confidence in his own judgment, and that confidence appears not to have been misplaced.

CHARLES DARWIN.[1]

CHARLES DARWIN died on the 19th of April last, a few months after the completion of his seventy-third year; and on the 26th, the mortal remains of the most celebrated man of science of the nineteenth century were laid in Westminster Abbey, near to those of Newton.

He was born at Shrewsbury, February 12, 1809, and was named Charles Robert Darwin. But the middle appellation was omitted from his ordinary signature and from the title-pages of the volumes which, within the last twenty-five years, have given such great renown to an already distinguished name. His grandfather, Dr. Erasmus Darwin, — who died seven years before his distinguished grandson was born, — was one of the most notable and original men of his age; and his father, also a physician, was a person of very marked character and ability. His maternal grandfather was Josiah Wedgwood, who, beginning as an artisan potter, produced the celebrated Wedgwood ware, and became a Fellow of the Royal Society and a man of much scientific mark. The importance of heritability, which is an essential part of Darwinism, would seem to have had a significant illustration in the person of its great expounder. He was educated at the Shrewsbury Grammar School and at Edinburgh University, where, following the example of his grandfather, he studied for two sessions, having the medical profession in view, and where, at the close of the year 1826, he made his first contribution to natural history in two papers (one of them on the ova of Flustra). Soon finding the medical profession not to his liking, he proceeded to the University of Cambridge, entering Christ's College, and took his bachelor's degree in

[1] Proceedings of the American Academy of Arts and Science, xvii. 449. (1882.)

1831; that of M. A. in 1837, after his return from South America.

It is said that Darwin was a keen fox-hunter in his youth, — not a bad pursuit for the cultivation of the observing powers. There is good authority for the statement — though it has nowhere been made in print — that at Cambridge he was disposed at one time to make the Church his profession, following the example of Buckland and of his teacher, Sedgwick. But in 1831, just as he was taking his bachelor's degree, Captain Fitzroy offered to receive into his own cabin any naturalist who was disposed to accompany him in the Beagle's surveying voyage round the world. Mr. Darwin volunteered his services without salary, with the condition only that he should have the disposal of his own collections. And this expedition of nearly five years — from the latter part of September, 1831, to the close of October, 1836 — not only fixed the course and character of the young naturalist's life-work, but opened to his mind its principal problems and suggested the now familiar solution of them. For he brought back with him to England a conviction that the existing species of animals and plants are the modified descendants of earlier forms, and that the internecine struggle for life in which these modifiable forms must have been engaged would scientifically explain the changes. The noteworthy point is that both the conclusion and the explanation were the legitimate outcome of real scientific investigation. It is an equally noteworthy fact, and a characteristic of Darwin's mind, that these pregnant ideas were elaborated for more than twenty years before he gave them to the world. Offering fruit so well ripened upon the bough, commending the conclusions he had so thoroughly matured by the presentation of very various lines of facts, and of reasonings close to the facts, unmixed with figments and *à priori* conceptions, it is not so surprising that his own convictions should at the close of the next twenty years be generally shared by scientific men. It is certainly gratifying that he should have lived to see it, and also have outlived most of the obloquy and dread which the promulgation of these opinions aroused.

Mr. Darwin lived a very quiet and uneventful life. In 1839 he married his cousin, Emma Wedgwood, who with five sons and two daughters survives him; he made his home on the border of the little hamlet of Down, in Kent, — " a plain but comfortable brick house in a few acres of pleasure-ground, a pleasantly old-fashioned air about it, with a sense of peace and silence ; " and here, attended by every blessing except that of vigorous health, he lived the secluded but busy life which best suited his chosen pursuits and the simplicity of his character. He was seldom seen even at scientific meetings, and never in general society; but he could welcome his friends and fellow-workers to his own house, where he was the most charming of hosts.

At his home, without distraction and as continuously as his bodily powers would permit, Mr. Darwin gave himself to his work. At least ten of his scientific papers, of greater or less extent, had appeared in the three years between his return to England and his marriage ; and in the latter year (1839) he published the book by which he became popularly known, namely, the " Journal of Researches into the Natural History and Geology of the Countries visited during the Voyage of the Beagle," which has been pronounced " the most entertaining book of genuine travels ever written," and it certainly is one of the most instructive. His work on " Coral Reefs " appeared in 1842, but the substance had been communicated to the Geological Society soon after his return to England ; his papers on " Volcanic Islands," on the " Distribution of Erratic Boulders and Contemporaneous Unstratified Deposits in South America," on the " Fine Dust which falls on Vessels in the Atlantic Ocean," and some other geological as well as zoölogical researches, were published previously to 1851. Between that year and 1855 he brought out his most considerable contributions to systematic zoölogy, his monographs on the Cirripedia and the fossil Lepadidæ.

We come to the first publication of what is now known as Darwinism. It consists of a sketch of the doctrine of Natural Selection, which was drawn up in the year 1839, and copied and communicated to Messrs. Lyell and Hooker in

1844, being a part of the manuscript of a chapter in his
"Origin of Species;" also of a private letter addressed to
the writer of this memorial in October, 1857, — the publication of which (in the Journal of the Proceedings of the Linnæan Society, Zoölogical Part, iii. 45–53, issued in the summer of 1858) was caused by the reception by Darwin himself
of a letter from Mr. Wallace, inclosing a brief and strikingly
similar essay on the same subject, entitled "On the Tendency
of Varieties to depart indefinitely from the Original Type."
Mr. Darwin's action upon the reception of this rival essay
was characteristic. His own work was not yet ready, and the
fact that it had been for years in preparation was known only
to the persons above mentioned. He proposed to have the
paper of Mr. Wallace (who was then in the Moluccas) published at once, in anticipation of his own leisurely prepared
volume; and it was only under the solicitation of his friends
cognizant of the case that his own early sketch and the cor-
roboratory letter were printed along with it.

The precursory essays of Darwin and Wallace, published
in the Proceedings of a scientific society, can hardly have been
read except by a narrow circle of naturalists. Most thoughtful investigating naturalists were then in a measure prepared
for them. But toward the close of the following year (in the
autumn of 1859) appeared the volume "On the Origin of
Species by means of Natural Selection, or the Preservation of
Favored Races in the Struggle for Life," the first and most
notable of that series of duodecimos which have been read
and discussed in almost every cultured language, and which
within the lifetime of their author have changed the face and
in some respect the character of natural history, — indeed
have almost as deeply affected many other lines of investigation and thought.

In this Academy, where the rise and progress of Darwinian
evolution have been attentively marked and its bearings critically discussed, and at this date, when the derivative origin of
animal and vegetable species is the accepted belief of all of us
who study them, it would be superfluous to give any explanatory account of these now familiar writings; nor, indeed,

would the pages which we are accustomed to consecrate to the memory of our recently deceased associates allow of it. Let us note in passing that the succeeding volumes of the series may be ranked in two classes, one of which is much more widely known than the other. One class is of those which follow up the argument for the origination of species through descent with modification, or which widen its base and illustrate the *modus operandi* of Natural Selection. Such are the two volumes on "Domesticated Animals and Cultivated Plants," illustrating Variation, Inheritance, Reversion, Interbreeding, etc.; the volume on the "Descent of Man, and Selection in Relation to Sex," — which extended the hypothesis to its logical limits — and that, "On the Expression of the Emotions in Man and the Lower Animals," published in 1872, which may be regarded as the last of this series. Since then Mr. Darwin appears to have turned from the highest to the lower forms of life, and to have entered upon the laborious cultivation of new and special fields of investigation, which, although prosecuted on the lines of his doctrine and vivified by its ideas, might seem to be only incidentally connected with the general argument. But it will be found that all these lines are convergent. Nor were these altogether new studies. The germ of the three volumes upon the Relation of Insects to Flowers and its far-reaching consequences, is a little paper, published in the year 1858, "On the Agency of Bees in the Fertilization of Papilionaceous Flowers, and on the Crossing of Kidney Beans;" the first edition of the volume on "The various Contrivances by which Orchids are Fertilized by Insects" appeared in 1862, thus forming the second volume of the whole series; and the two volumes "On the Effects of Cross and Self-Fertilization in the Vegetable Kingdom," and "The different Forms of Flowers on Plants of the same Species," which, along with the new edition of "The Fertilization of Orchids," were all published in 1876 and 1877, originated in two or three remarkable papers contributed to the "Journal of the Linnæan Society" in 1862 and 1863, but are supplemented by additional and protracted experiments. The volume on "Insectivorous Plants," and the noteworthy

conclusions in respect to the fundamental unity, and therefore
common source, of vegetable and animal life, grew out of an
observation which the author made in the summer of 1860,
when he " was surprised by finding how large a number of
insects were caught by the leaves of the common Sun-dew
(*Drosera rotundifolia*), on a heath in Sussex." Almost
everybody had noticed this; and one German botanist (Roth),
just a hundred years ago, had observed and described the
movement of the leaf in consequence of the capture. But
nothing came of it, or of what had been as long known of our
Dionæa, beyond a vague wonderment, until Mr. Darwin took
up the subject for experimental investigation. The precursor
of his volume on " The Movements and Habits of Climbing
Plants," published in 1875, as well as of the recent and larger
volume on " The Power of Movement in Plants," 1880, was
an essay published in the " Journal of the Linnæan Society "
in 1865; and this was instigated by an accidental but capital
observation made by a correspondent, in whose hands it was
sterile; but it became wonderfully fertile when touched by
Darwin's genius.[1] His latest volume, on " The Formation of
Vegetable Mould through the Action of Worms," is a development, after long years, of a paper which he read before the
Geological Society of London in 1837.

These subsidiary volumes are less widely known than those

---

[1] Mr. Darwin's quickness in divining the meaning of seemingly unimportant things is illustrated in his study of Dionæa. Noting that the
trap upon irritation closes at first imperfectly, leaving some room within
and a series of small interstices between the crossed spines, but after a
time, if there is prey within, shuts down close, he at once inferred that
this was a provision for allowing small insects to escape, and for retaining
only those large enough to make the long process of digestion remunerative. To test the surmise, he asked a correspondent to visit the habitat
of Dionæa at the proper season, and to ascertain by the examination of a
large number of the traps in action whether any below a certain considerable size were to be found in them. The result confirmed the inference,
a comparatively trivial but characteristic illustration of Darwin's confidence in the principle of utility, and a good example of the truth of the
dictum, which was by some thought odd when first made, namely, that
Darwin had restored teleology to natural history, from which the study
of morphology had dissevered it.

of the other class; but they are of no less interest, and they are very characteristic of the author's genius and methods, — characteristic also of his laboriousness. For the amount of prolonged observation, watchful care, and tedious experiment they have demanded is as remarkable as the skill in devising simple and effectual modes of investigation is admirable. That he should have had the courage to undertake and the patience to carry on new inquiries of this kind after he had reached his threescore and ten years of age, and after he had attained an unparalleled breadth of influence and wealth of fame, speaks much for his energy and for his devotion to knowledge for its own sake. Indeed, having directed the flow of scientific thought into the new channel he had opened, along which the current set quicker and stronger than he could have expected, he seems to have taken up with fresh delight studies which he had marked out in early years, or topics which from time to time had struck his acute attention. To these he gave himself, quite to the last, with all the spirit and curiosity of youth. Evidently all this amount of work was done for the pure love of it; it was all done methodically, with clear and definite aim, without haste, but without inter-mission.

It would confidently be supposed that in this case genius and industry were seconded by leisure and bodily vigor. Fortunately Darwin's means enabled him to control the disposition of his time. But the voyage of the Beagle, which was so advantageous to science, ruined his health. A sort of chronic sea-sickness, under which all his work abroad was performed, harassed him ever afterwards. The days in which he could give two hours to investigation or writing were counted as good ones, and for much of his life they were largely outnumbered by those in which nothing could be attempted. Only by great care and the simplest habits was he able to secure even a moderate amount of comfortable existence. But in this respect his later years were the best ones, and therefore the busiest. In them also he had most valuable filial aid. There was nothing to cause much anxiety until his seventy-third birthday had passed, or to excite alarm until the week before his death.

It may without exaggeration be said that no scientific man, certainly no naturalist, ever made an impression at once so deep, so wide, and so immediate. The name of Linnæus might suggest comparison; but readers and pupils of Linnæus over a century ago were to those of Darwin as tens are to thousands, and the scientific as well as the popular interest of the subjects considered were somewhat in the same ratio. Humboldt, who, like Darwin, began with research in travel, and to whom the longest of lives, vigorous health, and the best opportunities were allotted, essayed similar themes in a more ambitious spirit, enjoyed equal or greater renown, but made no deep impression upon the thought of his own day or of ours. As one criterion of celebrity, it may be noted that no other author we know of ever gave rise in his own active lifetime to a special department of bibliography. Dante-literature and Shakespeare-literature are the growth of centuries; but *Darwinismus* had filled shelves and alcoves and teeming catalogues while the unremitting author was still supplying new and ever novel subjects for comment. The technical term which he chose for a designation of his theory, and several of the phrases originated in explanation of it only twenty-five years ago, have already been engrafted into his mother tongue, and even into other languages, and are turned to use in common as well as in philosophical discourse, without sense of strangeness.

Wonderful indeed is the difference between the reception accorded to Darwin and that met with by his predecessor, Lamarck. But a good deal has happened since Lamarck's day; wide fields of evidence were open to Darwin which were wholly unknown to his forerunner; and the time had come when the subject of the origin and connection of living forms could be taken up as a research rather than as a speculation. Philosophizers on evolution have not been rare; but Darwin was not one of them. He was a scientific investigator, — a philosopher, if you please, but one of the type of Galileo. Indeed very much what Galileo was to physical science in his time, Darwin is to biological science in ours. This without reference to the fact that the writings of both conflicted with

similar prepossessions; and that the Darwinian theory, legitimately considered, bids fair to be placed in this respect upon
the same footing with the Copernican system.

An English poet wrote that he awoke one morning and
found himself famous. When this happened to Darwin, it
was a genuine surprise. Although he had addressed himself
simply to scientific men, and had no thought of arguing his
case before a popular tribunal, yet " The Origin of Species "
was too readable a book upon too sensitive a topic to escape
general perusal; and this, indeed, must in some sort have
been anticipated. But the avidity with which the volume was
taken up, and the eagerness of popular discussion which en-
sued, were viewed by the author — as his letters at the time
testify — with a sense of amused wonder at an unexpected
and probably transient notoriety.

The theory he had developed was presented by a working
naturalist to his fellows, with confident belief that it would
sooner or later win acceptance from the younger and more
observant of these. The reason why these moderate expectations were much and so soon exceeded are not far to seek,
though they were not then obvious to the world in general.
Although mere speculations were mostly discountenanced by
the investigating naturalists of that day, yet their work and
their thoughts were, consciously or unconsciously, tending in
the direction of evolution. Even those who manfully rowed
against the current were more or less carried along with it,
and some of them unwittingly contributed to its force. Most
of them in their practical studies had worked up to, or were
nearly approaching, the question of the relation of the past
inhabitants of the earth to the present, and of the present to
one another, in such wise as to suggest inevitably that, somehow or other, descent with modification was eventually to be
the explanation. This was the natural outcome of the line of
thought of which Lyell early became the cautious and fair-
minded expositor, and with which he reconstructed theoretical
geology. If Lyell had known as much at first hand of botany
or zoölogy as he knew of geology, it is probable that his celebrated chapter on the permanence of species in the " Prin-

ciples" would have been reconsidered before the work had passed to the ninth edition in 1853. He was convinced that species went out of existence one by one, through natural causes, and that they came in one by one, bearing the impress of their immediate predecessors; but he saw no way to connect the two through natural operations. Nor, in fact, had any of the evolutionists been able to assign real causes capable of leading on such variations as are of well-known occurrence to wider and specific or generic differences. Just here came Darwin. When upon the spot he had perceived that the animals of the Galapagos must be modified forms derived from the adjacent continent, and he soon after worked out the doctrine of natural selection. This supplied what was wanting for the condensation of opinions and beliefs, and the collocation of rapidly accumulating facts, into a consistent and workable scientific theory, under a principle which unquestionably could directly explain much, and might indirectly explain more.

It is not merely that Darwin originated and applied a new principle. Not to speak of Wallace, his contemporary, who came to it later, his countryman, Dr. Wells, as Mr. Darwin points out, "distinctly recognizes the principle of natural selection, and this is the first recognition which has been indicated; but he applied it only to the races of men, and to certain characters alone." Darwin, like the rest of the world, was unaware of this anticipation until he was preparing the fourth edition of his "Origin of Species," in 1866, when he promptly called attention to it, perhaps magnifying its importance. However this be, Darwin appears to have been first and alone in apprehending and working out the results which necessarily come from the interaction of the surrounding agencies and conditions under which plants and animals exist, including, of course, their action upon each other. Personifying the *ensemble* of these and the consequences, — namely, the survival only of the fittest in the struggle for life, — under the term of Natural Selection, Mr. Darwin, with the instinct of genius divined, and with the ability of a master worked out its pregnant and far-reaching applications. He not only saw its strong points, but he foresaw its limitations, indicated most of the objections in advance of his opponents, weighed them with judicial mind, and where he could not obviate them, seemed never disposed to underrate their force. Although naturally disposed to make the most of his theory, he distinguished between what he could refer to known causes and what thus far is not referable to them. Consequently, he kept clear of that common confusion of thought which supposes that natural selection originates the variations which it selects. He believed, and he has shown it to be probable, that external conditions induce the actions and changes in the living plant or animal which may lead on to the difference between one species and another; but he did not maintain that they produced the changes, or were sufficient scientifically to explain them. Unlike most of his contemporaries in this respect, he appears to have been thoroughly penetrated by the idea that the whole physiological action of the plant or animal is a response of the living organism to the action of the surroundings.

The judicial fairness and openness of Darwin's mind, his penetration and sagacity, his wonderful power of eliciting the meaning of things which had escaped questioning by their very commonness, and of discerning the great significance of causes and interactions which had been disregarded on account of their supposed insignificance, his method of reasoning close to the facts and in contact with the solid ground of nature, his aptness in devising fruitful and conclusive experiments, and in prosecuting nice researches with simple but effectual appliances, and the whole rare combination of qualities which made him *facile princeps* in biological investigation, — all these gifts are so conspicuously manifest in his published writings, and are so fully appreciated, that there is no need to celebrate them in an obituary memorial. The writings also display in no small degree the spirit of the man, and to this not a little of their persuasiveness is due. His desire to ascertain the truth, and to present it purely to his readers, is everywhere apparent. Conspicuous, also, is the absence of all trace of controversy and of everything like pretension; and this is remarkable, considering how censure and how praise were heaped upon him without stint. He does not teach didactically, but takes the reader along with him as his companion in observation and in experiment. And in the same spirit, instead of showing pique to an opponent, he seems always to regard him as a helper in his search for the truth. Those privileged to know him well will certify that he was one of the most kindly and charming, unaffected, simple-hearted, and lovable of men.

How far and how long the Darwinian theory will hold good, the future will determine. But in its essential elements, apart from *à priori* philosophizing, with which its author had nothing to do, it is an advance from which it is evidently impossible to recede. As has been said of the theory of the Conservation of Energy, so of this: "The proof of this great generalization, like that of all other generalizations, lies mainly in the fact that the evidence in its favor is continually augmenting, while that against it is continually diminishing, as the progress of science reveals to us more and more of the workings of the universe."

JOSEPH DECAISNE.[1]

JOSEPH DECAISNE, the oldest member of the Botanical Section on the foreign list, died at Paris, on the 8th of February last, in the seventy-fifth year of his age. He was elected into this Academy in August, 1846, along with Agassiz and De Verneuil. He was born at Brussels, March 11, 1807, the second of three brothers, one of whom became a distinguished painter, and the other the head of the medical department of the Belgian army. He came to Paris and entered the Jardin des Plantes when a lad of seventeen years, and in its service his whole subsequent life was passed. The young employé attracted the attention of Adrien de Jussieu, who, seeing his promise and unusual botanical knowledge, soon placed him at the head of the seed department, and in 1833 made him his *Aide-naturaliste*, thus giving the young gardener opportunity for the studies and researches by which he won a place among the foremost botanists of the time. For more than forty years the administration of the Jardin des Plantes and the duties of the chair of Culture at the Museum were in his hands, he having supplied the place of Mirbel through the closing years of the latter's life, and succeeded him as professor in the year 1851; and these duties he continued to fulfill to the last. He was elected a member of the Institute in 1847, in succession to Dutrochet; for forty years he was one of the editors of, and since the death of his colleague, Adolphe Brongniart, he was the sole editor of the botanical portion of the " Annales des Sciences Naturelles." In the Annales he had published some good botanical papers, the earliest in the year 1831. But his first distinction was gained by his anatomical and physiological researches upon the Madder-plant, a monograph containing the results of which appeared at Brussels in 1837, and was said to be " one of the most able memoirs that has ever been published on the physiological history of plants and their bearing on

[1] Proceedings American Academy of Arts and Science, xvii. 458. (1882.)

practical cultivation and manufactures." Two years later, in connection with the chemist Péligot, he published an investigation of the anatomical structure of the Sugar-beet. His classical memoir upon the structure and development of the Mistletoe appeared in 1840, and is of purely scientific interest. In the year 1841 he showed that the Corallines, which had been wrongly carried over to the animal kingdom with the Corals and their allies, were genuine Seaweeds, disguised by the incorporation of a great amount of lime into their tissues. And about this time, in connection with his friend and former pupil, Thuret, he discovered and illustrated the male organs of the *Fuci*, as well as the mode of impregnation and reproduction, thus initiating the investigations which in the hands of the late Thuret and others have revolutionized phycology.

Leaving these researches for his associate to complete and publish, thenceforth Decaisne turned all his attention to phanerogamous botany, morphological and systematic. Two orders were elaborated by him for De Candolle's "Prodromus," *Asclepiadaceæ* and *Plantaginaceæ*, the former demanding much minute research; he produced in 1868, in conjunction with Le Maout, that admirable text-book, the "Traité Générale de Botanique," profusely illustrated by his own facile pencil, which is well known in the original and in the English translation edited by Sir Joseph Hooker. But the works by which he will be most widely known, and which were connected especially with his directorship of the Jardin des Plantes, are that incomparable series of colored illustrations of fruits, together with descriptive text, known as "Le Jardin Fruitier du Muséum," and his subsidiary investigations and publications upon the *Pomaceæ* and their allies. These important publications began in the year 1858, and were completed only a year or two ago.

Decaisne never married: he lived his simple and devoted life in the house on Rue Cuvier in the Jardin des Plantes, where he died, regretted and beloved, the last of the line of illustrious botanists — such as Mirbel, Adrien de Jussieu, Gaudichaud, and Adolphe Brongniart — who were associated in the administration of this institution thirty or forty years ago.

GEORGE ENGELMANN.[1]

In the death of Dr. Engelmann, which took place on the 4th of February last, the American Academy has lost one of its very few Associate Fellows in the Botanical Section, and the science one of its most eminent and venerable cultivators. He was born at Frankfort-on-the-Main, February 2, 1809, and had therefore just completed his seventy-fifth year. His father, a younger member of the family of Engelmanns who for several generations served as clergymen at Bacharach on the Rhine, was also educated for the ministry, and was a graduate of the university of Halle, but he devoted his life to education. Marrying the daughter of George Oswald May, a somewhat distinguished portrait-painter, they established at Frankfort, and carried on for a time with much success, a school for young ladies, such as are common in the United States, but were then a novelty in Germany.

George Engelmann was the eldest of thirteen children born of this marriage, nine of whom survived to manhood. Assisted by a scholarship founded by "the Reformed Congregation of Frankfort," he went to the university of Heidelberg in the year 1827, where he had as fellow-students and companions Karl Schimper and Alexander Braun. With the latter he maintained an intimate friendship and correspondence, interrupted only by the death of Braun in 1877. The former, who manifested unusual genius as a philosophical naturalist, after laying the foundations of phyllotaxy, to be built upon by Braun and others, abandoned, through some singular infirmity of temper, an opening scientific career of the highest promise, upon which the three young friends, Agassiz, Braun, and Schimper, and in his turn Engelmann, had zealously entered.

[1] Proceedings American Academy of Arts and Science, xix. 516. (1884.)

Embarrassed by some troubles growing out of a political demonstration by the students at Heidelberg, Engelmann in the autumn of 1828 went to the university of Berlin for two years; and thence to Würzburg, where he took his degree of Doctor in Medicine in the summer of 1831. His inaugural dissertation, " De Antholysi Prodromus," which he published at Frankfort in 1832, testifies to his early predilection for botany, and to his truly scientific turn of mind. It is a morphological dissertation, founded chiefly on the study of monstrosities, illustrated by five plates filled with his own drawings. It was therefore quite in the line with the little treatise on the Metamorphosis of Plants, published forty years before by another and the most distinguished native of Frankfort, and it appeared so opportunely that it had the honor of Goethe's notice and approval. Goethe's correspondent, Madame von Willema, sent a copy to him only four weeks before his death. Goethe responded, making kind inquiries after young Engelmann, who, he said, had completely apprehended his ideas of vegetable morphology, and had shown such genius in their development that he offered to place in this young botanist's hands the store of unpublished notes and sketches which he had accumulated.

The spring and summer of 1832 were passed at Paris in medical and scientific studies, with Braun and Agassiz as companions, leading, as he records, " a glorious life in scientific union, in spite of the cholera." Meanwhile, Dr. Engelmann's uncles had resolved to make some land investments in the valley of the Mississippi, and he willingly became their agent. At least one of the family was already settled in Illinois, not far from St. Louis. Dr. Engelmann, sailing from Bremen for Baltimore in September, joined his relatives in the course of the winter, made many lonely and somewhat adventurous journeys on horseback in southern Illinois, Missouri, and Arkansas, which yielded no other fruits than those of botanical exploration; and finally he established himself in the practice of medicine at St. Louis, late in the autumn of 1835. St. Louis was then rather a frontier trading-post than a town, of barely eight or ten thousand inhabitants. He lived to see it become a metropolis of over four hundred thousand. He began in absolute poverty, the small means he had brought from Europe completely exhausted. In four years he had laid the foundations of success in his profession, and had earned the means for making a voyage to Germany, and, fulfilling a long-standing engagement, for bringing to a frugal home the chosen companion of his life, Dora Hartsmann, his cousin, whom he married at Kreuznach, on the 11th of June, 1840. On his way homeward, at New York, the writer of this memorial formed the personal acquaintance of Dr. Engelmann; and thus began the friendship and the scientific association which has continued unbroken for almost half a century.

Dr. Engelmann's position as a leading physician in St. Louis, as well among the American as the German and French population, was now soon established. He was even able in 1856, without risk, to leave his practice for two years, to devote most of the first summer to botanical investigation in Cambridge, and then, with his wife and young son, to revisit their native land, and to fill up a prolonged vacation in interesting travel and study. In the year 1868 the family visited Europe for a year, the son remaining to pursue his medical studies in Berlin. And lastly, his companion of nearly forty years having been removed by death in January, 1879, and his own robust health having suffered serious and indeed alarming deterioration, he sailed again for Germany in the summer of 1883. The voyage was so beneficial that he was able to take up some botanical investigations, which, however, were soon interrupted by serious symptoms. But the return voyage proved wonderfully restorative; and when, in early autumn, he rejoined his friends here, they could hope that the unfinished scientific labors, which he at once resumed with alacrity of spirit, might still for a while be carried on with comfort. So indeed they were, in some measure, after his return to his home, yet with increasing infirmity and no little suffering until the sudden illness supervened which, in a few days, brought his honorable and well-filled life to a close.

In the latter part of his life Dr. Engelmann was able to explore considerable portions of his adopted country, the mountains of North Carolina and Tennessee, the Lake Superior region, and the Rocky Mountains and contiguous plains in Colorado and adjacent territories, and so to study in place, and with the particularity which characterized his work, the *Cacti*, the *Coniferæ*, and other groups of plants which he had for many years been specially investigating. "In 1880 he made a long journey through the forests of the Pacific States, where he saw for the first time in the state of nature plants which he had studied and described more than thirty years before. Dr. Engelmann's associates [so one of them declares] will never forget his courage and industry, his enthusiasm and zeal, his abounding good-nature, and his kindness and consideration of every one with whom he came in contact." His associates, and also all his published writings, may testify to his acuteness in observation, his indomitable perseverance in investigation, his critical judgment, and a rare openness of mind which prompted him continually to revise old conclusions in the light of new facts or ideas.

In the consideration of Dr. Engelmann's botanical work — to which these lines will naturally be devoted — it should be remembered that his life was that of an eminent and trusted physician, in large and general practice, who even in age and failing health was unable, however he would have chosen, to refuse professional services to those who claimed them; that he devoted only the residual hours, which most men use for rest or recreation, to scientific pursuits, mainly to botany, yet not exclusively. He was much occupied with meteorology. On establishing his home at St. Louis, he began a series of thermometrical and barometrical observations, which he continued regularly and systematically to the last, when at home always taking the observations himself, — the indoor ones even up to the last day but one of his life. Even in the last week he was seen sweeping a path through the snow in his garden to reach his maximum and minimum thermometers. His latest publication (issued since his death by the St. Louis Academy of Sciences) is a digest and full representation of the thermometrical part of these observations for forty-seven years. He apologizes for not waiting the completion of the half-century before summing up the results, and shows that these could not after three more years be appreciably different.

A list of Dr. Engelmann's botanical papers and notes, collected by his friend and associate, Professor Sargent, and published in Coulter's "Botanical Gazette" for May, 1884, contains about one hundred entries, and is certainly not quite complete. His earliest publication, his inaugural thesis already mentioned (De Antholysi Prodromus), is a treatise upon teratology in its relations to morphology. It is a remarkable production for the time and for a mere medical student with botanical predilections. There is an interesting recent analysis of it in "Nature" for April 24, by Dr. Masters, the leading teratologist of our day, who compares it with Moquin-Tandon's more elaborate "Tératologie Végétale," published ten years afterwards, and who declares that, "when we compare the two works from a philosophical point of view, and consider that the one was a mere college essay, while the other was the work of a professed botanist, we must admit that Engelmann's treatise, so far as it goes, affords evidence of deeper insight into the nature and causes of the deviations from the ordinary conformation of plants than does that of Moquin."

Transferred to the valley of the Mississippi and surrounded by plants most of which still needed critical examination, Dr. Engelmann's avocation in botany and his mode of work were marked out for him. Nothing escaped his attention; he drew with facility; and he methodically secured his observations by notes and sketches, available for his own after-use and for that of his correspondents. But the lasting impression which he has made upon North American botany is due to his wise habit of studying his subjects in their systematic relations, and of devoting himself to a particular genus or group of plants (generally the more difficult) until he had elucidated it as completely as lay within his power. In this way all his work was made to tell effectively.

Thus his first monograph was of the genus Cuscuta (published in the American Journal of Science, in 1842), of which when Engelmann took it up we were supposed to have only one indigenous species, and that not peculiar to the United States, but which he immediately brought up to fourteen species without going west of the Mississippi Valley. In the year 1859, after an investigation of the whole genus in the materials scattered through the principal herbaria of Europe and this country, he published in the first volume of the St. Louis Academy of Sciences a systematic arrangement of all the *Cuscutæ*, characterizing seventy-seven species, besides others classed as perhaps varieties.

Mentioning here only monographical subjects, we should next refer to his investigations of the Cactus family, upon which his work was most extensive and important, as well as particularly difficult, and upon which Dr. Engelmann's authority is of the very highest. He essentially for the first time established the arrangement of these plants upon floral and carpological characters. This formidable work was begun in his "Sketch of the Botany of Dr. A. Wislizenus's Expedition from Missouri to Northern Mexico," in the latter's memoir of this tour, published by the United States Senate. It was followed up by his account (in the American Journal of Science, 1852) of the Giant Cactus on the Gila (*Cereus giganteus*) and an allied species; by his synopsis of the *Cactaceæ* of the United States, published in the "Proceedings of the American Academy of Arts and Sciences," 1856; and by his two illustrated memoirs upon the southern and western species, one contributed to the fourth volume of the series of Pacific Railroad Expedition Reports, the other to Emory's Report on the Mexican Boundary Survey. He had made large preparations for a greatly needed revision of at least the North American *Cactaceæ*. But, although his collections and sketches will be indispensable to the future monographer, very much knowledge of this difficult group of plants is lost by his death.

Upon two other peculiarly American groups of plants, very difficult of elucidation in herbarium specimens, Yucca and Agave, Dr. Engelmann may be said to have brought his work up to the time. Nothing of importance is yet to be added to what he modestly styles "Notes on the Genus Yucca," published in the third volume of the Transactions of the St. Louis Academy, 1873, and not much to the "Notes on Agave," illustrated by photographs, included in the same volume and published in 1875.

Less difficult as respects the material to work upon, but well adapted for his painstaking, precise, and thorough handling, were such genera as Juncus (elaborately monographed in the second volume of the Transactions of the St. Louis Academy, and also exemplified in distributed sets of specimens), Euphorbia (in the fourth volume of the Pacific Railroad Reports, and in the Botany of the Mexican Boundary), Sagittaria and its allies, Callitriche, Isoetes (of which his final revision is probably ready for publication), and the North American *Loranthaceæ*, to which Sparganium, certain groups of Gentiana, and some other genera, would have to be added in any complete enumeration. Revisions of these genera were also kindly contributed to Dr. Gray's Manual; and he was an important collaborator in several of the memoirs of his surviving associate and friend.

Of the highest interest, and among the best specimens of Dr. Engelmann's botanical work, are his various papers upon the American Oaks and the *Coniferæ*, published in the "Transactions of the St. Louis Academy," and elsewhere, the results of long-continued and most conscientious study. The same must be said of his persevering study of the North American Vines, of which he at length recognized and characterized a dozen species, — excellent subjects for his nice discrimination, and now becoming of no small importance to grape-growers, both in this country and in Europe. Nearly all that we know scientifically of our species and forms of Vitis is directly due to Dr. Engelmann's investigations. His first separate publication upon them, "The Grape Vines of Missouri," was published in 1860; his last, a reëlaboration of the American species, with figures of their seeds, is in the third edition of the Bushberg Catalogue, published only a few months ago.

Imperfect as this mere sketch of Dr. Engelmann's botanical authorship must needs be, it may show how much may be done for science in a busy physician's *horæ subsecivæ*, and in his occasional vacations. Not very many of those who could devote their whole time to botany have accomplished as much. It need not be said, and yet perhaps it should not pass unrecorded, that Dr. Engelmann was appreciated by his fellow-botanists both at home and abroad, that his name is upon the rolls of most of the societies devoted to the investigation of nature, that he was " everywhere the recognized authority in those departments of his favorite science which had most interested him," and that, personally one of the most affable and kindly of men, he was as much beloved as respected by those who knew him.

More than fifty years ago his oldest associates in this country — one of them his survivor — dedicated to him a monotypical genus of plants, a native of the plains over whose borders the young immigrant on his arrival wandered solitary and disheartened. Since then the name of Engelmann has, by his own researches and authorship, become unalterably associated with the Buffalo-grass of the plains, the noblest Conifers of the Rocky Mountains, the most stately Cactus in the world and with most of the associated species, as well as with many other plants of which perhaps only the annals of botany may take account. It has been well said by a congenial biographer, that " the western plains will still be bright with the yellow rays of *Engelmannia*, and that the splendid Spruce, the fairest of them all, which bears the name of Engelmann, will still, it is to be hoped, cover with noble forests the highest slopes of the Rocky Mountains, recalling to men, as long as the study of trees occupies their thoughts, the memory of a pure, upright, and laborious life."

OSWALD HEER.[1]

OSWALD HEER, the most eminent investigator of the fossil plants and insects of the tertiary period, died on the 27th of September last, shortly after he had entered upon the seventy-fifth year of his age.

He was born at the hamlet of Nieder-Utzwyl, in Canton St. Gallen, Switzerland, August 31, 1809, passed most of his youth at Matt, in Canton Glarus, where his father was the parish clergyman, pursued his academic and professional studies at the university of Halle, and was ordained as minister of the Gospel in the year 1831. The next year he went to Zurich, where he resided for the rest of his life. Here he studied medicine for a time, but soon devoted himself seriously to entomology and botany, of which he was fond from boyhood. In 1834 he became Privat-docent of these sciences; in 1852, when the university of Zurich was developed, he became its professor of botany, and in 1855 he took a similar chair in the Polytechnicum. Most of his earlier publications were entomological; and it was by the way of entomology that he entered upon his distinguished career as a palæontologist. His life-long friend, the eminent Escher von der Linth, appreciating his rare powers of observation, induced him to undertake the study of the fossil insects of the celebrated tertiary deposits of Oeningen. The results of his labors in this virgin field were published between the years 1847 and 1853. His attention had from the first been attracted to the plants associated with the insect remains. His first palæo-botanical paper appeared in 1851; the three volumes of his " Flora Tertiaria Helvetiæ " were issued between 1855 and 1859; in 1862 his memoir on the fossil flora of Bovey-Tracey (Eng-

[1] Proceedings of the American Academy of Arts and Science, xix. 556. (1884.)

land) was published in the Philosophical Transactions of the Royal Society, London. About the same time also appeared a paper in the Journal of the Geological Society on certain fossil plants of the Isle of Wight. For the benefit of his health, always delicate and then much impaired, he passed the winter of 1854–55 in Madeira, and on his return published a paper on the fossil plants of that island, and an article on the probable origin of the actual flora and fauna of the Azores, Madeira, and the Canaries. In this, and in his work, published in 1860, on " Tertiary Climates in their Relation to Vegetation " (which the next year appeared also in a French translation by his young friend Gaudin), Heer brought out his theory of a Miocene Atlantis. His more extensive and popular treatise upon past climates as illustrated by vegetable palæontology, his " Urwelt der Schweiz," — a vivid portraiture of the past of his native country, — appeared in 1865, and afterwards in a revised French edition, with his friend Gaudin (who died soon after) for collaborator as well as translator. There was also an English translation by Heywood, published in 1876, and, indeed, it is said to have been translated into six languages.

In 1877 Heer completed his " Flora Fossilis Helvetiæ," a square-folio volume, with seventy plates, which extended and supplemented his Tertiary Flora of that country, being devoted to the illustration of the fossil plants of the Carboniferous, the Triassic, the Jurassic, and the Cretaceous, as well as the Eocene formations.

The life-long delicacy of Heer's health prevented his making any extensive explorations in person. But materials for his investigation came to him in even embarrassing abundance, not only from his own country, — where, even before he was widely known (as his fellow-countryman and his distinguished fellow-worker in palæo-botany, Lesquereux, informs us), a lady opened upon her property near Lausanne quarries and tunnels expressly for the discovery and collection of fossil plants, and sent them by tons to Zurich, — but from all parts of the world collections were pressed upon him, and his whole time and strength were given to their study. In this way he became interested in the arctic fossil flora, of which he became the principal investigator and expounder. His first essay in the domain which he has made so peculiarly his own was in a paper on certain fossil plants of Vancouver's Island and British Columbia, published in 1865; and in 1868 he brought out the first of that most important series of memoirs upon the ancient floras of arctic America, Greenland, Spitzbergen, Nova Zembla, arctic and subarctic Asia, etc., which, collected, make up the seven quarto volumes of the "Flora Fossilis Arctica." The seventh volume of this monumental work was brought to a conclusion only a few months before the author's death.

Heer's researches into the fossil botany of the tertiary deposits were very important in their bearings. They made it certain that our actual temperate floras round the world had a common birthplace at the north, where the continents are in proximity; they essentially identified the direct or collateral ancestors of our existing forest-trees which flourished within the arctic zone when it enjoyed a climate resembling our own at present; and they leave the similarities and the dissimilarities of the temperate floras of the Old and the New World to be explained as simple consequences of established facts. Thus Heer himself did away with his own hypothesis of a continental Atlantis by bringing to light the facts which proved that there was no need of it. And while thus justifying the ideas which had been brought forward in one of the memoirs of the American Academy (in 1859) before these fossil data were known, he was not slow to adopt and to extend the tentative views which he had confirmed.

A list of Heer's scientific publications is given in the "Botanisches Centralblatt," No. 5, for 1884. They are seventy-seven in number, besides the seven quarto volumes of the "Flora Fossilis Arctica," which comprise a considerable number of independent memoirs. These works make an era in vegetable palæontology. Their crowning general interest is that they bring the vegetation of the past into direct connection with the present.

Although he lived to a good old age, and was never inactive, Heer was for most of his life an invalid, suffering from pulmonary disease. For the last twelve years his work was carried on at his bedside or from his bed, assisted by a devoted and accomplished daughter; he seldom left his house, except to pass the last two winters in the milder climate of Italy. Last summer, having finished his "Flora Fossilis Arctica," in the hope of recruiting his exhausted strength he was removed to the most sheltered spot on the shores of the Lake of Geneva, but without benefit. He died at Lausanne, at his brother's house, on the 27th of September, 1883. It has been well said of him, in a tribute which a personal friend and fellow-naturalist paid to his memory, that " a man more lovable, more sympathetic, and a life more laborious and pure, one could scarcely imagine."

Heer was elected into the Academy in May, 1877. He is botanically commemorated in a genus of beautiful Melastomaceous plants indigenous to Mexico.

GEORGE BENTHAM.[1]

GEORGE BENTHAM, one of the most distinguished botanists of the present century, and at the time of his death one of the oldest, was born at Stoke, a suburb of Portsmouth, September 22, 1800. He died at his house, No. 25 Wilton Place, London, on the 10th of September, 1884, a few days short of eighty-four years old. His paternal grandfather, Jeremiah Bentham, a London attorney or solicitor, had two sons, who both became men of mark, Jeremy and Samuel. The latter and younger had two sons, only one of whom, the subject of this memoir, lived to manhood. George Bentham's mother was a daughter of Dr. George Fordyce, a Scottish physician who settled in London, was a Fellow of the Royal Society, a lecturer on chemistry, and the author of some able medical works, also of a treatise upon Agriculture and Vegetation. It was from his mother that George Bentham early imbibed a fondness for botany.

The early part of his life and education was somewhat eventful and peculiar, and in strong contrast with the later. His father, General, subsequently Sir Samuel Bentham, was an adept in naval architecture. At the age of twenty-two he visited the arsenals of the Baltic for the improvement of his knowledge; thence he traveled far into Siberia. He became intimate with Prince Potemkin, by whom he was induced to enter the civil and afterwards the military service of the Empress Catharine. He took part in a naval action against the Turks on the Black Sea, and was rewarded with the command of a regiment stationed in Siberia, with which he traversed the country even to the frontiers of China. After ten years he returned to England, where his inventive skill and experience found a fitting field in the service of the Admiralty, in

[1] Proceedings American Academy of Arts and Science, xx. 527. (1884.)

which he attained the post of Inspector-General of Naval Works. Among the services he rendered was that of bringing to England the distinguished engineer Isambard Mark Brunel. In the year 1805, General Bentham was sent by the Admiralty to St. Petersburg, to superintend the building in Russia of vessels for the British navy. He took his family with him ; and there began the education of George Bentham, in the fifth year of his age, under the charge of a Russian lady who could speak no English, where he learned to converse fluently in Russian, French, and German, besides acquiring the rudiments of Latin as taught by a Russian priest. On the way back to England, two or three years later, the detention of a month or two in Sweden gave opportunity for learning enough of Swedish to converse in that language and to read it with tolerable ease in after life. Returning to England, the family settled at Hampstead, and the children pursued their studies under private tutors. In the years 1812–13, during the excitement produced by the French invasion of Russia and the burning of Moscow, our young polyglot " budded into an author, by translating (along with his brother and sister) and contributing to a London magazine a series of articles from the Russian newspapers, detailing the operations of the armies." In 1814, upon the downfall of Napoleon, the Bentham family crossed over to France, prepared for a long stay, remained in the country (at Tours, Saumur, and Paris) during the hundred days preceding Napoleon's final overthrow, and in 1816 Sir Samuel Bentham set out upon a prolonged and singular family tour, *en caravane*, through the western and southern departments of France. To quote from the published account from which most of these biographical details are drawn, and which were taken from Mr. Bentham's own memoranda [1] —

" The *cortége* consisted of a two-horse coach fitted up as a sleeping apartment ; a long, low, two-wheeled, one-horse spring van for General and Mrs. Bentham, furnished with a library and piano ; and another, also furnished, for his daughters and their governess. The plan followed was to travel by day

[1] An article in "Nature," Oct. 2, 1884, by Sir Joseph Dalton Hooker.

from one place of interest to another, bivouacking at night by the road, or in the garden of a friend, or in the precincts of the prefectures, to which latter he had credentials from the authorities in the capital. In this way he visited Orleans, Tours, Angoulême, Bordeaux, Toulouse, Montpellier, and finally Montauban, where a lengthened stay was made in a country house hired for the purpose. From Montauban (the *cortége* having broken down in some way) they proceeded still by private conveyances to Carcassonne, Narbonnes, Nîmes, Tarascon, Marseilles, Toulon, Hyères."

It was in the early part of this tour that young Bentham's attention was first turned to botany. Happening to take up De Candolle's edition of Lamarck's " Flore Française," which his mother, who was fond of the subject, had just purchased, he was struck with the methodical analytical tables, and he proceeded immediately to apply them to the first plant he could lay hold of. " His success led him to pursue the diversion of naming every plant he met with." During his long stay at Montauban he entered as a student in the Protestant theological school of that town, pursuing " with ardor the courses of mathematics, Hebrew, and comparative philology, the latter a favorite study in after life," and at home giving himself to music, in which he was remarkably gifted, to Spanish, to botany, and, with great relish, to society. Soon after, the family was established upon a property of 2000 acres, purchased by his father in the vicinity of Montpellier. Here he resumed the intimacy of his boyhood with John Stuart Mill, who was five years his junior, and whose life-long taste for botany was probably fixed during this residence of seven or eight months in the Bentham family in the year 1820. About this time Bentham occupied himself with ornithology and then with entomology, finding time, however, for another line of study ; for at the age of twenty he had begun a translation into French of his Uncle Jeremy's " Chrestomathia," which was published in Paris some years afterwards ; and he soon after translated also the essay on " Nomenclature and Classification." This was followed by his own " Essai sur la Nomenclature et Classification," published in Paris. This,

his original scientific production, was one of some mark, for it is praised by Stanley-Jevons in his recent "History of the Sciences."

On attaining his majority, his elder and only brother having died, he was placed in management of his father's Provençal estate, an employment which he took up with alacrity and prosecuted with success, turning to practical account his methodical habits, his indomitable industry, and his familiarity with Provençal country life and language. The latter he spoke like a native. A language always seemed to come to him without effort. Meanwhile his leisure hours were given to philosophical studies, his holidays to botanical excursions into the Cevennes and the Pyrenees. In the year 1853, a visit to England upon business relating to his father's French estate, where it seemed probable that he was to spend his life, was followed by circumstances which gave him back to his native country. He brought to his Uncle Jeremy a French translation of the latter's "Chrestomathia"; he made the acquaintance of Sir James Edward Smith, Robert Brown, Lambert, Don, and the other English botanists of the day; visited Sir William, then Professor Hooker, at Glasgow, and Walker Arnott in Edinburgh; took the latter with him the next summer to France, where the two botanists herborized together in Languedoc and the Pyrenees; and, returning to London, he accepted his uncle's pressing invitation to remain and devote a portion of his time to the preparation of the latter's manuscripts for the press, at the same time pursuing legal studies at Lincoln's Inn. He was in due time called to the bar, and in 1832 he held his first and last brief. In that year Jeremy Bentham died, bequeathing most of his property to his nephew. This was much less than was expected, owing to bad management on his uncle's part and to the extravagant sums spent by his executors in the publication of the philosopher's posthumous work. But it sufficed, in connection with the paternal inheritance, which fell to him in the year previous, for the modest independence which allowed of undistracted devotion to his favorite studies. These were for a time divided between botany, jurisprudence, and logic, not to speak of editorial work upon his father's papers relating to the management of the navy and the administration of the national dockyards.

The first publication was botanical, and was published in Paris, in the year 1826, — his "Catalogue des Plantes Indigènes des Pyrénées et du Bas Languedoc." To this is prefixed an interesting narrative of a botanical tour in the Pyrenees, and some remarks upon the mode of preparing such catalogues in order to bring out their greatest utility, — remarks which already evince the wisdom for which he was distinguished in after years. He also reformed and reëlaborated our difficult genera of the district, — Cerastium, Orobanche, Helianthemum, and Medicago. The next, perhaps, was an article upon codification — wholly disagreeing with his uncle — which attracted the attention of Brougham, Hume, and O'Connell; also one upon the laws affecting larceny, which Sir Robert Peel complemented and made use of, and another on the law of real property.

But his most considerable work of the period received scant attention at the time from those most interested in the subject, and passed from its birth into oblivion, from which only in these later years it has been rescued, yet without word or sign from its author. This work (of 287 octavo pages) was published in London in 1827, under the title of "Outline of a New System of Logic, with a critical examination of Dr. Whately's Elements of Logic." It was in this book that the quantification of the predicate was first systematically applied, in such wise that Stanley-Jevons[1] declares it to be "undoubtedly the most fruitful discovery made in abstract logical science since the time of Aristotle." Before sixty copies of the book had been sold, the publisher became bankrupt, and the whole impression of this work of a young and unknown author was sold for waste paper. One of the extant copies, however, came into the hands of the distinguished philosopher Sir William Hamilton, to whom the discovery of the quantification of the predicate was credited, and who, in claiming it, brought " an acrimonious charge of plagiarism " against Pro-

[1] In "Contemporary Review," xxi., 1873, p. 823.

fessor De Morgan upon this very subject. Yet this very book of Mr. Bentham is one of the ten placed by title at the head of Sir William Hamilton's article on logic in the Edinburgh Review for April, 1833, is once or twice referred to in the article, and, a dozen years later, in the course of the controversy with De Morgan, Sir William alluded to this article as containing the germs of his discovery. We may imagine the avidity with which De Morgan, injuriously attacked, would have seized upon Mr. Bentham's book if he had known of it. It is not so easy to understand how Mr. Bentham, although now absorbed in botanical researches, could have overlooked this controversy in the " Athenæum," or how, if he knew of it, he could have kept silence. It was only at the close of the year 1850 that Mr. Warlow sent from the coast of Wales a letter to the " Athenæum," in which he refers to Bentham's book as one which had long before anticipated this interesting discovery. Although Hamilton himself never offered explanation of his now unpleasant position (for the note obliquely referring to the matter in the second edition of his Discussions is not an explanation), Mr. Blaine did (in the " Athenæum for February 1, 1851) immediately endeavor to discredit the importance of Bentham's work, and again in 1873 ("Contemporary Review," xxi.), in reply to Herbert Spencer's reclamation of Bentham's discovery. To this Stanley-Jevons made reply in the same volume (pp. 821–824); and later, in his " Principles of Science " (ii. 387), this competent and impartial judge, in speaking of the connection of Bentham's work " with the great discovery of the quantification of the predicate," adds : —

" I must continue to hold that the principle of quantification is explicitly stated by Mr. Bentham ; and it must be regarded as a remarkable fact in the history of logic, that Hamilton, while vindicating in 1847 his own claims to originality and priority as against the scheme of De Morgan, should have overlooked the much earlier and more closely related discoveries of Bentham."

It must be that Hamilton reviewed Bentham's book without reading it through, or that its ideas did not at the time leave any conscious impression upon the reviewer's mind, yet may have fructified afterwards.

After his uncle's death in 1832, Mr. Bentham gave his undivided attention to botany. He became a Fellow of the Linnæan Society in 1828. Robert Brown soon after presented his name to the Royal Society, but withdrew it before the election, to mark the dissatisfaction on the part of scientific men with the management of the society when a royal duke was made president. Consequently he did not become F. R. S. until 1862. In 1829, when the Royal Horticultural Society was much embarrassed, he accepted the position of honorary secretary, with his friend Lindley as associate. Under their management it was soon extricated from its perilous condition, attained its highest prosperity and renown, and did its best work for horticulture and botany. In 1833 he married the daughter of Sir Harford Brydges, for many years British ambassador in Persia, and the next year he took up his residence in the house in Queen Square Place, Westminster, inherited from his uncle, in which Jeremy Bentham and his own paternal grandfather had dwelt for almost a century. The house no longer exists, but upon its site stands the western wing of the "Queen Anne Mansions." The summer of 1836 was passed in Germany, at points of botanical interest and wherever the principal herbaria are preserved, the whole winter in Vienna. Some account of this tour, and interesting memoranda of the botanists, gardens, and herbaria visited, communicated in familiar letters to Sir William Hooker, were printed at the time (without the author's name) in the second volume of the "Companion to the Botanical Magazine." Similar visits for botanical investigation, mingled with recreation, were made almost every summer to various parts of the continent; in one of them he revisited the scenes of his early boyhood in Russia, traveled with Mrs. Bentham to the fair at Nijni-Novgorod, and thence to Odessa, by the rude litter-like conveyances of the country.

In 1842 he removed with his herbarium to Pontrilas House in Herefordshire, an Elizabethan mansion belonging to his brother-in-law, and combined there the life of a country squire

with that of a diligent student, until 1854, when, returning to
London, he presented his herbarium and botanical library to
the Royal Gardens at Kew, where they were added to the
still larger collections of Sir William Hooker.　After a short
interval Mr. Bentham took up his residence at No. 25 Wilton
Place, between Belgrave Square and Hyde Park, which was
his home for the rest of his life.　Thence, autumn holidays
excepted, with perfect regularity for five days in the week he
resorted to Kew, pursued his botanical investigations from ten
to four o'clock, then, returning, he wrote out the notes of his
day's work before dinner, hardly ever breaking his fast in the
long interval.　With such methodical habits, with freedom
from professional or administrative functions which consume
the precious time of most botanists, with steady devotion to his
chosen work, and with nearly all authentic materials and
needful appliances at hand or within reach, it is not surprising
that he should have undertaken and have so well accomplished
such a vast amount of work ; and he has the crowning merit
and happy fortune of having completed all that he undertook.

Nor did he decline duties of administration and counsel
which could rightly be asked of him.　The presidency of the
Linnæan Society, which he accepted and held for eleven years
(1863 to 1874), was no sinecure to him ; for he is said to have
taken on no small part of the work of secretary, treasurer, and
botanical editor.　Somewhat to the surprise of his younger
associates, who knew him only as the recluse student, he made
proof in age of the fine talent for business and the conduct of
affairs which had distinguished his prime in the management
of the Horticultural Society ; and in his annual presidential
addresses, which form a volume of permanent value, his discussions of general as well as of particular scientific questions
and interests bring out prominently the breadth and fullness
of his knowledge and the soundness of his judgment.

The years which followed his retirement from the chair of
the Linnæan Society, at the age of seventy-three, were no less
laborious or less productive than those preceding ; at the age
of eighty (as the writer can testify) the diminution of bodily
strength had wrought no obvious abatement of mental power

and not much of facility; and he was able to finish in the spring of 1883 the great work upon which he was engaged. As was natural, his corporeal strength gave way when his work was done. After a year and a half of increasing debility he died simply of old age — the survivor of his wife for three or four years, the last of the Benthams, for he had no children, nor any collateral descendants of the name.

A large part of his modest fortune was bequeathed to the Linnæan Society, to the Royal Society, for its scientific relief fund, and in other trusts for the promotion of the science to which his long life was so perseveringly devoted.

The record of no small and no unimportant part of a naturalist's work is to be found in scattered papers, and those of George Bentham are quite too numerous for individual mention. The series begins with an article upon *Labiatæ*, published in " Linnæa " in 1831; it closes with one in the " Journal of the Linnæan Society," read April 19, 1883, indicating the parts taken by the two authors in the elaboration of the " Genera Plantarum," then completed. Counting from the date of the Catalogue of Pyrenean plants, 1826, there are fifty-seven years of authorship. His first substantial volume in botany was the " Labiatarum Genera et Species, or a description of the genera and species of plants of the order *Labiatæ*, with their general history, characters, affinities, and geographical distribution," an octavo of almost 800 pages, of which the first part was published in 1832, the last in 1836. He found even the European part of this large order in much confusion; his monograph left its seventeen hundred and more of species so well arranged (under 107 genera and in tribes of his own creation) that there was little to alter, except as to the rank of certain groups, when he revised them for the " Prodromus " in 1848, and finally revised the genera (now increased to 136, and with estimated species almost doubled) for the " Genera Plantarum " in 1876. Although the work of a beginner, it took rank as the best extant monograph of its kind, namely, one of a large natural order, without plates. In it Mr. Bentham first set the example in any large way, of consulting all the available herbaria for the inspection and determination of type-specimens. To this end he made journeys to the continent every year from 1830 to 1834, visiting nearly all the public and larger private herbaria. In the years during which the monograph of *Labiatæ* was in progress, Mr. Bentham elaborated and published the earlier of the papers which have particularly connected his name with North American botany. These are, first, the reports on some of the new ornamental plants raised in the Horticultural Society's Garden from seeds collected in western North America by Douglas, under the auspices of that society, by which were first made known to botanists and florists so many of the characteristic genera and species of Oregon and California, now familiar in gardens, Gilias and Nemophilas, Limnanthes, Phacelias, Brodiæas, Calochorti, Eschscholtzias, Collinsias, and the like; then the monograph of *Hydrophylleæ* (1834), followed the next year by that on Hosackia, and that on the *Eriogoneæ*, — all American and chiefly North American plants, — the first-fruits of a great harvest which even now has not wholly been gathered in; the field is so vast, though the laborers have not been few. Later, the "Plantæ Hartwegianæ," an octavo volume begun in 1839, but finished in 1857 with the Californian collections; and in 1844, the "Botany of the Voyage of the Sulphur," in quarto, the first part of which relates to Californian botany. The various papers upon South American botany are even more numerous; one of them being that in which Heliamphora, of British Guiana, a new genus of Pitcher Plants, of the Sarracenia family, was established.

Bentham's labors upon the great order *Leguminosæ* began early, with his "Commentationes de Leguminosarum Generibus," published in the Annals of the Vienna Museum, being the work of a winter's holiday (1836-7) passed in that capital in the herbarium then directed by Endlicher. This was followed by a series of papers, mostly monographs of genera, in Hooker's "Journal of Botany," in the "Journal of the Linnæan Society," and elsewhere, by the elaboration of the order for the imperial "Flora Brasiliensis," and later, by the "Revision of the Genus Cassia," and that of the suborder

Mimoseæ, in the "Transactions of the Linnæan Society," the latter (a quarto volume in size) published as late as the year 1875. Both are perfect models of monographical work.

An important series of monographs in another and more condensed form was contributed to De Candolle's "Prodromus," namely, the Tribe *Ericeæ* in the seventh volume, the *Polemoniaceæ* in the ninth, the *Scrophulariaceæ* in the tenth, the *Labiatæ* forming the greater part of the twelfth, and the *Eriogoneæ* in the fourteenth; these together filling 1133 pages according to the surviving editor. If not quite the largest collaborator of the De Candolles, as counted in pages, he was so in the number of plants described, and his work was of the best. It was also ready in time, which is more than can be said of the collaborators in general.

There are few parts of the world upon the botany of which Mr. Bentham has not touched — Tropical America, in the ample collections of Mr. Spruce, and those of Hartweg, distributed, and the former partly and the latter wholly determined by him, as also Hinds' collections made in the voyage of the Sulphur, besides what has already been adverted to; Polynesia, from Hinds' and Barclay's collections; Western Tropical Africa, in the Niger Flora, most of the "Flora Nigritiana" being from his hand; the "Flora Hongkongensis," in which he began the series of British colonial floras; and finally that vast work, the "Flora Australiensis," in seven volumes, which the author began when he was over sixty years old and finished when he was seventy-seven. Nor did he neglect the cultivation of the narrow and more exhausted field of British botany. His "Handbook of the British Flora," for the use of beginners and amateurs, published in 1858, has gone through four large editions. Its special object was to enable a beginner or a mere amateur, with little or no previous scientific knowledge and without assistance, to work out understandingly the characters by which the plants of a limited flora may be distinguished from each other, these being expressed as much as possible in ordinary language, or in such technical terms as could be fully explained in the book itself and easily apprehended by the learner. The immediate and continued popularity of this handy volume, bringing the light of full knowledge and sound methods to guide the beginner's way, illustrates the advantage of having elementary works prepared by a master of the subject, whenever the master will take the necessary pains. To the same end, the author prepared for this volume an excellent and terse introduction to structural and descriptive botany, which has been prefixed to all the Colonial Floras. In the first edition to this British Flora it was attempted to use or to give English names to the genera and species throughout. This could be done only in such a familiar and well-trodden field as Britain, where almost every plant was familiar; but even here it failed, and in later editions the popular names were relegated to a subordinate position.

It has been stated that Mr. Bentham was over sixty years old when he undertook the "Flora Australiensis," and he was seventy-seven when he brought this vast work to completion, assisted only in notes and preliminary studies by Baron von Mueller of Melbourne. About the same time he courageously undertook the still greater task of a new "Genera Plantarum," to be worked out, not, like that of Endlicher, mainly by the compilation of published characters into a common formula, but by an actual examination of the extant materials, primarily those of the Kew herbarium, — this work, however, in conjunction with his intimate associate, Sir Joseph Hooker. This work is the only "joint production" in which Mr. Bentham ever engaged. The relations and position of the two authors made the association every way satisfactory, and the magnitude of the task made it necessary. The training and the experience of the two associates were very different and in some ways complemental, one having the greatest herbarium knowledge of any living botanist, the other, the widest and keenest observer of vegetable life under "whatever climes the sun's bright circle warms," as well as of antarctic regions which it warms very little. It would be expected, on the principle "juniores ad labores," that the laboring oar would be taken by the younger of the pair. It was long and severe work for both; but the veteran was happily quite free from, and his companion heavily weighted by, onerous official duties and cares; and so it came to pass that about two thirds of the orders and genera were elaborated by Mr. Bentham. In April, 1883, the completion of the work (*i. e.*, of the genera of Phænogamous plants, to which it was limited) closed this long and exemplary botanical career; and the short account which he gave to the Linnæan Society on the nineteenth of that month, specifying the conduct of the work and the part of the respective authors, was his last publication.

In this connection mention should also be made of the essays (which he simply calls "Notes") upon some of the more important orders which he investigated for the "Genera Plantarum," — the *Compositæ*, the Campanulaceous and the Oleaceous orders, the *Monocotyledoneæ* as to classification, the *Euphorbiaceæ*, the Orchis family, the *Cyperaceæ* and the *Gramineæ*. These are not mere abstracts, issued in advance, but critical dissertations with occasional discussions of some general or particular question of terminology or morphology. When collected they form a stout volume, which, along with the volume made up of his anniversary addresses when president of the Linnæan Society, and the paper on the progress and state of systematic botany, read to the British Association for the Advancement of Science in 1874, should be much considered by those who would form a just idea of the largeness of Mr. Bentham's knowledge and the character of his work.

It will have been seen that Mr. Bentham confined himself to the Phænogamia, to morphological, taxonomical, and descriptive work, not paying attention to the Cryptogamia below the Ferns, nor to vegetable anatomy, physiology, or palæontology. He was what will now be called a botanist of the old school. Up to middle age and beyond he used rather to regard himself as an amateur, pursuing botany as an intellectual exercise. "There are diversities of gifts;" perhaps no professional naturalist ever made more of his, certainly no one ever labored more diligently, nor indeed more successfully over so wide a field, within these chosen lines. For extent and variety of good work accomplished, for an intuitive sense of method, for lucidity and accuracy, and for insight, George Bentham may fairly be compared with Linnæus, De Candolle, and Robert Brown.

His long life was a perfect and precious example, much needed in this age, of persevering and thorough devotion to science while unconstrained as well as untrammeled by professional duty or necessity. For those endowed with leisure, to " live laborious days " in her service, it is not a common achievement.

The tribute which the American Academy of Sciences pays to the memory of a deceased foreign honorary member might here fittingly conclude. But one who knew him long and well may be allowed to add a word upon the personal characteristics of the subject of this memorial; the more so that he is himself greatly indebted for generous help. For, long ago, when in special need of botanical assistance, Mr. Bentham invited him and his companion to his house at Pontrilas, and devoted the greater part of his time for two months to this service. Mr. Bentham's great reserve and dryness in general intercourse, and his avoidance of publicity, might give the impression of an unsympathetic nature; but he was indeed most amiable, warm-hearted, and even genial, " the kindest of helpmates," the most disinterested of friends.

AUGUSTUS FENDLER.[1]

AFTER Dr. Engelmann's death, the beginning of a notice of Mr. Fendler was found upon his table, from which it was learned that he had died at Trinidad, some time previous. Inquiries sent to the Port of Spain, where he had for several years resided, remain unanswered. An autobiographical account which he addressed to a correspondent (and which, with some of his letters, we hope will before long be printed) enables us to state that Mr. Fendler was born at Gumbinnen, on the easternmost borders of Prussia, January 10, 1813, lost his father in infancy, was sent to the gymnasium of the town when twelve years old, but was at sixteen apprenticed to the town clerk, where, perhaps, he perfected the neat and clear handwriting with which his correspondents are familiar. Having a fondness for mathematics and chemistry, he obtained in 1834, upon examination, a nomination to the Royal Polytechnic School at Berlin, but relinquished it after a year on account of delicate health. In 1836 he came from Bremen to Baltimore, " with a couple of dollars in his pocket," worked in a tan-yard in Philadelphia, then in a lamp factory in New York; in 1838 he traveled in the most economical way to St. Louis, which required thirty days, and was employed by a lamp-maker who made " spirit-gas " for lighting public-houses, coal-gas being then unknown so far west. Soon after, he made his way to New Orleans and to Texas, where he was witness to the ravages of yellow fever in the summer and fall of 1839. He returned to Illinois, broken in health and empty in purse, taught school for some time; then, the spirit of wandering and of solitude coming strongly upon him, he took possession of an uninhabited island in the Missouri River, about three hundred miles above St. Louis, where he enjoyed a hermit's life for six months, and until a great spring rise of the

[1] American Journal of Science and Arts, 3 ser., xxix. 169. (1885.)

river threatened to sweep away his cabin, when he took to his canoe, and dropped down the stream among the floating logs and masses of ice. In 1844 he returned to Old Prussia on a visit, at Königsberg made the acquaintance of Ernest Meyer, the professor of botany, and learned from him — what he would have been most glad to know before — that dried specimens of plants for the herbarium might be disposed of at a reasonable price. Returning to St. Louis, he began to collect plants in this view, took the botanical specimens to Dr. Engelmann, who gave him botanical assistance and encouragement. In 1846 Dr. Engelmann and the writer of this notice obtained permission for the transportation of Mr. Fendler and his luggage along with the body of United States troops which took possession of Santa Fé, New Mexico; there he remained for about a year, and made his well-known New Mexican collection, the first-fruits of the botany of that interesting district. In 1849 he attempted another western botanical expedition, this time with Salt Lake in view. But on the plains he lost all his drying-paper in a flood of the Little Blue River; and he returned to St. Louis, to find that all his collections, books, journals, and other possessions had been burnt in the great conflagration which had just devastated that city. He now sought a different climate, and, at the approach of winter, went to the Isthmus of Panama for four months, made at Chagres an interesting botanical collection, returned by way of New Orleans to Arkansas, and to Memphis on the Tennessee side of the river, where for three years he carried on the camphene-light business, botanizing in the vicinity when he could. In 1854, the introduction of gas having made his occupation unprofitable, and a craving for new scenes being strong upon him, he sailed for La Guayra, went up to Caraccas and thence to Colonia Tovar, 6500 feet above the sea, built his cabin on the mountain side, where he lived four or five years and amassed his large and fine Venezuelan collections of dried plants, so well known in the principal herbaria of the world. His principal companions were his thermometer and barometer, and his careful meteorological observations were published by the Smithsonian Institution, in the report of the year 1857. Returning to Missouri in 1864, he bought some wild land at Allenton, cultivated and lived on it for seven years (except one winter passed in the herbarium at Cambridge), having the companionship and assistance of a half-brother who had joined him, and whom, being rather feeble-minded, he took care of for the rest of his life. In 1871, having sold his place in Missouri, he returned again to Prussia, intending to remain in his native country. But he soon longed for the New World, to which he returned in 1873 ; he settled in Wilmington, Delaware, where, having the botanical companionship of Mr. Canby, he again interested himself in his favorite pursuits, — but now much more in speculative physics. For years the thoughts of his solitary hours had turned upon the cause of gravitation and its prob-able connection with other forces, and while at Wilmington he wrote (and unhappily printed at his own expense) a thin octavo volume, entitled " The Mechanism of the Universe." Repeated attacks of acute rheumatism constrained him to seek again a tropical climate, this time the island of Trini-dad. He and his brother landed at the Port of Spain in June, 1877, where he passed the remainder of his days, living mainly on the products of the small plot of land which he purchased, renewing his old interest and activity in making botanical observations and collections, especially among the Ferns, of which he sent to Professor Eaton collections worthy of his better days. But, having exhausted in this respect the field within his immediate reach, and lost the vigor needed for laborious excursions, little had been heard of him for the past few years, and it is only indirectly that the fact of his death has been made known to us.

It is needless to say that Fendler was a quick and keen observer and an admirable collector. He had much literary taste, and had formed a very good literary style in English, as his descriptive letters show. He was excessively diffident and shy, but courteous and most amiable, gentle, and deli-cately refined. Many species of his own discovery commem-orate his name, as also a well-marked genus, a Saxifragaceous shrub, which is winning its way into ornamental cultivation.

CHARLES WRIGHT.[1]

CHARLES WRIGHT died on the 11th of August, at Wethersfield, Connecticut, at the home where he was born on the 29th
of October, 1811, and where the early as well as the later
years of his life were passed. He received his education in
the grammar school of his native village and at Yale College,
which he entered in 1831, graduating in 1835. His fondness
for botany was developed while he was in college, although,
so far as we can learn, he had no teacher. The opportunity
of gratifying this predilection in an inviting region may have
determined his acceptance, almost immediately after graduation, of an offer to teach in a private family at Natchez, Mississippi. Within a year pecuniary reverses of his employer
terminated this engagement. At this time there was a flow
of immigration into Texas, then an independent republic;
and Mr. Wright joining in it, in the spring of 1837 made his
way from the Mississippi to the Sabine, and over the border,
chiefly on foot, botanizing as he went. Making his headquarters for two or three years at a place then called Zarvala,
on the Neches, he occupied himself with land-surveying, explored the surrounding country, " learned to dress deer-skins
after the manner of the Indians, and to make moccasins and
leggins," " became a pretty fair deer-hunter," and inured himself to the various hardships of a frontier life at that period.
When the business of surveying fell off he took again to
teaching; and in the year 1844 he opened a botanical correspondence with the present writer, sending an interesting collection of the plants of eastern Texas to Cambridge. In 1845
he went to Rutersville in Fayette County, and for a year or
two he was a teacher in a so-called college at that place, or in
private families there and at Austin, devoting all his leisure

[1] American Journal of Science and Arts, 3 ser., xxxi. 12. (1886.)

to his favorite avocation. In the summer of 1847–8 he had an opportunity of carrying his botanical explorations farther south and west. His friend, Dr. Veitch, whom he had known in eastern Texas, raised a company of volunteers for the Mexican war, then going on (Texas having been annexed to the United States), and gave Mr. Wright a position with moderate pay and light duties. This took him to Eagle Pass on the Mexican frontier, where he botanized on both sides of the river. He returned to the north in the autumn of that year, with his botanical collections, and passed the ensuing winter in Connecticut and at Cambridge.

In the spring of 1849, Mr. Wright returned to Texas, and, at the beginning of the summer, with some difficulty obtained leave to accompany the small body of United States troops which was sent across the unexplored country from San Antonio to El Paso on the Rio Grande. Notwithstanding some commendatory letters from Washington, no other assistance was afforded than the conveyance of his trunk and collecting paper. He made the whole journey on foot, boarded with one of the messes of the transportation train, and endured many privations and hardships. The return to the sea-board, in autumn, was by rather a more northerly route and under somewhat less untoward conditions. The interesting collection thus made first opened to our knowledge the botany of the western part of Texas. It was published, as to the *Polypetalœ* and *Compositœ*, in the third volume of the " Smithsonian Contributions to Knowledge," as " Plantæ Wrightianæ," Part I, in 1852.

A year and more was then passed in the central portion of Texas, awaiting the opportunity for other distant explorations, supporting himself in part by teaching a small school. At length, in the spring of 1851, he joined the party under Colonel Graham, one of the commissioners for surveying and determining the United States and Mexican boundary from the Rio Grande to the Pacific, accepting a position partly as botanist, partly as one of the surveyors, which assured a comfortable maintenance and the wished for opportunity for botanical exploration in an untouched field. Attached only to Colonel Graham's party, he returned with him without reaching farther westward than about the middle of what is now the territory of Arizona, and in the summer of 1852 he returned with his extensive collections to San Antonio, and thence to St. Louis, to deliver his *Cactaceæ* to Dr. Engelmann, and with the remainder to Cambridge. These collections were the basis of the second part of "Plantæ Wrightianæ," published in 1853, and, in connection with those of Dr. Parry, Professor Thurber, and Dr. J. M. Bigelow, etc., of the Botany of the Mexican Boundary Survey, published in 1859. As Mr. Wright collected more largely than his associate botanists, and divided his collections into sets, his specimens are incorporated into a considerable number of herbaria, at home and abroad, and are the types of many new species and genera. No name is more largely commemorated in the botany of Texas, New Mexico, and Arizona than that of Charles Wright. It is an Acanthaceous genus of this district, of his own discovery, that bears the name of Carlowrightia. Surely no botanist ever better earned such scientific remembrance by entire devotion, acute observation, severe exertion, and perseverance under hardship and privation.

Mr. Wright's next expedition was made under more pleasant conditions. It was a long one around the world, as botanist to the North Pacific Exploring Expedition, fitted out under Captain Ringgold, who was during the cruise succeeded by Commander John Rodgers. After passing the winter of 1852-3 at his home in Connecticut and at Cambridge, he joined this expedition in the spring, and sailed in the United States ship "Vincennes" from Norfolk, Virginia, on the 11th of June. The collections made when touching at Madeira and Cape Verde were of course unimportant; but at Simon's Bay, just round the Cape of Good Hope, a stay of six weeks resulted in a very considerable collection of about 800 species within a small area, the Cape being wonderfully crowded with all kinds of plants. The voyage was thence to Sydney and through the Coral Sea to Hongkong, which was reached about the middle of March, 1854. The collection of over 500 species of Phænogamous plants, which was made during that spring and summer upon this little island, and supplemented in the spring of 1855, was in part the basis of Bentham's " Flora Hongkongensis." In the autumn of 1854, interesting collections were made on the Bonin and Loo Choo Islands, and later upon the islands between the latter and Japan. Still more extensive and important were the collections made in Japan, especially those of the northern island, although the stay was brief. Also those made in Behring's Straits, mainly on Kiene or Arakamtchetchene Island, on the verge of the Polar Sea, where the scientific members of the expedition passed the month of August and a part of September, 1855. Reaching San Francisco in October, the season being unpropitious for botany, Mr. Wright was detached from the expedition, and came home by way of San Juan del Sur and Nicaragua, botanizing for a few weeks upon an island in the Lake, and thence by way of Greytown to New York.

In the following autumn (of 1856) Mr. Wright began his prolific exploration of the botany of Cuba. Landing at Santiago de Cuba, on the southeastern part of the island, he passed the winter of 1856–7 and the greater part of the ensuing summer in that nearly virgin district, most hospitably entertained by his countryman, Mr. George Bradford, and among the caffetals of the mountains by M. Lescaille, returning home with his rich collections early in the autumn. A year later he revisited Cuba, was again received by his devoted friends, extended his botanical explorations to the northern coast, and also farther westward, exchanging the dense virgin forest for open Pine-woods, like those of the Atlantic southern States, stopping at various *hatos* or cattle-farms on his route, but reaching better accommodations at Bayamo, when his kind host, Dr. Don Manuel Yero, assisted him in making some profitable mountain excursions. In the winter and spring of 1861 he was again domiciled with the Lescailles at Monte Verde and at the other coffee-plantations of this kind family; and from thence he was able to extend his herborizations to the eastern coast from Baracoa to Cape Maysi. The next winter he made his way westward to near the centre of the island, making headquarters at the sugar-plantation of the hospitable Don Simon de Cardenas, thence visiting the Sienaga de Zapata, a great marshy tract toward the south coast. In the early summer he transferred his indefatigable operations to the Vuelt-abajo, as it is called, or that part of Cuba westward of Habana, making his home at Balestena, a cattle-farm at the southern base of the mountains opposite Bahia Honda, where he was long most hospitably entertained by Don Jose Blain and Don F. A. Sauvalle. From thence he pushed his explorations nearly to the southwestern extremity of the island at Cape San Antonio. In the summer of 1864 he came home with his large collections, remaining there and at Cambridge for about a year.

In the autumn of 1865 he went again and for the last time to Cuba, again traversed the Vuelt-abajo in various directions, proceeded by steamer to Trinidad, and botanized in the mountains behind that town ; thence by way of Santiago he revisited the scenes of his earlier explorations and the surviving friends who had efficiently promoted them. The oldest and best of them, the elder Lescaille, was now dead. In the month of July, 1867, our persevering explorer came home.

Mr. Wright's Cuban botanical collections, from time to time distributed into sets, with numbers, were acquired by several of the principal herbaria, the fullest sets of the Phænogamous and vascular Cryptogamous plants, by the herbarium of Cambridge, and by the late Professor Grisebach of Göttingen. Professor Grisebach was in these years engaged with his "Flora of the British West Indies"; so that he gladly undertook the determination of the plants of Cuba. They were accordingly mainly published in Grisebach's two papers, "Plantæ Wrightianæ e Cuba Orientali," in the "Memoirs of the American Academy of Arts and Sciences," 1860 and 1862, and in his "Catalogus Plantarum Cubensium exhibens collectionem Wrightianam aliasque minores ex Insula Cuba missas," an 8vo volume, published in Leipsic in 1866. The latter work enumerates the Ferns and their allies, but those for the earlier part were published in 1860 by Professor Eaton, in his "Filices Wrightianæ et Fendlerianæ," a paper in the eighth volume of the "Memoirs of the American Acad-

emy." The later collections were incompletely published in
the "Flora Cubana," a volume issued by F. A. Sauvalle at
Habana, in 1873 and later, a revision of Grisebach's Catalogue (without the references, but with Spanish vernacular
names attached) which was made by Mr. Wright, who added
the descriptions of a good many new species. The only other
direct publication by Mr. Wright is his "Notes on Jussiæa,"
in the tenth volume of the Linnæan Society's Journal. As to
the lower Cryptogams, Mr. Wright's very rich collections
were distributed in sets and published by specialists: the
Fungi, by Berkeley and the late Dr. Curtis; the *Musci*, by
the late Mr. Sullivant; the *Lichenes*, by Professor Tuckerman in large part, and certain tribes quite recently by Müller
of Geneva. So Mr. Wright's name is deeply impressed upon
the botany of the Queen of the Antilles.

There was a prospect that he might do some good work
for the botany of San Domingo; for in 1871 a government
vessel was sent to make some exploration of that island, and
Mr. Wright went with it. It was in winter, the dry season,
and the excursion across the country was hurried and unsat-
isfactory; so that the small collection made in this, his last
distant botanizing, was not of much account.

Mr. Wright's botanizing days were now essentially over.
He made, indeed, a visit to the upper part of Georgia in the
spring of 1875. But this was mainly for recuperation from
the effects of a transient illness, and for seeing again a relative
and companion of his youth, from whom he had long been
separated. A large part of several years was passed at Cambridge, taking a part of the work of the Gray Herbarium;
and one winter was passed at the Bussey Institution, in aiding
his associate of the South Pacific cruise, Professor Storer.
Of late there fell to him the principal charge of the family at
Wethersfield, consisting of a brother who had become an
invalid, and of two sisters in feeble health, all unmarried and
aging serenely together. By degrees his own strength was
sapped by some organic disease of the heart, which had given
him serious warning; and on the 11th of August he suddenly succumbed, while making his usual round at evening to

look after the domestic animals of the homestead. Not returning when expected, he was sought for; the body was found as if in quiet repose, but the spirit had departed.

Mr. Wright was a person of low stature and well-knit frame, hardy rather than strong, scrupulously temperate, a man of simple ways, always modest and unpretending, but direct and downright in expression, most amiable, trusty, and religious. He accomplished a great amount of useful and excellent work for botany in the pure and simple love of it; and his memory is held in honorable and grateful remembrance by his surviving associates.

GEORGE W. CLINTON.[1]

GEORGE W. CLINTON died, at Albany, on the 7th of September last, in the seventy-eighth year of his age. He was the son of DeWitt Clinton, one of the most distinguished governors, and the grand-nephew of George Clinton, the first governor of the State of New York. He was born on the 13th of April, 1807, whether in the city of New York or in the home on Long Island is uncertain. He became a student in Albany Academy in the year 1816, when his father entered upon his first tenure of office as governor, entered Hamilton College in 1821, was graduated in 1825, was led by his early scientific tastes to the study of medicine, which he pursued for a year or two; at least he attended two courses of lectures at the then flourishing country medical school at Fairfield, New York. There his acquaintance with Professor James Hadley further developed his fondness for chemistry and botany, as it did that of the writer of this notice a few years afterwards. He also came under the instruction or companionship of Dr. Lewis C. Beck, a younger brother of his medical preceptor Dr. T. Romeyn Beck, attended a course of private lectures on botany given by Dr. William Tully, entered into correspondence with Rafinesque, Torrey, etc., and so bid fair to give himself to scientific studies, as we may suppose with the approval of his father, who, it is well known, had a decided scientific bent. But Governor Clinton's death in February, 1828, wrought a change in his prospects and in the course of his life. Acting upon the advice of his father's friend, Ambrose Spencer, the distinguished chief justice of the State, he took up the study of law, attended the law lectures of Judge Gould at Litchfield, Connecticut, and continued his studies in Canandaigua, New York, in the office of John C. Spencer, whose daughter he

[1] American Journal of Science and Arts, 3 ser., xxxi. 17. (1886.)

married. Admitted to the bar in 1831, he established himself at Buffalo in 1836, and practised his profession most
acceptably until the year 1854, when he became judge of the
superior court of that city. This honorable position he continued to hold with entire approbation until January, 1878,
when he retired under the provision of the constitution upon
attaining the age of seventy years. Then he resumed the
practice of the law for two or three years ; but at length he
took up his residence in Albany, partly for the more convenient rendering of his service as a regent of the university
of the State, and its vice-chancellor, but mainly for investigating and editing the papers and writings of his great-uncle
George Clinton. On the afternoon of the 7th of September he
took an accustomed walk in the Rural Cemetery of Albany,
and there he died, probably quite instantaneously ; for when
his body was found, two or three hours later, some withered
sprays of White Melilot, which he had gathered, were still
clasped in his hand.

Judge Clinton's professional life need not here be considered. I did not know him, but knew of him, as a botanist
in his younger days. About the year 1860, after buying a
botanical book for his daughter, the turning over its pages
revived an almost forgotten delight ; and when his attention
was again given to the flowers he had so long neglected, we
soon came into correspondence. "I might have become a
respectable naturalist," he writes, "but was torn from it in
my youth. . . . To become a botanist is now hopeless ; I am,
and must remain a mere collector. But then I collect for
my friends and for the Buffalo Society of the Natural Sciences. If I can please my friends and help the Society it
pleases me. I want it to succeed. Money I cannot give it,
and I give it all I can, the benefit of my example and pleasant
labors." An instructive and pleasant, and on his part a
sprightly correspondence it has been, and most ardent and
successful were his efforts in the development of the Society
of the Natural Sciences over which he presided, and especially of its herbarium which he founded. In the spring of
1864 he wrote : "To-morrow I believe I shall be able to mail

you my 'Preliminary List of the Plants of Buffalo.' And I demand that immediately upon its reception, you write me, saying 'pretty well for you.' I do feel gratified that I have at last made the mitiest mite of a contribution to science. I know how extremely minute it is. I would not be so exacting but for the fact that my letter-book is just full, and I want to commence a new one with a letter from you, I mean with a note from you, a letter is too ambitious."

As this modest Preliminary List exemplifies the beginning, so the full and critically prepared "Catalogue of the Native and Naturalized Plants of the City of Buffalo and its Vicinity," (pp. 215), published in 1882–3, marks the conclusion and shows the fruits of Judge Clinton's work upon the flora of the district around Buffalo. This catalogue was indeed prepared and published by his near friend and associate Mr. Day, with a thoroughness and judgment which have been much commended. But the collection and elaboration of the materials, the critical determination of the species, and the preparation of the "Clinton Herbarium," as it is now appropriately called, were essentially his own work in the *horæ subsecivæ* of a busy professional life. If during middle life and while making his way in his chosen vocation he abandoned his early scientific avocation, he took it up again when he had achieved a position which allowed some well-earned leisure, and he pursued it with an added zeal, energy, and acumen, which should give his name a place among the botanical worthies, — to be remembered after those who knew and appreciated and loved him have passed away. A little Scirpus specifically bears his name; but I never see the modest Liliaceous plant of our northern woods, called Clintonia in honor of the father, without associating it with the son.

Judge Clinton's contributions to the literature of the legal profession consisted mainly of his "Digest of the Decisions of the Law and Equity Courts of the State of New York," in three stout volumes. But he was a not unfrequent and a fascinating writer in the newspapers of the city, an occasional lecturer upon historical as well as scientific topics, and an organizer or promoter of every good civic work. He was a person of marked and distinct individuality. It had been said of him that "he was not like anybody else, did not look like anybody else, and did not talk like anybody else." But his ways and conversation were peculiarly winning and delightful. Of a rather large family of children, four survive, two of them sons, and a goodly number of grandchildren.

EDMOND BOISSIER.[1]

EDMOND BOISSIER died on the 25th of September, at his country residence in Canton Vaud, Switzerland, at the age of seventy-five years and three months. Having known him personally almost from the beginning of his botanical career, which has been so honorable and distinguished, it is a melancholy satisfaction as well as a duty, to pay this passing tribute to his memory.

Boissier came from one of those worthy families which were lost to France and gained to Geneva by the revocation of the Edict of Nantes, — a family that has proved its talents and high character in more than one of its members. Madame the Countess de Gasparin is a sister next to him in age, and the two had their education very much in common. He was born at Geneva, May 25, 1810, brought up and educated there, except that the summers were passed at his father's place at Valeyres, which he in time inherited, and where his life was closed. From his youth he was fond of natural history and of travel. It was not in his disposition, nor of the Genevese spirit of that day, to lead an aimless life ; so, when he came to choose what may be called his profession, it was natural that, at Geneva, in the days of the elder De Candolle, he took to botany. He showed his great good sense by his early judicious choice of a field and by his unbroken devotion to it. To the Mediterranean region, to southern Spain, and the Orient most of his work relates. After a year or two of careful preparation he went to Spain, in 1837, explored especially Granada and the eastern Pyrenees, and between 1839 and 1845 brought out his "Voyage Botanique dans le midi de l'Espagne," in two large quarto volumes, the first of narrative and plates, one hundred and eighty in number, the

second of descriptive matter relating to the Granadan flora.
Among the species he brought to light was the *Abies Pin-sapo*, the beautiful Fir-tree now so well known in cultivation. His narrative, besides its botanical interest, is charming reading.

In 1842, after his marriage to his cousin, of the De la Rive family, he traveled with his wife in Greece, Anatolia, Syria, and Egypt. It was to his dear companion that he dedicated two of their joint discoveries, *Omphalodes Luciliæ* and *Chionodoxa Luciliæ*. In 1849 he experienced the great sorrow of his life in her death from typhoid fever, during a second journey in the south of Spain. Between 1842 and 1854 he published the first series of his " Diagnoses Plantarum Orientalium Novarum," filling two volumes, and in 1855 the second series of almost equal extent ; in 1848 he completed his monograph of the *Plumbaginaceæ* ; in 1862 he promptly finished his conscientious elaboration of the great genus Euphorbia for De Candolle's " Prodromus," and in 1866 brought out the " Icones Euphorbiarum," of one hundred and twenty folio plates from outline drawings by Heyland. In 1881 he made a trip to Norway with his associate Reuter. Not to mention other journeys, he was again in Spain and adjacent countries in 1877, and lastly in 1881, his eighth visit, — then in wretched health. Passing by scattered papers of his, we come to his great work, the " Flora Orientalis," in five octavo volumes. It comprehends Greece and Turkey up to Dalmatia and the Balkans ; the Crimea ; Egypt up to the first cataracts ; northern Arabia down to the tropical line ; Asia Minor, Armenia, Syria, and Mesopotamia ; Turkestan up to 45° of latitude, Persia, Afghanistan, and Beloochistan — that is, up to the borders of India. The first volume was published in 1867 ; the fifth, in 1884, brings the work down to its conclusion with the *Pteridophytes ;* and the manuscript for a supplementary volume, for recent discoveries and some reëlaboration, was about half finished when he laid down his pen under an attack seemingly no worse than the many he had recovered from, but which now terminated his earthly life.

It was a noble life, shadowed by an early bereavement, and in later years worn by painful disease, — the manly life of one who lived simply and wrought industriously where many others with his independent fortune would have lived idly and luxuriously; and he was no less a loyal and public-spirited citizen. Upon an occasion when, long ago, we met him at Geneva, he had no time for botanical parlance, for he was doing duty in the ranks of the federal army. Later, at a time of commotion at Geneva, he helped to quell a revolutionary riot, and received a painful bayonet wound in the service. True to his ancestry, he was a devoted Protestant Christian, a trusted member of the synod of the Free Church in Canton Vaud, where he lived when not in winter residence at Geneva, and where his assiduous attentions to the poor and sick will be remembered. He was a man of fine presence, and till past middle life of much bodily vigor. As a botanist he gave himself to systematic work only, for which he had a fine tact, and, like the school in which he was bred, perhaps a faculty of excessive discrimination. No man living knew the Europeo-Caucasian plants so well, or could describe them better; and his herbarium must be, with possibly one rival, the most extensive and valuable private collection in Europe. He loved living flowers as well, and rejoiced in his choice conservatory collections at Rivage, on the shores of the Leman, and in his well-stocked rock-works of alpine plants which adorn his grounds at Valeyres.

A charming biographical notice by one who knew him well through his whole life, M. De Candolle, is contained in the "Archives des Science" of the "Bibliothèque Universelle de Geneva" for October last.

JOHANNES AUGUST CHRISTIAN RŒPER.[1]

JOHANNES AUGUST CHRISTIAN RŒPER died on the 17th of March, 1884, at the age of eighty-four. He had been for some time the oldest botanist we know of, at least the oldest botanical author ; for his first work, a monograph of the German species of Euphorbia, was published in 1824. He was director of the Botanic Garden at Basle in 1828, when he published his classical paper " De Organis Plantarum," and he may have been so in 1826, when he contributed to Seringe's " Mélanges Botaniques " his paper on the nature of flowers and inflorescences, which first put the latter upon a scientific basis and essentially established the present nomenclature. He was botanical professor there in 1830, when he published his tract " De Floribus et Affinitatibus Balsaminearum." In these essays he gave the promise of being one of the foremost morphological botanists of the age. Some time before the year 1840 he was translated to Rostock, where he held the botanical professorship for more than forty years, but without fulfilling the promise of his youth by additional contributions to the science of any considerable importance. There are, however, some articles from him in the " Botanische Zeitung," and other German periodicals, the latest in the year 1859. In 1851 he was chosen a Foreign Member of the Linnæan Society of London. We find no record of the time or place of his nativity, but we infer from a statement in the preface of his work on Euphorbia, which was published at Göttingen, that he was a German and not a Swiss. He is said to have been most amiable, and of deep religious convictions.

[1] American Journal of Science and Arts, 3 ser., xxxi. 22. (1886.)

LOUIS AGASSIZ.

THERE is no need to give an abstract of the contents of these fascinating volumes,[1] for everybody is reading them. Most are probably wishing for more personal details, especially of the American life; but the editorial work is so deftly and delicately done, and " the story of an intellectual life marked by rare coherence and unity " is so well arranged to tell itself and make its impression, that we may thankfully accept what has been given us, though the desired " fullness of personal narrative " be wanting.

Twelve years have passed since Agassiz was taken from us. Yet to some of us it seems not very long ago that the already celebrated Swiss naturalist came over in the bloom of his manly beauty to charm us with his winning ways, and inspire us with his overflowing enthusiasm, as he entered upon the American half of that career which has been so beneficial to the interests of natural science. There are not many left of those who attended those first Lowell Lectures in the autumn of 1846, — perhaps all the more taking for the broken English in which they were delivered, — and who shared in the delight with which, in a supplementary lecture, he more fluently addressed his audience in his mother-tongue.

In these earliest lectures he sounded the note of which his last public utterance was the dying cadence. For, as this biography rightly intimates, his scientific life was singularly entire and homogeneous, — if not uninfluenced yet quite unchanged by the transitions which have marked the period. In a small circle of naturalists, almost the first that was assembled to greet him on his coming to this country, and of which the writer is the sole survivor, when Agassiz was inquired of as to

[1] "Louis Agassiz, his Life and Correspondence." (The Andover Review, January, 1886, p. 39.)

his conception of "species," he sententiously replied : "A species is a thought of the Creator." To this thoroughly theistic conception he joined the scientific deduction which he had already been led to draw, that the animal species of each geological age, or even stratum, were different from those preceding and following, and also unconnected by natural derivation. And his very last published works reiterated his steadfast conviction that "there is no evidence of a direct descent of later from earlier species in the geological succession of animals." Indeed, so far as we know, he would not even admit that such "thoughts of the Creator" as these might have been actualized in the natural course of events. If he had accepted such a view, and if he had himself apprehended and developed in his own way the now wellnigh assured significance of some of his early and pregnant generalizations, the history of the doctrine of development would have been different from what it is, a different spirit and another name would have been prominent in it, and Agassiz would not have passed away while fighting what he felt to be — at least for the present — a losing battle. It is possible that the "whirligig of time" may still "bring in his revenges," but not very probable.

Much to his credit, it may be said that a good share of Agassiz's invincible aversion to evolution may be traced to the spirit in which it was taken up by his early associate Vogt, and, indeed, by most of the German school then and since, which justly offended both his scientific and his religious sense. Agassiz always "thought nobly of the soul," and could in no way approve either materialistic or agnostic opinions. The idealistic turn of his mind was doubtless confirmed in his student days at Munich, whither he and his friend Braun resorted after one session at Heidelberg, and where both devotedly attended the lectures of Schelling, — then in his later glory, — and of Oken, whose "Natur-Philosophie" was then in the ascendant. Although fascinated and inspired by Oken's *à priori* biology (built upon morphological ideas which had not yet been established but had, in part, been rightly divined), the two young naturalists were not carried away by it, — probably because they were such keen and conscientious observers, and were kept in close communion with work-a-day Nature. As Agassiz intimates, they had to resist " the temptation to impose one's own ideas upon Nature, to explain her mysteries by brilliant theories rather than by patient study of the facts as we find them," and that " overbearing confidence in the abstract conceptions of the human mind as applied to the study of nature; " although, indeed, he adds, " the young naturalist of that day who did not share, in some degree, the intellectual stimulus given to scientific pursuits by physio-philosophy would have missed a part of his training." That training was not lost upon Agassiz. Although the adage in his last published article, " A physical fact is as sacred as a moral principle," was well lived up to, yet ideal prepossessions often had much to do with his marshaling of the facts.

Another professor at Munich, from whom Agassiz learned much, and had nothing to unlearn, was the anatomist and physiologist Döllinger. He published little; but he seems to have been the founder of modern embryological investigation, and to have initiated his two famous pupils, first Von Baer, and then Agassiz, into at least the rudiments of the doctrine of the correspondence between the stages of the development of the individual animal with that of its rank in the scale of being, and the succession in geological time of the forms and types to which the species belongs : a principle very fertile for scientific zoölogy in the hands of both these naturalists, and one of the foundations of that theory of evolution which the former, we believe, partially accepted, and the other wholly rejected.

The botanical professor, the genial Von Martius, should also be mentioned here. He found Agassiz a student, barely of age; he directly made him an author, and an authority in the subject of his predilection. Dr. Spix, the zoölogical companion of Martius in Brazilian exploration, died in 1826 ; the fishes of the collection were left untouched. Martius recognized the genius of Agassiz, and offered him, and indeed pressed him to undertake their elaboration. Agassiz brought

out the first part of the quarto volume on the Fishes of the
Brazilian Expedition of Spix and Martius before he took
his degree of Doctor of Philosophy, and completed it before
he proceeded to that of Doctor in Medicine in 1830. The
work opened his way to fame, but brought no money. Still,
as Martius defrayed all the expenses, the net result compared quite favorably with that of later publications. Moreover, out of it possibly issued his own voyage to Brazil in
later years, under auspices such as his early patron never
dreamed of.

This early work also made him known to Cuvier; so that
when he went to Paris, a year afterwards, to continue his
medical and scientific studies, — the one, as he deemed, from
necessity, the other from choice, — he was received as a fellow-
savant. Yet at first with a certain reserve, probably no more
than was natural in view of the relative age and position of
the two men ; but Agassiz, writing to his sister, says : " This
extreme but formal politeness chills you instead of putting
you at your ease ; it lacks cordiality, and to tell the truth, I
would gladly go away if I were not held fast by the wealth of
material of which I can avail myself." But only a month
later he writes — this time to his uncle — that, while he was
anxious lest he " might not be allowed to examine, and still
less to describe, the fossil fishes and their skeletons in the
Museum, . . . knowing that Cuvier intended to write a work
on this subject," and might naturally wish to reserve the materials for his own use, and when the young naturalist, as he
showed his own sketches and notes to the veteran, was faintly
venturing to hope that, on seeing his work so far advanced,
he might perhaps be invited to share in a joint publication,
Cuvier relieved his anxiety and more than fulfilled his half-
formed desires.

" He desired his secretary to bring him a certain portfolio of
drawings. He showed me the contents : they were drawings
of fossil fishes, and notes which he had taken in the British
Museum and elsewhere. After looking it through with me,
he said he had seen with satisfaction the manner in which I
had treated this subject ; that I had, indeed, anticipated him,

since he had intended at some future time to do the same thing; but that, as I had given it so much attention, and had done my work so well, he had decided to renounce his project, and to place at my disposition all the materials he had collected and all the preliminary notes he had taken."

Within three months Cuvier fell under a stroke of paralysis, and shortly died. The day before the attack he had said to Agassiz, " Be careful, and remember that work kills." We doubt if it often kills naturalists, unless when, like Cuvier, they also become statesmen.

But to live and work, the naturalist must be fed. It was a perplexing problem how possibly to remain a while longer in Paris, which was essential to the carrying on of his work, and to find the means of supplying his very simple wants. And here the most charming letters in these volumes are, first, the one from his mother, full of tender thoughtfulness, and making the first suggestion about Neuchâtel and its museum, as a place where the aspiring naturalist might secure something more substantial than "brilliant hopes " to live upon; next, that from Agassiz to his father, who begs to be told as much as he can be supposed to understand of the nature of this work upon fossil fishes, which called for so much time, labor, and expense; and, almost immediately, Agassiz's letter to his parents, telling them that Humboldt had, quite spontaneously and unexpectedly, relieved his present anxieties by a credit of a thousand francs, to be increased, if necessary. Humboldt had shown a friendly interest in him from the first, and had undertaken to negotiate with Cotta, the publisher, in his behalf; but, becoming uneasy by the delay, and feeling that " a man so laborious, so gifted, and so deserving of affection . . . should not be left in a position where lack of serenity disturbs his power of work," he delicately pressed the acceptance of this aid as a confidential transaction between two friends of unequal age.

Indeed, the relations between the " two friends," one at that time sixty-three, and the other twenty-five, were very beautiful, and so continued, as the correspondence shows. Humboldt's letters (we wish there were more of them) are particularly delightful, are full of wit and wisdom, of almost paternal solicitude, and of excellent counsel. He enjoins upon Agassiz to finish what he has in hand before taking up new tasks (this is in 1837), not to spread his intellect over too many subjects at once, nor to go on enlarging the works he had undertaken; he predicts the pecuniary difficulties in which expansion would be sure to land him, bewails the glacier investigations, and closes with "a touch of fun, in order that my letter may seem a little less like preaching. A thousand affectionate remembrances. No more ice, not much of echinoderms, plenty of fish, recall of ambassadors *in partibus*, and great severity toward booksellers, an infernal race, two or three of which have been killed under me."

The ambassadors *in partibus* were the artists Agassiz employed and sent to England or elsewhere to draw fossil fishes for him in various museums, at a cost which Humboldt knew would be embarrassing. The ice, which he would have no more of, refers to the glacier researches upon which Agassiz was entering with ardor, laying one of the solid foundations of his fame. Curiously enough, both Humboldt and Von Buch, with all their interest in Agassiz, were quite unable to comprehend the importance of an inquiry which was directly in their line, and, indeed, they scorned it; while the young naturalist, without training in physics or geology, but with the insight of genius, at once developed the whole idea of the glacial period, with its wonderful consequences, upon his first inspection of the phenomena shown him by Charpentier in the valley of the Rhône.

It is well that Humboldt's advice was not heeded in this regard. Nevertheless, he was a wise counselor. He saw the danger into which his young friend's enthusiasm and boundless appetite for work was likely to lead him. For of Agassiz it may be said, with a variation of the well-known adage, that there was nothing he touched that he did not aggrandize. Everything he laid hold of grew large under his hand, — grew into a mountain threatening to overwhelm him, and would have overwhelmed any one whose powers were not proportionate to his aspirations. Established at Neuchâtel, and giving himself with ardor to the duties of his professorship, it was surely enough if he could do the author's share in the production of his great works on the fossil and the fresh-water fishes, without assuming the responsibilities and cares of publication as well, and even of a lithographic establishment which he set up mainly for his own use. But he carried on, *pari passu,* or nearly so, his work on fossil Mollusca, — a quarto volume with nearly a hundred plates, — his monographs of Echinoderms, living and fossil, his investigations of the embryological development of fishes, and that laborious work, the "Nomenclator Zoölogicus," with the "Bibliographia," later published in England by the Ray Society. Moreover, of scattered papers, those of the Royal Society's Catalogue which antedate his arrival in this country are more than threescore and ten. He had help, indeed; but the more he had the more he enlarged and diversified his tasks, Humboldt's sound advice about his zoölogical undertakings being no more heeded than his fulminations against the glacial theory.

In the midst of all this, Agassiz turned his glance upon the glaciers, and the "local phenomenon" became at once a cosmic one. So far a happy divination; but he seems to have believed quite to the last that not only the temperate zones, but whole intertropical continents — at least the American — had been sheeted with ice. The narrative in the first volume will give the general reader a vivid but insufficient conception of the stupendous work upon which he so brilliantly labored for nearly a decade of years.

Cœlum non animum mutant who come with such a spirit to a wider and, scientifically, less developed continent. First as visitor, soon as denizen, and at length as citizen of the American republic, Agassiz rose with every occasion to larger and more various activities. What with the Lowell Institute, the college in Charleston, South Carolina, and Cornell University, in addition to Harvard, he may be said to have held three or four professorships at once, none of them sinecures. He had not been two months in the country before a staff of assistants was gathered around him and a marine zoölogical laboratory was in operation. The rude shed on the shore and the small wooden building at Cambridge developed under his hand into the Museum of Zoölogy, — if not as we see it now, yet into one of the foremost collections. Who can say what it would have been if his plans and ideas had obtained full recognition, and " expenditure " had seemed to the trustees, as it seemed to him, " the best investment," or if efficient filial aid, not then to be dreamed of, had not given solid realization to the high paternal aspirations! In like manner grew large under his hand the Brazilian exploration so generously provided for by a Boston citizen and fostered by an enlightened emperor ; and on a similar scale was planned, and partly carried out, the " Contributions to the Natural History of the United States," as the imperial quarto work was modestly entitled, which was to be published " at the rate of one volume a year, each volume to contain about three hundred pages and twenty plates," with simple reliance upon a popular subscription ; — and so, indeed, of everything which this large-minded man undertook.

While Agassiz thus was a magnanimous man, in the literal as well as the accepted meaning of the word, he was also, as we have seen, a truly fortunate one. Honorable assistance came to him at critical moments, such as the delicate gift from Humboldt at Paris, which perhaps saved him to science ; such as the Wollaston prize from the Geological Society in 1834, when he was struggling for the means of carrying on the Fossil Fishes. The remainder of the deficit of this undertaking he was able to make up from his earliest earnings in America. For the rest, we all know how almost everything he desired — and he wanted nothing except for science — was cheerfully supplied to his hand by admiring givers. Those who knew the man during the twenty-seven years of his American life can quite understand the contagious enthusiasm and confidence which he evoked. The impression will in some degree be transmitted by these pleasant and timely volumes, which should make the leading lines of the life of Agassiz clear to the newer generation, and deepen them in the memory of an older one.

EDWARD TUCKERMAN.[1]

ON the 15th of March last, the Academy lost one of the older and more distinguished members of the botanical section, the lichenologist, Edward Tuckerman.

He was born in Boston, December 7, 1817, was the eldest son of a Boston merchant of the same name and of Sophia (May) Tuckerman. He was prepared for college at the Boston Latin School, whence, in obedience to his father's choice rather than of his own, he went to Union College at Schenectady. Entering as a sophomore, he took his B. A. degree in 1837. He then entered the Harvard Law School, took his degree in 1839, and remained in residence in Cambridge for a year or two longer. In the year 1841 he went to Germany and Scandinavia, going as far north as Upsala, devoting himself, as in a subsequent visit, to philosophical, historical, and botanical studies. On his return, in September, 1842, he made, with the writer of this notice, a botanical excursion to the White Mountains of New Hampshire, with which he was already familiar. At the close of that or early in the following year he took up his residence at Union College, proceeded to the M. A. degree, and there prepared and privately published one of the smaller, but noteworthy, of his botanical papers.

In the year 1844 or 1845 he returned to Cambridge, and in the autumn of 1846, in his twenty-ninth year, he became again an undergraduate. Applying for admission to the incoming senior class, he remarked to President Quincy that his father had broken the family tradition by sending him to another college, and that he proposed to correct the mistake. To the suggestion, that, being already an alumnus of the Law School as well as of Union, the University would willingly concede to him the earlier degrees he sought, he replied that

[1] Proceedings of the American Academy of Arts and Sciences, new ser. xiii., 539. (1886.)

he proposed to receive them in the ordinary way. He accordingly passed the regular examinations, took the whole routine of the studies of his class, and so was graduated with distinction in the class of 1847, — a unique but characteristic illustration of a loyal spirit, becoming " small by degrees and beautifully less."

His passion for university study was not yet quite satiated. For, two or three years later, he entered the Harvard Divinity School, passed through its course of study and prescribed exercises, — among them the delivery of a sermon in one of the Cambridge churches, — and so, in the year 1852, he became for the third time an alumnus of Harvard.

In May, 1854, he married in Boston Sarah Eliza Sigourney Cushing, who survives him, without offspring. Removing that year to Amherst, he built with excellent taste, upon a beautiful site, the house which has ever since been their abode. Although mainly devoted to botanical investigations, his first official connection with Amherst College was that of lecturer in history, then that of professor of oriental history, down to the year 1858, when he was collated to the chair of botany, which he held to the end of his life, although of late years relieved from the duty of class instruction. The college did itself the honor to confer upon its professor the degree of LL. D.

We cannot say when or how Professor Tuckerman became a botanist. But at an early period he was intimate with Dr. Harris, then University librarian, and with the ardent William Oakes of Ipswich, upon whom, through Dr. Osgood of Danvers, descended the mantle of Manasseh Cutler, of Essex County, the earliest New England botanist.

He must have been attracted to the Lichens almost from the beginning; for his first publications were upon Lichens of New England, largely those of his own collecting in the White and Green Mountains, in two papers, one communicated to the Boston Natural History Society, in 1838 or 1839, the other in 1840. These were soon followed by papers on Phænogamous botany, namely: one " On Oakesia a new Genus of the Order Empetreæ," a contribution made while he was abroad, in the summer of 1842, to Hooker's "London Journal of Botany." Unfortunately, the interesting plant which he thus dedicated to his botanical associate, William Oakes, who well deserved such commemoration, proved to be a second species of Corema. In 1843, at Schenectady, he privately printed and issued his "Enumeratio Methodica Caricum quarundam" (pp. 21, 8vo), in which he displayed not only his critical knowledge of the large and difficult genus Carex, but also his genius as a systematizer; for this essay was the first considerable, and a really successful, attempt to combine the species of this genus into natural groups. It is wholly in Latin, which he much affected for scientific disquisition as well as for technical characters, and used with facility and elegance. In the same year also appeared, in the American Journal of Science, the first of his "Observations on some interesting Plants of New England." This was followed in 1848 by a second, and in 1849 by a third paper in the same Journal; these containing, *inter alia*, his elaboration of our species of Potamogeton, then for the first time critically studied. These papers — with one or two in Hovey's Magazine and elsewhere, at about the same date — may be said to have ended his work in Phænogamous botany, although his interest in the subject never died out; for when he accepted the chair of botany at Amherst he began the preparation of "A Catalogue of Plants growing without cultivation within thirty miles of Amherst College," which he published in the year 1875, the late Mr. Charles Frost of Brattleborough contributing the lower Cryptogamia other than the Lichens. In matter and form, as well as in typography (in which Professor Tuckerman had exquisite taste), this catalogue is one of the very best.

But it was to Lichenology that his strength, as indeed almost his whole life, was most assiduously devoted. When, in his youth, the active members of the newly organized Natural History Society of Boston divided among themselves the work of making better known the animals, plants, and minerals of Massachusetts, the study of the Lichens either was assigned to him or he volunteered to undertake it. From this came those earliest papers which have already been mentioned. Also his " Synopsis of the Lichens of New England, the other Northern States, and British America," communicated to this Academy in the autumn of 1847, which is the most considerable botanical contribution to the first volume of the Proceedings. The fourth, fifth, sixth, and seventh volumes contain other of his Lichenological papers, of wholly original matter and critical character, — largely upon collections which had begun to come to him from the Rocky Mountain region, from Texas, the Pacific coast, the Sandwich Islands, and especially from the rich materials gathered in Cuba and elsewhere by the late Charles Wright. In these years, too, he much helped the study of his favorite plants by the preparation and issue of his " Lichenes Americæ Septentrionalis Exsiccati," in six fasciculi, or three volumes, highly valued by those who fortunately possess them. Equally fortunate are the herbaria which possess the " Lichenes Caroli Wrightii Cubæ curante E. Tuckerman," which authenticate his thorough work upon that portion of Mr. Wright's Cuban collections that he undertook to elaborate.

Passing without notice various subsidiary contributions both to journals and to the reports of exploring expeditions, we come to a pamphlet which he independently published at Amherst, in 1866, entitled " Lichens of California, Oregon, and the Rocky Mountains, so far as yet known," which, small though it be (pp. 35, 8vo), is particularly noteworthy; for in this he lays down the principles and matured opinions which he had adopted, and which he firmly adhered to, for the taxonomy and classification of Lichens. These are fully exemplified in the two systematic works to which Professor Tuckerman's later years and maturest powers were persistently devoted, — works which, partly from their publication somewhat out of the ordinary channels, are by no means so well known as they should be, but which surely secure to their author the position of a master in his department, — in which, indeed, we suppose he has left behind him no superior. These works are, first, the " Genera Lichenum, an Arrangement of the North American Lichens " (pp. 283, 8vo), published at Amherst in the year 1872 ; second, the "Synopsis of the North American Lichens," Part I, comprising the *Parmeliacei, Cladoniei,* and *Cœnogoniei,* published in Boston in 1882. It is hoped, but it is not yet certain, that some portions of the remainder, relating to the less conspicuous but more difficult tribes, may have been substantially made ready for the printer. The loss, we fear, is irreparable ; for the work cannot be completed by other hands upon quite the same lines, nor in our day with the same knowledge and insight ; and Professor Tuckerman's mode of exposition is inimitable.

That which Professor Tuckerman did accomplish, however, suffices to show the wide reach and remarkable precision of his knowledge, his patience and thoroughness in investigation, his sagacity in detecting affinities, and his philosophical and rather peculiar turn of mind. He wrote in a style which — though perhaps founded on that of his botanical model, Fries, for succinctness, and that of his favorite German philosophical masters for involution — was yet all his own, and which was the more pronounced in advancing years, when, owing to increasing deafness and delicate health, he led a more secluded life. In disquisition, the long and comprehensive sentences which he so carefully constructs are unmistakably clear to those who will patiently plod their way through them, and his choice even of unusual words is generally felicitous ; but sometimes the statements are so hedged about and interpenetrated by qualifications or reservations, and so pregnant with subsidiary although relevant considerations, that they are far from easy reading. Like nests of pill-boxes, they are packed into least bulk ; but for practical use they need to be taken apart.

That Professor Tuckerman could write idiomatic and clear-flowing English upon occasion, the delightful introduction to his edition of Josselyn's "New England's Rarities" demonstrates ; and in the framing of botanical descriptive phrases, Latin or English, in which clearness and brevity with just order and proportion are desiderata, he had hardly a superior.

As has been said, his botanical model was Elias Fries. He

had visited him at Upsala, and he kept up a correspondence with him to the end of the venerable botanist's life. He caught from Fries, or he developed independently, and cultivated to perfection, that sense of the value of the indefinable something which botanists inadequately express by the term "habit," which often enables the systematist to divine much further than he can perceive in the tracing of relationships. Upon this, in direct reference to Fries, and with a use of the term that seems to correlate it with "insight," Tuckerman remarks: "So great is the value of Habit in minds fully qualified to apprehend and appreciate its subtleties, that such minds may not only anticipate what the microscope is to reveal, but help us to understand its revelations." It should be remembered, however, that when Fries did the best of his work there were no microscopes of much account; and it is probable that Tuckerman would have done more, and perhaps have reached some different conclusions, if he had earlier and more largely used the best instrumental appliances of the time. One advantage, however, of his way of study, and his philosophical conception of an ideal connection of forms which are capable of a wide play of variation, was that he took broad views of genera and species. So he was quite unlike that numerous race of specialists who, in place of characterizing species, describe specimens, and to whom "genus" means the lowest recognizable group of species.

As to the vexed question in Lichenology, which came to him rather late and seemed to threaten the stability of his work, it was most natural that, at his time of life, he did not take kindly to the Algo-fungal notion of Lichens, and that he was convinced of its falsity by questionable evidence.

Professor Tuckerman was much more than an excellent specialist. Happily, he did not become such until he had laid a good foundation, for the time, in general systematic botany; and his early studies show that he was a man of scholarly culture over an unusually wide range. He was at home in the leading modern languages; he wrote Latin with reasonable facility, and botanical Latin remarkably well; he had given serious attention to law, divinity, philosophy, and history; and he was fond of antiquarian and genealogical researches. He privately published (without date) a handsome edition of Josselyn's "New England's Rarities Discovered," with copious critical annotations, of 134 pages, including an introduction of 27 pages, which contains a biography of Josselyn and a sketch of the earlier sources of our knowledge of New England plants and of some of the people who made them known.[1] Among them is a biographical notice of Manasseh Cutler, one of the very first elected Fellows of this Academy, the earliest botanical contributor to its Memoirs, — pastor, naturalist, and statesman, the builder of New England in Ohio, probably the originator of the Dane Resolutions in Congress, — a man whose name deserves larger remembrance than it has yet received.

Professor Tuckerman was elected into this Academy in May, 1845. He was one of the corporate members of the National Academy of Sciences at Washington, and an honorary member of several of the learned societies and academies of Europe. He was still young when Nuttall dedicated to him the genus Tuckermania, founded upon one of the handsomer of the Californian *Compositæ*, which is retained as a subgenus. For one who did not attain the age of sixty-seven, his publications span a remarkably wide interval. It is said that he contributed several short articles on antiquarian topics to the "Mercantile Journal," in the year 1832 ; also that, in 1832 and 1833, he assisted the late Mr. Samuel G. Drake in the preparation of his "Book of the Indians" and "Indian Wars." Then, between 1834 and 1841, he contributed to the "New York Churchman" no less than fifty-four articles, under the title of "Notitia Literaria" and "Adversaria," upon points in history, biography, and theology. His latest botanical article was contributed to the "Bulletin of the Torrey Botanical Club" in 1884. A little later, possibly, are some of his contributions to the "Church Eclectic," mostly pseudonymous, — critical notices of recent theological works. He was a keen critic, and very independent in his judgments. He had

[1] It appears that this was a contribution to the fourth volume of the "Archæologia Americana," published in 1860.

sounded in his time the depths of various opinions. But as he was born into, so he died, as he had lived, devoutly, in the communion of the Protestant Episcopal Church. With some interruptions, and of late under increasing infirmities, he yet continued his Lichenological studies until within a few weeks before the end. Living for a long while in comparative seclusion, few of our younger botanists can have known him personally, or much by correspondence ; and most of his old associates and near friends, who knew him best and prized him highly for his sterling character, have gone before him.

INDEX.

Terminology of æstivation, 181.

Tertiary arctic forests, 160; vegetation of Greenland, 227.

Thalictrum clavatum, 39.

Thalictrum Richardsonii, 39.

Tilia heterophylla, 37.

Torrey, John, sketch of, 359.

Torrey and Gray's Flora, 249.

Torreya, distribution of, 149; genealogy of, 161; pilgrimage to, 189.

Tortworth, the Chestnut of, 94.

Trautvetteria palmata, 44.

Trees, absence of, from the pampas, 263; manner of determining their age discussed, 84; manner of growth of, described, 80; of Canada and Great Britain contrasted, 266; the longevity of, 71.

Troglodytes Gorilla, Wyman's paper on, 386.

Trons, the Sycamore-Maple of, 89.

Trumbull, J. H., letter on the Jerusalem Artichoke, 198.

Tuckerman, Edward, sketch of, 491.

Vaccinium Constablæi, 65.

Vaccinium erythrocarpum, 45.

Vaccinium macrocarpum, 167.

Varieties, do they wear out or tend to wear out? 174.

Varieties, gender of names of, 257.

Vegetable kingdom, history of, discussed, 140.

Vegetation, aspects of North American, 143, 261.

Veratrum parviflorum, 48.

Vienna, herbarium at, 18.

Wadsworth Oak, 97.

Walker-Arnot, George A., sketch of, 347.

Ward, Nathaniel Bagshaw, sketch of, 349.

Webb, P. Barker, herbarium of, 16.

Weed, definition of a, 234.

Weeds of California, 241; of eastern North America, 241; origin of American, 235; pertinacity and predominance of, 234.

Welwitsch, Frederick, sketch of, 358.

White Pine, the, 107.

Wight, Robert, sketch of, 356.

Wistaria, 154.

Wright, Charles, sketch of, 468.

Wyman, Jeffries, sketch of, 377.

Yew-trees, ancient, 103; of Fountains' Abbey, 104.

Zigadenus glaucus, 49.